Staff

Larry Loeppke	*Managing Editor*
Nichole Altman	*Developmental Editor*
Lori Church	*Permissions Coordinator*
Maggie Lytle	*Cover*
Tara McDermott	*Designer*
Kari Voss	*Typesetting Supervisor/Co-designer*
Jean Smith	*Typesetter*
Sandy Wille	*Typesetter*
Karen Spring	*Typesetter*

Sources for Statistical Reports

U.S. State Department *Background Notes* (2003)

C.I.A. *World Factbook* (2002)

World Bank *World Development Reports* (2002/2003)

UN *Population and Vital Statistics Reports* (2002/2003)

World Statistics in Brief (2002)

The Statesman's Yearbook (2003)

Population Reference Bureau *World Population Data Sheet* (2002)

The World Almanac (2003)

The Economist Intelligence Unit (2003)

Copyright

Cataloging in Publication Data
Main entry under title: Global Studies: Japan and the Pacific Rim. 8th ed.
  1. East Asia—History—20th century–. 2. East Asia—History—Politics and government—20th century–.
I. Title: Japan and the Pacific Rim. II. Collinwood, Dean W., *comp*.
ISBN 0–07–311219–4 ISSN 1059-5988

Eighth Edition

Printed in the United States of America 1234567890BAHBAH54 Printed on Recycled Paper

Japan and the Pacific Rim

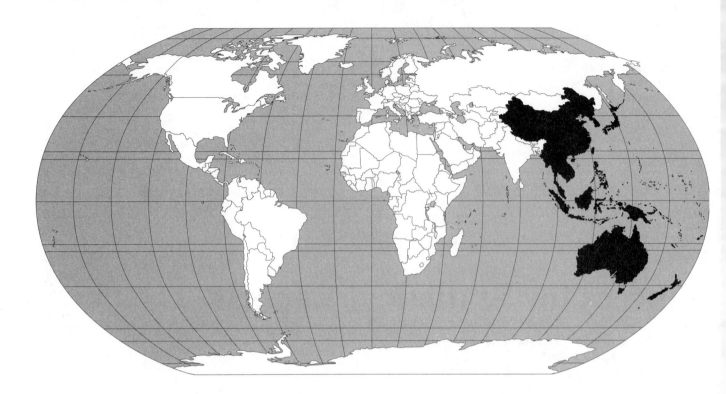

AUTHOR/EDITOR

Dr. Dean W. Collinwood

The author/editor of *Global Studies: Japan and the Pacific Rim* teaches Asian politics courses in the Political Science Department at the University of Utah in Salt Lake City. He also heads GLOBUS Inc., an international consulting firm with activities in Japan and China. He received his Ph.D. from the University of Chicago where he studied in both the political science and education departments. His M.Sc. in international relations is from the University of London, and his B.A. in political science with a Japanese minor is from Brigham Young University. Dr. Collinwood was a Fulbright Scholar at Tokyo University and Tsuda College in Japan and has conducted research throughout Asia and the Pacific. He is Chair of the Board of Trustees of the Asia Pacific Council, an international non-profit organization sponsoring internships in Asia, and he is past President of the Western Conference of the Association for Asian Studies. He has authored many books and articles on Japan, Korea, and several other countries, and some of his books are used by the U.S. State Department to train new Foreign Service officers.

GLOBAL STUDIES

JAPAN AND THE PACIFIC RIM

EIGHTH EDITION

Dr. Dean W. Collinwood
University of Utah

OTHER BOOKS IN THE GLOBAL STUDIES SERIES
- Africa
- China
- Europe
- India and South Asia
- Latin America
- The Middle East
- Russia, the Eurasian Republics, and
  Central/Eastern Europe

McGraw-Hill/Dushkin Company
2460 Kerper Boulevard, Dubuque, Iowa 52001
Visit us on the Internet—http://www.dushkin.com

Contents

Articles from the World Press

Using *Global Studies: Japan and the Pacific Rim*

THE GLOBAL STUDIES SERIES

The Global Studies series was created to help readers acquire a basic knowledge and understanding of the regions and countries in the world. Each volume provides a foundation of information—geographic, cultural, economic, political, historical, artistic, and religious—that will allow readers to better assess the current and future problems within these countries and regions and to comprehend how events there might affect their own well-being. In short, these volumes present the background information necessary to respond to the realities of our global age.

Each of the volumes in the Global Studies series is crafted under the careful direction of an author/editor—an expert in the area under study. The author/editors teach and conduct research and have traveled extensively through the regions about which they are writing.

MAJOR FEATURES OF THE GLOBAL STUDIES SERIES

The Global Studies volumes are organized to provide concise information on the regions and countries within those areas under study. The major sections and features of the books are described here.

Regional Essays

For *Global Studies: Japan and the Pacific Rim*, the author/editor has written two essays focusing on the religious, cultural, sociopolitical, and economic differences and simliarities of the countries and peoples in the region: "The Pacific Rim: Diversity and Interconnection," and "The Pacific Islands: Opportunities and Limits." Detailed maps accompany each essay.

Country Reports

Concise reports are written for each of the countries within the region under study. These reports are the heart of each Global Studies volume. *Global Studies: Japan and the Pacific Rim, Eighth Edition,* contains 20 country reports, including a lengthy special report on Japan.

The country reports are composed of five standard elements. Each report contains a detailed map visually positioning the country among its neighboring states; a summary of statistical information; a current essay providing important historical, geographical, political, cultural, and economic information; a historical timeline, offering a convenient visual survey of a few key historical events; and four "graphic indicators," with summary statements about the country in terms of development, freedom, health/welfare, and achievements.

A Note on the Statistical Reports

The statistical information provided for each country has been drawn from a wide range of sources. (The most frequently referenced are listed on page iii.) Every effort has been made to provide the most current and accurate information available. However, sometimes the information cited by these sources differs to some extent; and, all too often, the most current information available for some countries is somewhat dated. Aside from these occasional difficulties, the statistical summary of each country is generally quite complete and up to date. Care should be taken, however, in using these statistics (or, for that matter, any published statistics) in making hard comparisons among countries. We have also provided comparable statistics for the United States and Canada, which can be found on pages x and xi.

World Press Articles

Within each Global Studies volume is reprinted a number of articles carefully selected by our editorial staff and the author/editor from a broad range of international periodicals and newspapers. The articles have been chosen for currency, interest, and their differing perspectives. There are 24 articles in *Global Studies: Japan and the Pacific Rim, Eighth Edition.*

WWW Sites

An extensive annotated list of selected World Wide Web sites can be found on the facing page (ix) in this edition of *Global Studies: Japan.* In addition, the URL addresses for country-specific Web sites are provided on the statistics page of most countries. All of the Web site addresses were correct and operational at press time. Instructors and students alike are urged to refer to those sites often to enhance their understanding of the region and to keep up with current events.

Glossary, Bibliography, Index

At the back of each Global Studies volume, readers will find a glossary of terms and abbreviations, which provides a quick reference to the specialized vocabulary of the area under study and to the standard abbreviations used throughout the volume.

Following the glossary is a bibliography that lists general works, national histories, and current-events publications and periodicals that provide regular coverage on Japan.

The index at the end of the volume is an accurate reference to the contents of the volume. Readers seeking specific information and citations should consult this standard index.

Currency and Usefulness

Global Studies: Japan and the Pacific Rim, like the other Global Studies volumes, is intended to provide the most current and useful information available necessary to understand the events that are shaping the cultures of the region today.

This volume is revised on a regular basis. The statistics are updated, regional essays and country reports revised, and world press articles replaced. In order to accomplish this task, we turn to our author/editor, our advisory boards, and—hopefully—to you, the users of this volume. Your comments are more than welcome. If you have an idea that you think will make the next edition more useful, an article or bit of information that will make it more current, or a general comment on its organization, content, or features that you would like to share with us, please send it in for serious consideration.

Selected World Wide Web Sites for Japan and the Pacific Rim

(Some Web sites continually change their structure and content, so the information listed here may not always be available. Check our Web site at: http://www.dushkin.com/online/ —Ed.)

GENERAL SITES

CNN Online Page
http://www.cnn.com

A U.S. 24-hour video news channel. News is updated every few hours.

C-SPAN ONLINE
http://www.c-span.org

See especially C-SPAN International on the Web for International Programming Highlights and archived C-SPAN programs.

I-Trade International Trade Resources & Data Exchange
http://www.i-trade.com

Monthly exchange-rate data, U.S. Document Export Market Information (GEMS), U.S. Global Trade Outlook, and the CIA WorldFACT Book are available. This website is available in either English of Japanese.

Social Science Information Gateway (SOSIG)
http://sosig.esrc.bris.ac.uk

The project of the Economic and Social Research Council (ESRC) is presented here. It catalogs 22 subjects and lists developing-countries' URL addresses.

U.S. Agency for International Development (USAID)
http://www.info.usaid.gov

Graphically presented U.S. trade statistics related to Japan, China, Taiwan, and other Pacific Rim countries are available at this site.

U.S. Central Intelligence Agency Home Page
http://www.cia.gov

This site includes publications of the CIA, current Worldfact Book, and maps.

U.S. Department of State Home Page
http://www.state.gov/index.html

On this Web site Country Reports, Human Rights, International Organizations, and other features are organized alphabetically.

World Bank Group
http://www.worldbank.org

Find news (i.e., press releases, summary of new projects, speeches), publications, topics in development, countries and regions on this Web site. It links to other financial organizations. This site is available in English, Chinese, Spanish, and French.

World Health Organization (WHO)
http://www.who.int

Maintained by WHO's headquarters in Geneva, Switzerland, it is possible to uses Excite search engine to conduct keyword searches from here.

World Trade Organization
http://www.wto.org

The Web site's topics include legal frameworks, trade and environmental policies, recent agreements, etc. Available in English, Spanish, and French.

WWW Virtual Library Database
http://conbio.net/vl/database/

Easy search for country-specific sites that provide news, government, and other information is possible from this site.

United Nations
http://www.un.org

Offical site of the United Nations with reports on international programs in Asia financed by the UN. Available in English, Arabic, Chinese, French, Russian, and Spanish.

JAPAN

Japan Ministry of Foreign Affairs
http://www.mofa.go.jp

"What's New" lists events, policy statements, press releases on this Web site. The Foreign Policy section has speeches, archive, and information under Countries and Region, Friendship. Available in English and Japanese.

Japan Policy Research Institute (JPRI)
http://www.jpri.org

Headings on this site include "What's New" and Publications before 1996.

The Japan Times Online
http://www.japantimes.co.jp

This daily online newspaper is offered in English and contains late-breaking news.

THE PACIFIC RIM

Asia Gateway
http://www.asiagateway.com

Access country profiles, including lifestyles, business, and other data from this site. Look in "What's New" for news highlights.

Asia-Yahoo
http://www.yahoo.com/Regional/Regions/Asia/

This specialized Yahoo search site permits keyword search on Asian events, countries, or topics.

Inside China Today
http://www.insidechina.com

The European Information Network is organized under Headline News, Government, and Related Sites, Mainland China, Hong Kong, Macau, and Taiwan. Requires membership to access data.

Internet Guide for China Studies
http://sun.sino.uni-heidelberg.de/igcs/index.html

Coverage of news media, politics, legal and human rights information, as well as China's economy, philosophy and religion, society, arts, culture, and history may be found here. Available in German and English.

NewsDirectory.com
http://www.newsd.com

This site, a Guide to English-Language Media Online, lists over 7,000 actively updated papers and magazines.

Signposts to Asia and the Pacific
http://www.signposts.uts.edu.au

This Australian site contains databases, news, key country contacts, articles, and links to other Pacific Rim sites.

South-East Asia Information
http://sunsite.nus.edu.sg/asiasvc.html

A gateway for country-specific research is presented here. Information on Internet Providers and Universities in Southeast Asia is available as well as links to Asian online services.

See individual country report pages for additional Web sites.

The United States (United States of America)

GEOGRAPHY

Area in Square Miles (Kilometers): 3,717,792 (9,629,091) (about 1/2 the size of Russia)

Capital (Population): Washington, DC (3,997,000)

Environmental Concerns: air and water pollution; limited freshwater resources, desertification; loss of habitat; waste disposal; acid rain

Geographical Features: vast central plain, mountains in the west, hills and low mountains in the east; rugged mountains and broad river valleys in Alaska; volcanic topography in Hawaii

Climate: mostly temperate, but ranging from tropical to arctic

PEOPLE

Population

Total: 280,563,000

Annual Growth Rate: 0.89%

Rural/Urban Population Ratio: 24/76

Major Languages: predominantly English; a sizable Spanish-speaking minority; many others

Ethnic Makeup: 77% white; 13% black; 4% Asian; 6% Amerindian and others

Religions: 56% Protestant; 28% Roman Catholic; 2% Jewish; 4% others; 10% none or unaffiliated

Health

Life Expectancy at Birth: 74 years (male); 80 years (female)

Infant Mortality: 6.69/1,000 live births

Physicians Available: 1/365 people

HIV/AIDS Rate in Adults: 0.61%

Education

Adult Literacy Rate: 97% (official)

Compulsory (Ages): 7–16; free

COMMUNICATION

Telephones: 194,000,000 main lines

Daily Newspaper Circulation: 238/1,000 people

Televisions: 776/1,000 people

Internet Users: 165,750,000 (2002)

TRANSPORTATION

Highways in Miles (Kilometers): 3,906,960 (6,261,154)

Railroads in Miles (Kilometers): 149,161 (240,000)

Usable Airfields: 14,695

Motor Vehicles in Use: 206,000,000

GOVERNMENT

Type: federal republic

Independence Date: July 4, 1776

Head of State/Government: President George W. Bush is both head of state and head of government

Political Parties: Democratic Party; Republican Party; others of relatively minor political significance

Suffrage: universal at 18

MILITARY

Military Expenditures (% of GDP): 3.2%

Current Disputes: various boundary and territorial disputes; "war on terrorism"

ECONOMY

Per Capita Income/GDP: $36,300/$10.082 trillion

GDP Growth Rate: 0%

Inflation Rate: 3%

Unemployment Rate: 5.8%

Population Below Poverty Line: 13%

Natural Resources: many minerals and metals; petroleum; natural gas; timber; arable land

Agriculture: food grains; feed crops; fruits and vegetables; oil-bearing crops; livestock; dairy products

Industry: diversified in both capital and consumer-goods industries

Exports: $723 billion (primary partners Canada, Mexico, Japan)

Imports: $1.148 trillion (primary partners Canada, Mexico, Japan)

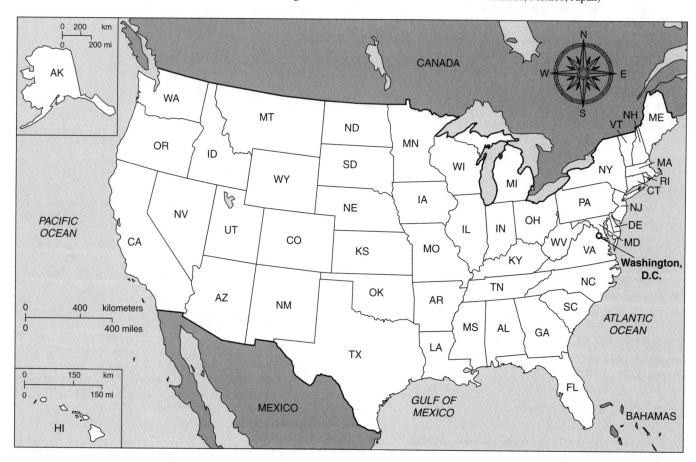

Canada

GEOGRAPHY

Area in Square Miles (Kilometers):
3,850,790 (9,976,140) (slightly larger than the United States)
Capital (Population): Ottawa (1,094,000)
Environmental Concerns: air and water pollution; acid rain; industrial damage to agriculture and forest productivity
Geographical Features: permafrost in the north; mountains in the west; central plains; lowlands in the southeast
Climate: varies from temperate to arctic

PEOPLE

Population

Total: 31,903,000
Annual Growth Rate: 0.96%
Rural/Urban Population Ratio: 23/77
Major Languages: both English and French are official
Ethnic Makeup: 28% British Isles origin; 23% French origin; 15% other European; 6% others; 2% indigenous; 26% mixed
Religions: 46% Roman Catholic; 36% Protestant; 18% others

Health

Life Expectancy at Birth: 76 years (male); 83 years (female)
Infant Mortality: 4.95/1,000 live births
Physicians Available: 1/534 people

HIV/AIDS Rate in Adults: 0.3%

Education

Adult Literacy Rate: 97%
Compulsory (Ages): primary school

COMMUNICATION

Telephones: 20,803,000 main lines
Daily Newspaper Circulation: 215/1,000 people
Televisions: 647/1,000 people
Internet Users: 16,840,000 (2002)

TRANSPORTATION

Highways in Miles (Kilometers): 559,240 (902,000)
Railroads in Miles (Kilometers): 22,320 (36,000)
Usable Airfields: 1,419
Motor Vehicles in Use: 16,800,000

GOVERNMENT

Type: confederation with parliamentary democracy
Independence Date: July 1, 1867
Head of State/Government: Queen Elizabeth II; Prime Minister Jean Chrétien
Political Parties: Progressive Conservative Party; Liberal Party; New Democratic Party; Bloc Québécois; Canadian Alliance
Suffrage: universal at 18

MILITARY

Military Expenditures (% of GDP): 1.1%
Current Disputes: maritime boundary disputes with the United States

ECONOMY

Currency ($U.S. equivalent): 1.39 Canadian dollars = $1
Per Capita Income/GDP: $27,700/$875 billion
GDP Growth Rate: 2%
Inflation Rate: 3%
Unemployment Rate: 7%
Labor Force by Occupation: 74% services; 15% manufacturing; 6% agriculture and others
Natural Resources: petroleum; natural gas; fish; minerals; cement; forestry products; wildlife; hydropower
Agriculture: grains; livestock; dairy products; potatoes; hogs; poultry and eggs; tobacco; fruits and vegetables
Industry: oil production and refining; natural-gas development; fish products; wood and paper products; chemicals; transportation equipment
Exports: $273.8 billion (primary partners United States, Japan, United Kingdom)
Imports: $238.3 billion (primary partners United States, European Union, Japan)

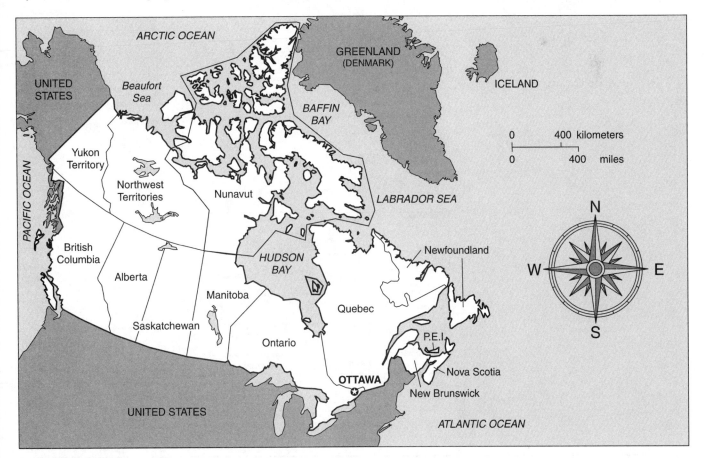

GLOBAL STUDIES

This map is provided to give you a graphic picture of where the countries of the world are located, the relationship they have with their region and neighbors, and their positions relative to major trade and power blocs. We have focused on certain areas to illustrate these crowded regions more clearly. The Japan and Pacific Rim region is shaded for emphasis.

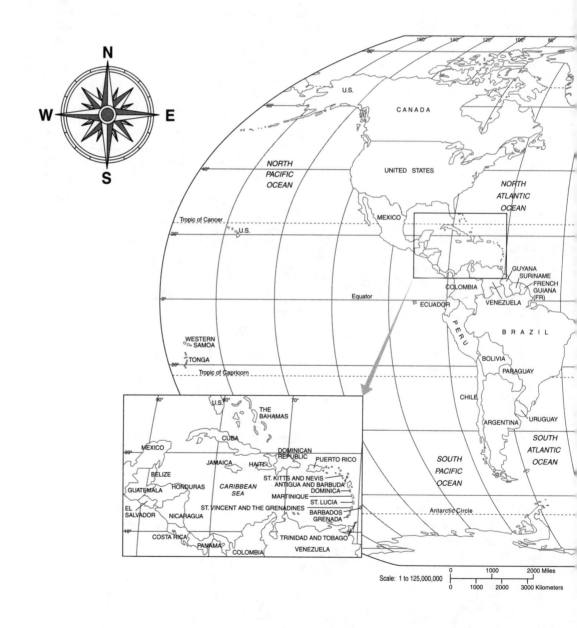

Pacific Rim

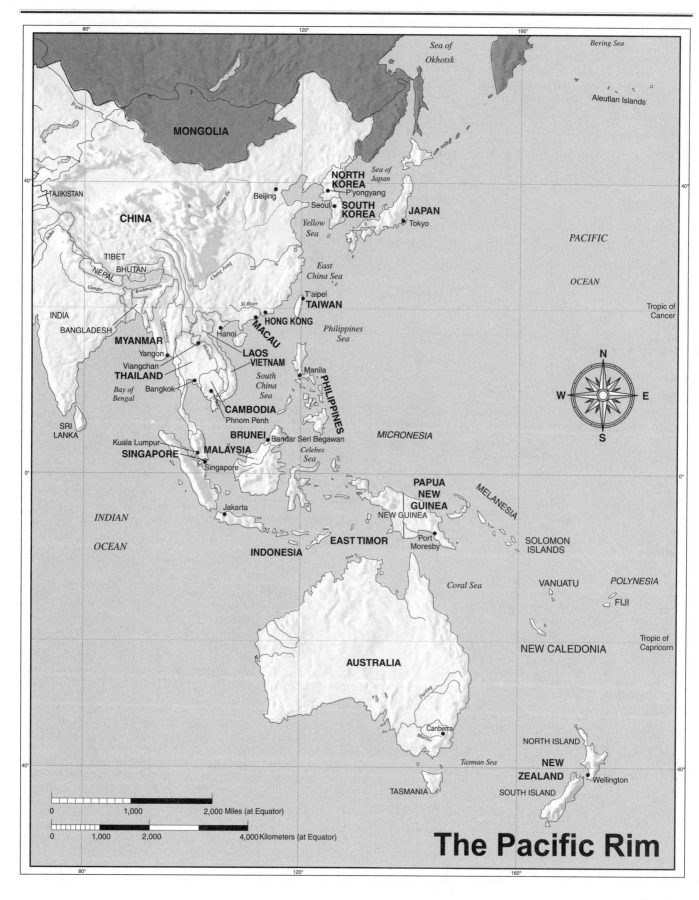

The Pacific Rim

The Pacific Rim: Diversity and Interconnection

WHAT IS THE PACIFIC RIM?

The term *Pacific Rim,* as used in this book, refers to 21 countries or administrative units along or near the Asian side of the Pacific Ocean, plus the numerous islands of the Pacific. Together, they are home to approximately 30 percent of the world's population and produce about 20 percent of the world's gross national product (GNP).

It is not a simple matter to decide which countries to include in a definition of the Pacific Rim. For instance, if we were thinking geographically, we might include Mexico, Chile, Canada, the United States, Russia, and numerous other countries that border the Pacific Ocean, while eliminating Myanmar (Burma) and Laos, since they are not technically on the rim of the Pacific. But our definition, and hence our selected inclusions, stem from fairly recent developments in economic and geopolitical power that have affected the countries of Asia and the Pacific in such a way that these formerly disparate regions are now being referred to by international corporate and political leaders as a single bloc.

Most people living in the region that we have thus defined do not think of themselves as "Pacific Rimmers." In addition, many social scientists, particularly cultural anthropologists and comparative sociologists, would prefer not to apply a single term to such a culturally, politically, and sociologically diverse region. Some countries, it is true, share certain cultural influences, such as the Confucian family and work ethic in China, Japan, and Korea, and Theravada Buddhism and rice cultivation in Southeast Asia. But these commonalities have not prevented the region from fracturing into dozens of societies, often very antagonistic toward one another.

However, for more than 20 years, certain powerful forces have been operating in the Rim that have had the effect of pulling all of the countries toward a common philosophy. Indeed, it appears to be the case that most Pacific Rim countries are attempting to implement some version of free-market capitalism (as opposed to communism, socialism, asceticism, and other practices) and are rapidly acquiring the related values of materialism and consumerism. For most, although not all, of the Pacific Rim countries, a common awareness appears to be developing of the value of peaceful interdependence, rather than military aggression, for the improvement and maintenance of a high standard of living.

What are the powerful forces that are fueling lifestyle convergence in the Rim? There are many, including nationalism and rapid advances in global communications. Japanese investment money is another; it has perhaps been the most important factor in jump-starting the formerly sleepy economies of many Pacific Rim nations. The Japanese yen, in the form of direct investment, loans, and development aid, has saturated the region for years. Using the infusion of capital, many countries have now copied Japan's export-oriented market strategy and are producing economic growth never before even dreamed of. Similarly, with massive investments from Europe, Japan, and the United States, the Chinese economy has become a power-house, and Chinese companies, with the support of the government, are now looking for other places in the Pacific Rim to invest. Today, a complex web of intra-regional investment by many countries is sustaining dramatic growth, even in the face of some recent cutbacks by the Japanese.

It is true that by the mid-1990s the Japanese government, worried about the loss of jobs at home, began to urge private businesses to pull back on investment in other countries in the region. The Japanese government itself cut overseas development assistance (ODA) by 10 percent in 2002; but even as recently as 1999, Japan's ODA was $11 billion, as compared to only $8.4 billion for the United States. More cuts in overseas investments may be forthcoming (despite pleas from world organizations for Japan not to abandon the region) as Japan continues to struggle with a decade-long recession caused by sloppy lending and bad loan management by Japan's banks. But dire predictions of Japanese economic weakness notwithstanding, Japan is not likely to withdraw from active involvement in the economies of the region. To do so would only exacerbate its problems. Japan continues to be an economic powerhouse; some companies have never made more money than they have during the current recession. Toyota, for example, now has a larger share of the U.S. auto market than ever. Economic aid to Pacific Rim countries, such as the US$1.56 billion aid package granted to Indonesia in 2000 and the US$1.5 billion program for Mekong River projects in Thailand, Vietnam, Cambodia, Laos, and Myanmar, will continue. Thus, the homogenizing economic lifestyle trends will also likely continue.

In the 1960s, when the Japanese economy had completely recovered from the devastation of World War II, the Japanese looked to North America and Europe for markets for their increasingly high-quality products. Japanese business continues to seek out markets and resources globally; but, in the 1980s, in response to the movement toward a truly common European economic community as well as in response to free trade agreements among North American countries, Japan began to invest more heavily in countries nearer its own borders. The Japanese hoped to guarantee themselves market and resource access should they find their products frozen out of the emerging European and North American economic blocs. Now China, Taiwan, Singapore, and others are doing the same thing. The unintended, but not unwelcome, consequences of this policy were the revitalization of many Asia–Pacific economies and the solidification of lines of communication between governments and private citizens within the region. Recognizing this interconnection has prompted many people to refer to the countries we treat in this book as a single unit, the Pacific Rim.

TROUBLES IN THE RIM

Twenty years ago, television images of billionaire Japanese businesspeople buying up priceless artworks at auction houses, and filthy-rich Hong Kong Chinese driving around in Rolls-Royces overshadowed the reality of the region: a place, where, for the most part, life is hard. For most of recorded history, most Pacific

(Photo by Lisa Clyde Nielsen)

The small island of Macau was acknowledged by China as a Portuguese settlement in 1557. The Portuguese influence is evident in the architecture of the downtown plaza pictured above. Today Macau is a gambling mecca, drawing an enormous number of avid fans from Hong Kong. This last outpost of European colonial power returned to Chinese control on December 20, 1999, after 442 years under Portugal.

Rim countries have not met the needs of their peoples. Whether it is the desire of rice farmers in Myanmar (formerly called Burma) for the right to sell their grain for personal profit, or of Chinese students to speak their minds without repression—in these and many other ways, the Pacific Rim has failed its peoples. In Vietnam, Myanmar, Laos, and Cambodia, for example, life is so difficult that thousands of families have risked their lives to leave their homelands. Some have swum across the wide Mekong River on moonless nights to avoid detection by guards, while others have sailed into the South China Sea on creaky and overcrowded boats (hence the name commonly given such refugees: "boat people"), hoping that people of goodwill, rather than marauding pirates, will find them and transport them to a land of safety. That does not always happen; in 2004, for example, Indonesia reported some 70 pirate attacks in its waters, and Malaysia reported 30. With the cut-off of refugee-support funds from the United Nations (UN), thousands of refugees remain unrepatriated, languishing in camps in Thailand, Malaysia, and other countries. Thousands of villagers driven from their homes by the

Myanmar Army await return. Meanwhile, the number of defectors from North Korea has been increasing steadily.

Between 1975 and 1994, almost 14,000 refugees reached Japan by boat, along with 3,500 Chinese nationals who posed as refugees in hopes of being allowed to live outside China. In 1998, the Malaysian government, citing its own economic problems, added to the dislocation of many people when it began large-scale deportations of foreign workers, many from Indonesia. Many of these individuals had lived in Malaysia for years. This "Operation Get Out" was expected to affect at least 850,000 people. These examples, and many others not mentioned here, stand as tragic evidence of the social and political instability of many Pacific Rim nations and of the intense ethnic rivalries that divide the people of the Rim.

Warfare

Of all the Rim's troubles, warfare has been the most devastating. In Japan and China alone, an estimated 15.6 million people died as a result of World War II. Not only have there been wars in which foreign powers like Britain, the United States, France, and the former Soviet Union have been involved, but there have been and continue to be numerous battles between local peoples of different tribes, races, and religions.

The potential for serious conflict remains in most regions of the Pacific Rim. Despite international pressure, the military dictators of Myanmar continue to wage war against the Karens and other ethnic groups within its borders. Japan remains locked in a dispute with Russia over ownership of islands to the north of Hokkaido. Taiwan and China still lay claim to each other's territory, as do the two Koreas; and both Taiwan and Japan lay claim to the Senkaku Island chain. The list of disputed borders, lands, islands, and waters in the Pacific Rim is very long; indeed, there are some 30 unresolved disputes involving almost every country of Asia and some of the Pacific Islands.

One example is the Spratly (sometimes spelled Spratley) Islands dispute. When the likelihood of large oil and cobalt deposits in the 340,000-square-mile ocean near the Spratlys was announced in the 1970s, China, Taiwan, Vietnam, the Philippines, Malaysia, and Brunei instantly laid claim to the area. By 1974, the Chinese Air Force and Navy were bombing a South Vietnamese settlement on the islands; by 1988, Chinese warships were attacking Vietnamese transport ships in the area. And by 1999, the Philippine Navy was sinking Chinese fishing boats off Mischief Reef. Both China and Vietnam have granted nearby oil-drilling concessions to different U.S. oil companies, so the situation remains tense, especially because China claims sovereignty, or at least hegemony, over almost the entire South China Sea and has been flexing its muscles in the area by stopping, boarding, and sometimes confiscating other nations' ships in the area. Some agreements among the claimants on protocol were reached in 2001, but as long as each—especially China—claims ownership, the dispute will continue to simmer.

In addition to these cross-border disputes, internal ethnic tensions are sometimes severe. Most Asian nations are composed of scores of different ethnic groups with their own languages, religions, and worldviews. In Fiji, it is the locals versus the immigrant Indians; in Southeast Asia, it is the locals versus the Chinese or the Muslims versus the Christians; in China, it is the Tibetans and most other ethnic groups versus the Han Chinese.

(UN photo by Shaw McCutcheon)

The number of elderly people in China will triple by the year 2025. Even though—and, ironically, because—it limits most couples to only one child, China will be faced with the increasing need of caring for retirement-age citizens. This group of elders in a village near Chengdu represents just the tip of an enormous problem for the future.

With the end of the Cold War in the late 1980s and early 1990s, many Asian nations found it necessary to seek new military, political, and economic alliances. For example, South Korea made a trade pact with Russia, a nation that, during the Soviet era, would have dealt only with North Korea; and, forced to withdraw from its large naval base in the Philippines, the United States increased its military presence in Singapore. The United States also began encouraging its ally Japan to assume a larger military role in the region. However, the thought of Japan re-arming itself causes considerable fear among Pacific Rim nations, almost all of which suffered defeat at the hands of the Japanese military only six decades ago. Nevertheless, Japan has acted to increase its military preparedness, within the narrow confines of its constitutional prohibition against re-armament. It now has the second-largest military budget in the world (it spends only 1 percent of its budget on defense, but because its economy is so large, the actual absolute expenditure is huge).

In response, China has increased its purchases of military equipment, especially from cash-hungry Russia. As a result, whereas the arms industry is in decline in some other world regions, it is big business in Asia. Four of the nine largest armies in the world are in the Pacific Rim. Thus, the tragedy of warfare, which has characterized the region for so many centuries, could continue unless governments manage conflict very carefully and come to understand the need for mutual cooperation.

In some cases, mutual cooperation is already replacing animosity. Thailand and Vietnam are engaged in sincere efforts to resolve fishing-rights disputes in the Gulf of Thailand and water-rights disputes on the Mekong River; North and South Korea have agreed to allow some cross-border visitation and are cooperating on a mammoth industrial park just inside North Korea; and even Taiwan and China have amicably settled issues relating to fisheries, immigration, and hijackings. Yet greed and ethnic and national pride are far too often just below the surface; left unchecked, they could catalyze a major confrontation in the region.

Population Imbalances

Another serious problem in the Pacific Rim is population imbalance. In some cases, there are too many people; in others, there will be too few in the future. At the moment, there are well over 2 billion people living in the region. Of those, approximately 1.3 billion are Chinese. Even though China's government has implemented the strictest family-planning policies in world history, the country's annual growth rate is such that more than 1 million inhabitants are added *every month*. This means that more new Chinese are born each year than make up the entire population of Australia! China's efforts to reduce its out-of-control population have produced some unwanted results: a huge imbalance in the ratio of boys to girls. With births limited to just one for

3

most families, and with the traditional Chinese preference for boys, many couples resort to selective abortions. As a result, there are now 120 boys born for every 100 girls, and as the boys reach marriageable age, they will find there are simply not enough females to go around; they will be, as the Chinese call them, "bare branches" with no off-spring.

Despite this extreme case, most Pacific Rim countries continue to promote family planning, and it is working. The World Health Organization (WHO) reports that about 217 million people in East Asia use contraceptives, as compared to only 18 million in 1965. Couples in some countries, including Japan, Taiwan, and South Korea, have been voluntarily limiting family size. Other states, such as China and Singapore, have promoted family planning through government incentives and punishments. The effort is paying off. The United Nations now estimates that the proportion of the global population living in Asia will remain relatively unchanged between now and the year 2025, and China's share will decline. A drop in birth rates is characteristic of almost the entire region: Birth rates started to drop in Japan and Singapore in the 1950s; Hong Kong, South Korea, the Philippines, Brunei, Taiwan, Malaysia, Thailand, and China in the 1960s; and Indonesia and Myanmar in the 1970s. In fact, in some countries, especially Japan, South Korea, and Thailand, single-child families and an aging population are creating problems in their own right as the ratio of productive workers to the overall population declines. The birth rate in Tokyo, for example, is 1.01 children per couple—far below replacement level. At this rate, Japan's population will be half its current size by the year 2100. Some experts are calling the situation a "demographic collapse." The low births are caused, in part, by the number of Japanese who do not marry—approximately 14 percent of all males over age 40 and 6 percent of females over 40 are not married. Many of these unmarrieds continue to live at their parents' home, coining the phrase "parasite singles." It is likely that these downward trends in Japan will soon be commonplace all over the Pacific Rim, some countries excepted.

The *lack* of young workers will be a problem for the next generation, but for now, so many children have already been born that Pacific Rim governments simply cannot meet their needs. For these new Asians, schools must be built, health facilities provided, houses constructed, and jobs created. This is not an easy challenge for many Rim countries. Moreover, as the population density increases, the quality of life decreases. By way of comparison, think of New York City, where the population is about 1,100 per square mile. Residents there, finding the crowding to be too uncomfortable, frequently seek more relaxed lifestyles in the suburbs. Yet in Tokyo, the density is approximately 2,400 per square mile; and in Macau, it is 57,576! Today, many of the world's largest cities are in the Pacific Rim: Shanghai, China, has about 12.9 million people (in the wider metropolitan area); Jakarta, Indonesia, has more than 11.4 million; Manila, the Philippines, is home to over 10 million; while Bangkok, Thailand, has about 7.5 million residents. And migration to the cities continues despite miserable living conditions for many (in some Asian cities, 50 percent of the population live in slum housing). One incredibly rapid-growth country is the Philippines; home to only about 7 million in 1898, when it was acquired by the United States, it is projected to have 130 million people in the year 2020.

TYPES OF GOVERNMENT IN SELECTED PACIFIC RIM COUNTRIES

PARLIAMENTARY DEMOCRACIES
Australia*
Fiji
New Zealand*
Papua New Guinea

CONSTITUTIONAL MONARCHIES
Brunei
Japan
Malaysia
Thailand

REPUBLICS
Indonesia
The Philippines
Singapore
South Korea
Taiwan

SOCIALIST REPUBLICS
China
Laos
Myanmar (Burma)
North Korea
Vietnam

OVERSEAS TERRITORIES/COLONIES
French Polynesia
New Caledonia

Australia and New Zealand have declared their intention to become completely independent republics.

Absolute numbers alone do not tell the whole story. In many Rim countries, 40 percent or more of the population are under age 15. Governments must provide schooling and medical care as well as plan for future jobs and housing for all these children. Moreover, as these young people age, they will require increased medical and social care. Scholars point out that, between 1985 and 2025, the numbers of old people will double in Japan, triple in China, and quadruple in Korea. In Japan, where replacement-level fertility was achieved in the 1960s, government officials are already concerned about the ability of the nation to care for the growing number of retirement-age people while paying the higher wages that the increasingly scarce younger workers are demanding.

Political Instability

One consequence of the overwhelming problems of population imbalances, urbanization, and continual military or ethnic conflict is disillusionment with government. In many countries of the Pacific Rim, people are challenging the very right of their governments to rule or are demanding a complete change in the political philosophy that undergirds government.

For instance, despite the risk of death, torture, or imprisonment, many college students in Myanmar have demonstrated against the current military dictatorship, and rioting students and workers in Indonesia were successful in bringing down the corrupt government of President Suharto. In some Rim countries, opposition groups armed with sophisticated weapons obtained from foreign nations roam the countryside, capturing towns and military installations. In less than a decade, the government of the Philippines endured six coup attempts; elite military dissidents have wanted to impose a patronage-style government like that of former president Ferdinand Marcos, while armed rural insurgents have wanted to install a Communist government. Thousands of students have been injured or killed protesting the governments of South Korea and China. Thailand has been beset by numerous military coups, the former British colony of Fiji recently endured two coups, and half a million residents of Hong Kong took to the streets to oppose Great Britain's decision to turn over the territory to China in 1997. Military takeovers, political assassinations, and repressive policies have been the norm in most of the countries in the region. Millions have spent their entire lives under governments they have never agreed with, and unrest is bound to continue, because people are showing less and less patience with imposed government.

Identity Confusion

A related problem is that of confusion about personal and national identity. Many nation-states in the Pacific Rim were created in response to Western pressure. Before Western influences came to be felt, many Asians, particularly Southeast Asians, did not identify themselves with a nation but, rather, with a tribe or an ethnic group. National unity has been difficult in many cases, because of the archipelagic nature of some countries or because political boundaries have changed over the years, leaving ethnic groups from adjacent countries inside the neighbor's territory. The impact of colonialism has left many people, especially those in places like Singapore, Hong Kong, and the Pacific islands, unsure as to their roots; are they European or Asian/Pacific, or something else entirely?

Indonesia illustrates this problem. People think of it as an Islamic country, as overall, its people are 87 percent Muslim. But in regions like North Sumatra, 30 percent are Protestant; in Bali, 94 percent are Hindu. In the former Indonesian territory and now independent country of East Timor, 49 percent are Catholic, while 51 percent are animist. Such stark differences in religious worldview, combined with the tendency of people of similar faith to live near each other and separate themselves geographically from others, make it difficult for a clear national identity to form. It is no surprise that Indonesia is currently grappling with several separationist movements. The Philippines is another example. With 88 different languages spoken, its people spread out over 12 large islands, and a population explosion (the average age is just 16), it is a classic case of psychological (and economic and political) fragmentation. Coups and countercoups rather than peaceful political transitions seem to be the norm, as people have not yet developed a sense of unified nationalism.

Uneven Economic Development

While millionaires in Singapore, Hong Kong, and Japan wrestle with how best to invest their wealth, others worry about how they will obtain their next meal. Such disparity illustrates another major problem afflicting the Pacific Rim: uneven economic development. The United Nations' Food and Agricultural Organization (FAO) estimates that 20 percent of the people in the Pacific Rim (the data exclude China) are undernourished. And while some people in, say, Tokyo and Singapore have a lifestyle similar to people in Europe and North America, others, for instance in Papua New Guinea, have a lifestyle not that different from that believed to have existed in the Stone Age. (It is also true that Asia overall consumes far less energy per person—0.99 ton of oil—than do Canada and the United States—8.67 and 8.77 tons, respectively. If energy consumption is equated to quality of life, then the gap between Asia and the developed world must stand as painful evidence of general worldwide equality.)

Many Asians, especially those in the Northeast Asian countries of Japan, Korea, and China, are finding that rapid economic change seems to render the traditions of the past meaningless. Moreover, economic success has produced a growing Japanese interest in maximizing investment returns, with the result that Japan (and, increasingly, South Korea, Taiwan, Singapore, and Hong Kong) is successfully searching out more ways to make money, while resource-poor regions like the Pacific islands lag behind. Indeed, with China receiving "normal trade relations" status from the United States in late 2000 and acceding to membership in the World Trade Organization (WTO) in 2001, many smaller countries, such as the Philippines, worry that business will move away from them and toward China, where labor costs are considerably lower. Indeed, the so-called "China Price," or the cost to make and sell goods made by low-wage Chinese laborers, is putting downward pressure on customary revenue streams in all Rim countries (and in Europe and North America as well).

The *developed nations* are characterized by political stability and long-term industrial success. Their per capita incomes are comparable to those of Canada, Northern Europe, and the United States, and they have achieved a level of economic sustainability. These countries are closely linked to North America economically. Japan, for instance, exports one third of its products to the United States.

The *newly industrializing countries* (NICs) are currently capturing world attention because of their rapid growth. Hong Kong, for example, has exported more manufactured products per year for the past decade than did the former Soviet Union and Central/Eastern Europe combined. Taiwan, famous for cameras and calculators, has had the highest average economic growth in the world for the past 20 years. South Korea is tops in shipbuilding and steel manufacturing and is the tenth-largest trading nation in the world.

The *resource-rich developing nations* have tremendous natural resources but have been held back economically by political and cultural instability and by insufficient capital to develop a sound economy. Malaysia is an example of a country attempting to overcome these drawbacks. Ruled by a coalition government representing nearly a dozen ethnic-based parties, Malaysia is richly endowed with tropical forests and large oil and natural-gas reserves. Developing these resources has taken years (the oil and gas fields began production as recently as 1978) and has required massive infusions of investment monies from Japan and other countries. By the mid-1990s, more than 3,000 foreign companies were doing business in Malaysia, and the country was moving into the ranks of the world's large exporters.

Command economies lag far behind the rest, not only because of the endemic inefficiency of the system but because military dictatorships and continual warfare have sapped the strength of the people, as in Laos, for example. Yet significant changes in some of these countries are now emerging. China and Vietnam, in particular, are eager to modernize their economies and institute market-based reforms. It is estimated that about half of China's economy is now privatized, although substantial government regulation remains throughout the system. Historically having directed its trade to North America and Europe, Japan is now finding its Asian/Pacific neighbors—especially the socialist-turning-capitalist ones—to be convenient recipients of its powerful economic and cultural influence.

Many of the *less developed countries* (LDCs) are the small micro-states of the Pacific with limited resources and tiny internal markets. Others, like Papua New Guinea, have only recently achieved political independence and are searching for a comfortable role in the world economy.

Environmental Destruction

Environmental destruction in the Pacific Rim is a problem of mammoth proportions. For more than 20 years, the former Soviet Union dumped nuclear waste into the Sea of Japan; China's use of coal in industrial development has produced acid rain over Korea and Japan; deforestation in Thailand, Myanmar, and other parts of Southeast Asia and China has destroyed many thousands of acres of watershed and wildlife habitat. On the Malaysian island of Sarawak, for example, loggers work through the night, using floodlights, to cut timber to satisfy the demands of international customers, especially Japanese. The forests there are disappearing at a rate of 3 percent a year. Highway and dam construction in many countries in Asia has seriously altered the natural environment. But environmental damage is perhaps most noticeable in the cities. Mercury pollution in Jakarta Bay has led to brain disorders among children in Indonesia's capital city; air pollution in Manila and Beijing ranks among the world's worst, while not far behind are Bangkok and Seoul; water pollution in Hong Kong has forced the closure of many beaches. The World Resources Institute of the U.S. Environmental Protection Agency (EPA) reports that greenhouse-gas emissions (annual emissions in billions of tons) are 5.7 for the United States, 6.2 for Canada, but in excess of 8 for Asia (of which half is from China alone).

An environmentalist's nightmare came true in 1997 and 1998 in Asia, when thousands of acres of timber went up in a cloud of smoke. Fueled by the worst El Niño–produced drought in 30 years and started by farmers seeking an easy way to clear land for farming, wildfires in Malaysia, Indonesia, and Brunei covered much of Southeast Asia in a thick blanket of smoke for months. Singapore reported its worst pollution-index record ever. All countries in the region complained that Indonesia, in particular, was not doing enough to put out the fires. Foreign-embassy personnel, citing serious health risks, left the region

until rains—or the lack of anything more to burn—extinguished the flames. Airports had to close, hundreds of people complaining of respiratory problems sought help at hospitals, and many pedestrians and even those inside buildings donned face masks. With valuable timber becoming more scarce all the time, many people around the world reacted with anger at the callous disregard for the Earth's natural resources.

While conservationists are raising the alarm about the region's polluted air and declining green spaces, medical professionals are expressing dismay at the speed with which serious diseases such as AIDS are spreading in Asia. The Thai government, despite impressive gains in the fight against HIV/AIDS, now believes that more than 2.4 million Thais are HIV-positive. World Health Organization data show that the AIDS epidemic is growing faster in Asia and Africa than anywhere else in the world.

In late 2002, a new and deadly coronavirus emerged among the population of Guangdong province in southern China. Called SARS (severe acute respiratory syndrome), the flu-like illness, for which there was no cure, quickly spread to Hong Kong, Singapore, Vietnam, Taiwan, Canada, and other countries, infecting thousands and killing hundreds. Not only did SARS create a serious health threat, it also damaged the reputation of China's leaders, who did not act quickly to handle the crisis and then tried to downplay the severity of the problem. Initially, China even refused to allow World Health Organization officials to visit Taiwan, claiming that Taiwan was not an independent country and could not invite such organizations inside its borders without China's permission. Although China relented on this point, the damage to its reputation had already been done. What is not yet clear is what damage the SARS outbreak will have on the Pacific Rim's economy. Various health officials issued advisories against travel to infected regions, and the tourism sector of the region had already suffered losses in excess of $11 billion by early 2003. Hotels and restaurants were reducing their staff, and airlines, already impacted by the fallout of the September 11, 2001 terrorist attacks on the New York City World Trade Center, were struggling with yet more layoffs and more red ink.

The region was just starting to recover from the SARS problem when another virus emerged: the bird flu. Affecting ducks, chickens, and even some non-fowl animals, the disease can only be eradicated by "culling" or killing all flocks where an infected bird is found. Millions of birds have been thus killed in Vietnam, Thailand, Malaysia, Laos, Indonesia, and other countries, with the resulting impact on people's incomes. Apparently, humans can also contract the virus, and with some 40 people felled by the disease in the Pacific Rim in 2004, some epidemiologists were predicting a world pandemic in which the bird flu would rival or exceed the human deaths attributed to the worldwide Spanish Flu epidemic in 1918. One beneficial outcome of this crisis has been increased communication and cooperation among Southeast Asian nations.

Natural Disasters

In December 2004, the world watched in horror as a mammoth tsunami or tidal wave generated by a 9.0 earthquake off the coast of Indonesia devastated the shores of 11 countries and killed more than 150,000. Billions of dollars of aid was sent to Indonesia, Malaysia, Myanmar, Thailand and other affected countries to help the survivors. While this event was truly cataclysmic, it must be remembered that similar events have always been part of the Pacific Rim. The chart below shows some other natural disasters in the region:

SELECTED NATURAL DISASTERS IN THE PACIFIC RIM 1920–2004

Year	Place	Event	Deaths
1920	Gansu, China	8.6 earthquake	200,000
1923	Kanto, Japan	7.9 earthquake	143,000
1932	Gansu, China	7.6 earthquake	70,000
1927	Tsinghai, China	7.9 earthquake	200,000
1976	Tangshan, China	7.5 earthquake	255,000
2004	South Asia	9.0 earthquake	150,000

Source: U.S. Geological Survey

In the year 2004 alone, the following natural disasters (a selected sample only) hit the region, causing loss of life and severe economic hardship for both individuals and governments: a tropical storm followed by a typhoon in the Philippines killed 566 people and left hundreds more missing or homeless; flooding in New Zealand killed 2 and caused US$200 million in damage; landslides in the southern Philippines left 13 dead and 150 missing; a 6.4 magnitude earthquake killed 22 people and caused dozens of buildings to collapse in southern Indonesia; the Mount Awu volcano in Indonesia erupted and forced thousands from their homes; six people were killed by flash floods in Vietnam; a cyclone with 105 mph winds hit Myanmar killing 140 people and leaving 18,000 people homeless. It might be expected that such disasters would hit a region known to be in one of the world's most seismically active zones, the "ring of fire," but expected or not, such shocks to the affected peoples and countries are extremely destabilizing and continually impede economic progress.

GUARDED OPTIMISM

Warfare, population imbalances, political instability, identity confusion, uneven development, and environmental and natural disasters would seem to be an irresolvable set of problems for the peoples of the Pacific Rim, but there is reason for guarded optimism about the future of the region. Unification talks between North and South Korea have finally resulted in real breakthroughs. For instance, in 2000, a railway between the demilitarized zone (DMZ) separating the two antagonists was being reconnected, after years of disuse. President Vladimir Putin of Russia agreed to reopen discussion with Japan over the decades-long Northern Territories dispute. Other important issues are also under discussion all over the region, and the UN peacekeeping effort in Cambodia seems to have paid off—at least there is a legally elected government in place, and most belligerents have put down their arms.

Until the Asian financial and currency crises of 1998–1999, the world media carried glowing reports on the burgeoning economic strength of many Pacific Rim countries. Typical was the *CIA World Factbook 1996–1997,* which reported high growth in gross national product per capita for most Rim countries: South Korea, 9.0 percent; Hong Kong, 5.0 percent; Indonesia, 7.5 percent; Japan (due to recession), 0.3 percent; Malaysia, 9.5 percent; Singapore, 8.9 percent; and Thailand, 8.6 percent. By comparison, the U.S. GNP growth rate was 2.1 percent; Great Britain, 2.7 percent; and Canada, 2.1 percent. Other reports on the Rim compared 1990s investment and savings percentages with those of 20 years earlier; in almost every case, there had been a tremendous improvement in the economic capacity of these countries.

Throughout the 1980s and most of the 1990s, the rate of economic growth in the Pacific Rim was indeed astonishing. In 1987, for example, the rate of real gross domestic product growth in the United States was 3.5 percent over the previous year. By contrast, in Hong Kong, the rate was 13.5 percent; in Taiwan, 12.4 percent; in Thailand, 10.4 percent; and in South Korea, 11.1 percent. In 1992, GDP growth throughout Asia averaged 7 percent, as compared to only 4.8 percent for the rest of the world. But recession in Japan (Japan had suffered zero or even negative growth in 1999 and 2000 and sluggish growth for 10 years before that), near financial collapse in Indonesia, and problems in other Asian countries have slowed the growth rates throughout the region. Still, Singapore and many other Pacific Rim economies are expected to grow faster than European and North American economies, and even politically chaotic Indonesia believes that its economy will stabilize very soon.

Ranking of Pacific Rim Countries by per Capita GDP 2004 (purchasing power parity; US dollar)

Comparisons: United States: $37,800; Canada: $29,800

1. Australia	$29,000	11. Thailand	$7,400
2. Hong Kong	$28,800	12. China	$5,000
3. Japan	$28,200	13. Philippines	$4,600
4. Singapore	$23,700	14. Indonesia	$3,200
5. Taiwan	$23,400	15. Vietnam	$2,500
6. New Zealand	$21,600	16. Papua New Guinea	$2,200
7. Macau	$19,400	17. Myanmar	$1,800
8. Burnei	$18,600	18. Laos	$1,700
9. South Korea	$17,800	19. North Korea	$1,300
10. Malaysia	$9,000	20. East Timor	$500

Source: CIA World Fact Book 2004

The significance of high growth rates, in addition to improvements in the standard of living, is the shift in the source of development capital, from North America to Asia. Historically, the economies of North America were regarded as the engine behind Pacific Rim growth; and yet today, growth in the United

ECONOMIC DEVELOPMENT IN SELECTED PACIFIC RIM COUNTRIES

Economists have divided the Rim into five zones, based on the level of development, as follows:

DEVELOPED NATIONS

Australia

Japan

New Zealand

NEWLY INDUSTRIALIZING COUNTRIES (NICs)

Hong Kong

Singapore

South Korea

Taiwan

RESOURCE-RICH DEVELOPING ECONOMIES

Brunei

Indonesia

Malaysia

The Philippines

Thailand

COMMAND ECONOMIES*

Cambodia

China

Laos

Myanmar (Burma)

North Korea

Vietnam

LESS DEVELOPED COUNTRIES (LDCs)

Papua New Guinea

Pacific Islands

*China, Vietnam, and, to a much lesser degree, North Korea are moving toward free-market economies.

talist enclaves of the People's Republic of China, Hong Kong, and Taiwan. Copying Japanese strategy and aided by a common written language and culture, this region has the potential of exceeding even the mammoth U.S. economy in the future. For now, however, and despite recent sluggish growth, Japan will remain the major player in the region.

Japan has been investing in the Asia/Pacific region for several decades. However, growing protectionism in its traditional markets as well as changes in the value of the yen and the need to find cheaper sources of labor (labor costs are 75 percent less in Singapore and 95 percent less in Indonesia) have raised Japan's level of involvement so high as to give it the upper hand in determining the course of development and political stability for the entire region. This heightened level of investment started to gain momentum in the mid-1980s. Between 1984 and 1989, Japan's overseas development assistance to the ASEAN countries amounted to $6.1 billion. In some cases, this assistance translated to more than 4 percent of a nation's annual national budget and nearly 1 percent of GDP. Private Japanese investment in ASEAN countries plus Hong Kong, Taiwan, and South Korea was $8.9 billion between 1987 and 1988. In more recent years, the Japanese government or Japanese business invested $582 million in an auto-assembly plant in Taiwan, $5 billion in an iron and steel complex in China, $2.3 billion in a bullet-train plan for Malaysia, and $530 million in a tunnel under the harbor in Sydney, Australia. Japan is certainly not the only player in Asian development (Japan has "only" about 20 projects under way in Vietnam, for example, as compared to 80 for Hong Kong and 39 for Taiwan), but the volume of Japanese investment is staggering. In Australia alone, nearly 900 Japanese companies are now doing business. Throughout Asia, Japanese is becoming a major language of business (along with Chinese which is also in demand as the Chinese economy grows in influence).

Although Japan works very hard at globalizing its markets and its resource suppliers, it has also developed closer ties with its nearby Rim neighbors. In a recent year, out of 20 Rim countries, 13 listed Japan as their first- or second-most-important trading partner, and several more put Japan third. Japan receives 42 percent of Indonesia's exports and 26 percent of Australia's; in return, 23 percent of South Korea's imports, 29 percent of Taiwan's, 30 percent of Thailand's, 24 percent of Malaysia's, and 23 percent of Indonesia's come from Japan. Pacific Rim countries are clearly becoming more interdependent—but simultaneously more dependent on Japan—for their economic success.

JAPANESE INFLUENCE, PAST AND PRESENT

This is certainly not the first time in modern history that Japanese influence has swept over Asia and the Pacific. A major thrust began in 1895, when Japan, like the European powers, started to acquire bits and pieces of the region. By 1942, the Japanese were in control of Taiwan, Korea, Manchuria and most of the populated parts of China, and Hong Kong; what are now Myanmar, Vietnam, Laos, and Cambodia; Thailand; Malaysia; Indonesia; the Philippines; part of New Guinea; and dozens of Pacific islands. In effect, by the 1940s, the Japanese were the dominant force in precisely the area that they are influencing now and that we are calling the Pacific Rim.

States and Canada trails many of the Rim economies. This anomaly can be explained, in part, by the hard work and savings ethics of Pacific Rim peoples and by their external-market–oriented development strategies. But, without venture capital and foreign aid, hard work and clever strategies would not have produced the rapid economic improvement that Asia has experienced over the past several decades. Japan's contribution to this improvement, through investments, loans, and donations, and often in much larger amounts than other investor nations such as the United States, cannot be overstated.

Some subregions are also emerging. There is, of course, the Association of Southeast Asian Nations (ASEAN) regional trading unit. In 2001, the ASEAN countries and China agreed to create a free-trade area simlar to the North American Free Trade Association (NAFTA) within 10 years. But the grouping that is gaining world attention is the informal region that people are calling "Greater China," consisting of the emerging capi-

(UN photo by Nichiro Gyogyo)

These men work on a Japanese factory ship, a floating cannery that processes salmon harvested from the Pacific.

The similarities do not end there, for, while many Asians of the 1940s were apprehensive about or openly resistant to Japanese rule, many others welcomed the Japanese invaders and even helped them to take over their countries. This was because they believed that Western influence was out of place in Asia and that Asia should be for Asians. They hoped that the Japanese military would rid them of Western rule—and it did: After the war, very few Western powers were able to regain control of their Asian and Pacific colonies.

Today, many Asians and Pacific islanders are concerned about Japanese financial and industrial influence in their countries, but they welcome Japanese investment anyway because they believe that it is the best and cheapest way to rid their countries of poverty and underdevelopment. So far, they are right—by copying the Japanese model of economic development, and thanks to Japanese trade, foreign aid, and investment, the entire region—some countries excepted—has increased its wealth and positioned itself to be a major player in the world economy for the foreseeable future.

It is important to note, however, that many Rim countries, such as China, Taiwan, Hong Kong, and South Korea, are strong challengers to Japan's economic dominance; in addition, Japan has not always felt comfortable about its position as head of the pack, for fear of a backlash. For example, Japan's higher

regional profile has prompted complaints against the Japanese military's World War II treatment of civilians in Korea and China and forced Japan to pledge $1 billion to various Asian countries as a symbolic act of apology.

Why have the Japanese re-created a modern version of the old Greater East Asian Co-Prosperity Sphere of the imperialistic 1940s? We cannot find the answer in the propaganda of wartime Japan, because fierce devotion to the emperor and the nation, and belief in the superiority of Asians over all other races are no longer the propellants in the Japanese economic engine. Rather, Japan courts Asia and the Pacific today to acquire resources to sustain its civilization. Japan is about the size of California, but it has five times as many people and not nearly as much arable land. Much of Japan is mountainous; many other parts are off limits because of active volcanoes (one tenth of all the active volcanoes in the world are in Japan); and, after 2,000-plus years of intensive and uninterrupted habitation, the natural forests have long since been consumed (though they have been replanted with new varieties), as are most of the other natural resources—most of which were scarce to begin with.

In short, Japan continues to extract resources from the rest of Asia and the Pacific because it is the same Japan as before—environmentally speaking, that is. Take oil. In the early 1940s, Japan needed oil to keep its industries (as well as its military machine) operating, but the United States wanted to punish Japan for its military expansion in Asia, so it shut off all shipments to Japan of any kind, including oil. That may have seemed politically right to the policymakers of the day, but it did not change Japan's resource environment; Japan still did not have its own oil, and it still needed as much oil as before. So Japan decided to capture a nearby nation that did have natural reserves of oil; in 1941, it attacked Indonesia and obtained by force the resource it had been denied through trade.

Japan has no more domestic resources now than it did half a century ago, and yet its needs—for food, minerals, lumber, paper—are greater. Except for fish, you name it—Japan does not have it. A realistic comparison is to imagine trying to feed half the population of the United States solely from the natural output of the U.S. state of Montana. As it happens, however, Japan sits next to the continent of Asia, which is rich in almost all the materials it needs. For lumber, there are the forests of Malaysia; for food, there are the farms and ranches of New Zealand and Australia; and for oil, there are Indonesia and Brunei, the latter of which sells about half of its exports to Japan. In sum, the quest for resources to maintain its quality of life is the reason Japan flooded its neighbors with Japanese yen in recent decades and why it will continue to maintain an active engagement with all Pacific Rim countries well into the future.

Catalyst for Development

In addition to the need for resources, Japan has turned to the Pacific Rim in an attempt to offset the anti-Japanese import or protectionist policies of its historic trading partners. Because so many import tariffs are imposed on products sold directly from Japan, Japanese companies find that they can avoid or minimize tariffs if they cooperate on joint ventures in Rim countries and have products shipped from there. The result is that both Japan and its host countries are prospering as never before. Sony Corporation, for example, assembles parts made in both Japan and

Singapore to construct videocassette recorders at its Malaysian factory, for export to North America, Europe, and other Rim countries. Toyota Corporation intends to assemble its automobile transmissions in the Philippines and its steering-wheel gears in Malaysia, and to assemble the final product in whichever country intends to buy its cars.

So helpful has Japanese investment been in spawning indigenous economic powerhouses that many other Rim countries are now reinvesting in the region. In particular, Hong Kong, Singapore, Taiwan, and South Korea are now in a position to seek cheaper labor markets in Indonesia, Malaysia, the Philippines, and Thailand. In recent years, they have invested billions of dollars in the resource- and labor-rich economies of Southeast Asia, increasing living standards and adding to the growing interconnectivity of the region. An example is a Taiwanese company that has built the largest eel-production facility in the world—in Malaysia—and ships its entire product to Korea and Japan.

Eyed as a big consumer as well as a bottomless source of cheap labor is the People's Republic of China. Many Rim countries, such as South Korea, Taiwan, Hong Kong, and Japan, are working hard to increase their trade with China. For over a decade, annual trade between Taiwan and China and between Hong Kong and China has been in the billions of dollars, despite political tensions. Japan was especially eager to resume economic aid to China in 1990 after temporarily withholding aid to China because of the Tiananmen Square massacre in Beijing. For its part, China is establishing free-enterprise zones that will enable it to participate more fully in the regional and world economy. China's new membership in the World Trade Organization will force it to be even more engaged in world trade. Already the Bank of China is the second-largest bank in Hong Kong.

Japan and a handful of other economic powerhouses of the Rim are not the only big players in regional economic development. The United States and Canada are major investors in the Pacific Rim (in computers and automobiles, for example), and Europe maintains its historical linkages with the region (such as oil). But there is no question that Japan has been the main catalyst for development. As a result, Japan itself has become wealthy. The Japanese stock market rivals the New York Stock Exchange, and there is a growing number of Japanese billionaires: Nobutada Saji worth US$6.9 billion; Yasuo Takie worth US$6.2 billion; Fukuzo Iwasaki worth US$5.7 billion; and many, many more. Loans secured by now-deflated land prices have damaged the Japanese banking industry in recent years, but it appears that the banks are recovering and that Japan will continue to play an active role in the economic development of Asia.

Not everyone is pleased with the way Japan has been giving aid and making loans in the region. Money invested by the Japan International Development Organization (JIDO) has usually been closely connected to the commercial interests of Japanese companies. For instance, commercial-loan agreements have often required that the recipient of low-interest loans purchase Japanese products.

Nevertheless, it is clear that many countries would be a lot worse off without Japanese aid. In a recent year, JIDO aid around the world was $10 billion. Japan is the dominant supplier of foreign aid to the Philippines and a major investor; in Thailand, where U.S. aid recently amounted to $20 million,

Japanese aid was close to $100 million. Some of this aid, moreover, gets recycled from one country to another within the Rim. Thailand, for example, receives more aid from Japan than any other country, but in turn, it supplies major amounts of aid to other nearby countries. Thus we can see the growing interconnectivity of the region, a reality now recognized formally by the establishment of the Asia-Pacific Economic Cooperation Group (APEC).

During the militaristic 1940s, Japanese dominance in the region produced antagonism and resistance. However, it also gave subjugated countries new highways, railways, and other infrastructural improvements. Today, while host countries continue to benefit from infrastructural advances, they also receive quality manufactured products. Once again, Northeast Asian, Southeast Asian, and South Pacific peoples have begun to talk about Japanese domination (recently tempered somewhat by the rise of China as a competitor). The difference this time is that few seem upset about it; many countries no longer believe that Japan has military aspirations against them, and they regard Japanese investment as a first step toward becoming economically strong themselves. Many people are eager to learn the Japanese language; in some cities, such as Seoul, Japanese has displaced English as the most valuable business language.

ASIAN FINANCIAL CRISIS

All over Asia, but especially in Thailand, Indonesia, Malaysia, South Korea, and the Philippines, business leaders and government economists found themselves scrambling in 1998 and 1999 to minimize the damage from the worst financial crisis in decades, a crisis that exploded in late 1997 with currency devaluations in Southeast Asia. With their banks and major corporations in deep trouble, governments began shutting down some of their expensive overseas consular offices, canceling costly public-works projects, and enduring abuse heaped on them by suddenly unemployed citizens.

For years, Southeast Asian countries copied the Japanese model: They stressed exports, and they allowed their governments to decide which industries to develop. This economic-development approach worked very well for a while, but governments were not eager to let natural markets guide production. So, even when profits from government-supported industries were down, the governments believed that they should continue to maintain these industries through loans from Japan and other sources. But banks can loan only so much, especially if it is "risky" money, and eventually the banks' creditworthiness was called into question. Money became tighter. Currencies were devalued, making it harder still to pay off loans and forcing many companies and banks into bankruptcy. Stock markets nose-dived. Thousands of workers were laid off, and many of the once-booming Asian economies hit hard times.

Japan's own economic sluggishness meant that it was not able to carry the same weight in solving the crisis as it might have done in the 1980s. One after another, the affected countries requested bailout assistance from the International Monetary Fund (IMF), a pool of money donated by some 183 nations. IMF funding halted the flow of red ink, and many countries are nearly back to normal rates of growth.

The crisis revealed how interconnected are the economies of the region and how much Asia has depended on Japan to stimu-

PACIFIC RIM BILLIONAIRES
in the Top 150 Wealthiest People In the World—2004

Tadashi Yanai	Japan	Kunio Busujima	Japan
C. Chearavanont	Thailand	Robert Kuok	Malaysia
Stanley Ho	Hong Kong	Michael Kadoorie	Hong Kong
Li Ka-shing	Hong Kong	Kun-Hee Lee	South Korea
The Kwok brothers	Hong Kong	Ananda Krishnan	Malaysia
Nobutada Saji	Japan	Yoshiaki Tsutsumi	Japan
Lee Shau Kee	Hong Kong	Terry Gou	Taiwan
Yasuo Takei	Japan	Y.C. Wang	Taiwan
Fukuzo Iwasaki	Japan	Rachman Halim	Indonesia
Eitaro Itoyama	Japan	Akira Mori	Japan
Tsai Wan Lin	Taiwan	Masayoshi Son	Japan
Yoshitaka Fukuda	Japan	Cheng Yu-tung	Hong Kong
Kyosuke Kinoshita	Japan	Tsai Wan Tsai	Taiwan
		Nina Wang	Hong Kong

Source: Forbes

late growth. It is no wonder why IMF officials say that Japan is the key; it must first fix its own problems before the rest of Asia will be able to count on it to stabilize future problems.

POLITICAL AND CULTURAL CHANGES

Although economic issues are important to an understanding of the Pacific Rim, political and cultural changes are also crucial. The new, noncombative relationship between the United States and the former Soviet bloc means that special-interest groups and governments in the Rim will be less able to rely on the strength and power of those nations to help advance or uphold their positions. Communist North Korea, for instance, can no longer rely on the Soviet bloc for trade and ideological support. North Korea may begin to look for new ideological neighbors or, more significantly, to consider major modifications in its own approach to organizing society.

Similarly, ideological changes are afoot in Myanmar, where the populace are tiring of life under a military dictatorship. The military can no longer look for guaranteed support from the former socialist world.

In the case of Hong Kong, the British government shied away from extreme political issues and agreed to the peaceful annexation in 1997 of a capitalist bastion by a Communist nation, China. It is highly unlikely that such a decision would have been made had the issue of Hong Kong's political status arisen during the anti-Communist years of the cold war. One must not get the impression, however, that suddenly peace has arrived in the Pacific Rim. But outside support for extreme ideological positions seems to be giving way to a pragmatic search for peaceful solutions. This should have a salutary effect throughout the region.

The growing pragmatism in the political sphere is yielding changes in the cultural sphere. Whereas the Chinese formerly looked upon Western dress and music as decadent, most Chi-

nese now openly seek out these cultural commodities and are finding ways to merge these things with the Communist polity under which they live. It is also increasingly clear to most leaders in the Pacific Rim that international mercantilism has allowed at least one regional country, Japan, to rise to the highest ranks of world society, first economically and now culturally and educationally. The fact that one Asian nation has accomplished this impressive achievement fosters hope that others can do so also.

Rim leaders also see, however, that Japan achieved its position of prominence because it was willing to change traditional mores and customs and accept outside modes of thinking and acting. Religion, family life, gender relations, recreation, and many other facets of Japanese life have altered during Japan's rapid rise to the top. Many other Pacific Rim nations—including Thailand, Singapore, and South Korea—seem determined to follow Japan's lead in this regard. Therefore, we are witnessing in certain high-growth Rim economies significant cultural changes: a reduction in family size, a secularization of religious impulses, a desire for more leisure time and better education, and a move toward acquisition rather than "being" as a determinant of one's worth. That is, more and more people are likely to judge others' value by what they own rather than by what they believe or do. Buddhist values of self-denial, Shinto values of respect for nature, and Confucian values of family loyalty are giving way slowly to Western-style individualism and the drive for personal comfort and monetary success. Formerly close-knit communities, such as those in the South Pacific, are finding themselves struggling with drug abuse, AIDS, and gang-related violence, just as in the more metropolitan countries. These changes in political and cultural values are at least as important as economic growth in projecting the future of the Pacific Rim.

The Pacific Islands Map

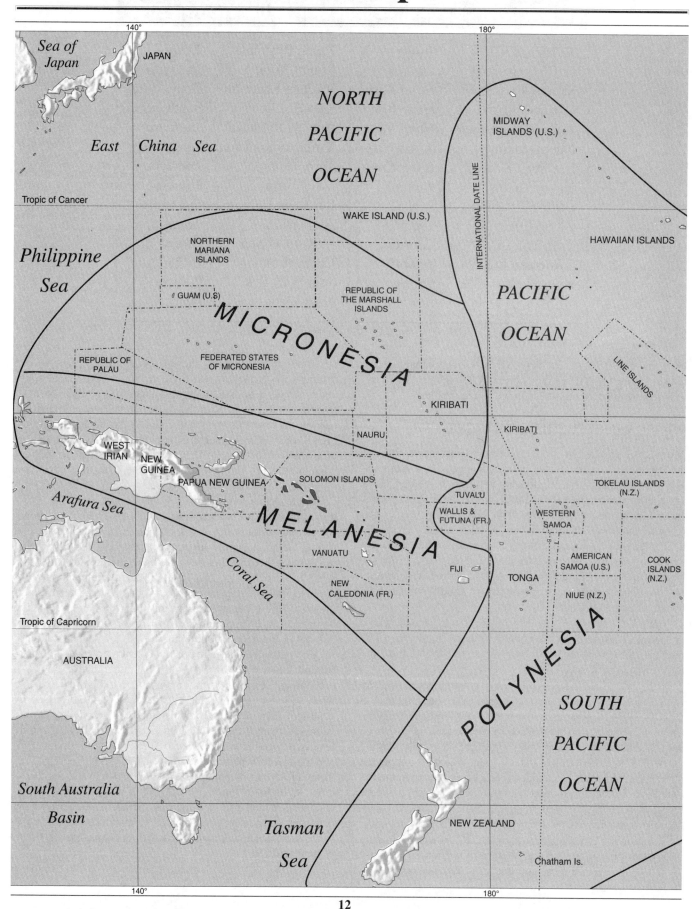

140° 180°

Sea of Japan

JAPAN

NORTH PACIFIC OCEAN

MIDWAY ISLANDS (U.S.)

East China Sea

INTERNATIONAL DATE LINE

Tropic of Cancer

WAKE ISLAND (U.S.)

HAWAIIAN ISLANDS

Philippine Sea

NORTHERN MARIANA ISLANDS

PACIFIC OCEAN

GUAM (U.S)

REPUBLIC OF THE MARSHALL ISLANDS

MICRONESIA

REPUBLIC OF PALAU

FEDERATED STATES OF MICRONESIA

LINE ISLANDS

KIRIBATI

KIRIBATI

NAURU

WEST IRIAN

NEW GUINEA

PAPUA NEW GUINEA

SOLOMON ISLANDS

TOKELAU ISLANDS (N.Z.)

TUVALU

Arafura Sea

MELANESIA

WALLIS & FUTUNA (FR.)

WESTERN SAMOA

VANUATU

AMERICAN SAMOA (U.S.)

COOK ISLANDS (N.Z.)

Coral Sea

FIJI

TONGA

NEW CALEDONIA (FR.)

NIUE (N.Z.)

Tropic of Capricorn

AUSTRALIA

POLYNESIA

SOUTH PACIFIC OCEAN

South Australia Basin

Tasman Sea

NEW ZEALAND

Chatham Is.

140° 180°

The Pacific Islands: Opportunities and Limits

PLENTY OF SPACE, BUT NO ROOM

There are about 30,000 islands in the Pacific Ocean. Most of them are found in the South Pacific and have been classified into three mammoth regions: *Micronesia*, composed of some 2,000 islands with such names as Palau, Nauru, and Guam; *Melanesia*, where 200 different languages are spoken on such islands as Fiji and the Solomon Islands; and *Polynesia*, comprised of Hawaii, Samoa, Tahiti, and other islands.

Straddling both sides of the equator and divided by the International Dateline, these territories are characterized as much by what is *not* there as by what *is*—that is, between any two tiny islands there might lie hundreds or even thousands of miles of open ocean. A case in point is the Cook Islands, in Polynesia. Associated with New Zealand, this 15-island group contains only 92 square miles of land but is spread over 714,000 square miles of open sea. So expansive is the space between islands that early explorers from Europe and the Spanish lands of South America often unknowingly bypassed dozens of islands that lay just beyond view in the vastness of the 64 million square miles of the Pacific—the world's largest ocean.

However, once the Europeans found and set foot on the islands, they inaugurated a process that irreversibly changed the history of island life. Their goals in exploring the Pacific were to convert islanders to Christianity and to increase the power and prestige of their homelands (and themselves) by obtaining resources and acquiring territory. They thought of themselves and European civilization as superior to others and often treated the "discovered" peoples with contempt. An example is the "discovery" of the Marquesas Islands (from whence came some of the Hawaiian people) by the Peruvian Spaniard Alvaro de Mendana. Mendana landed in the Marquesas in 1595 with some women and children—and, significantly, 378 soldiers. Within weeks, his entourage had planted three Christian crosses, declared the islands to be the possession of the king of Spain, and killed 200 islanders. Historian Ernest S. Dodge describes the brutality of the first contacts:

> The Spaniards opened fire on the surrounding canoes for no reason at all. To prove himself a good marksman one soldier killed both a Marquesan and the child in his arms with one shot as the man desperately swam for safety.... The persistent Marquesans again attempted to be friendly by bringing fruit and water, but again they were shot down when they attempted to take four Spanish water jars. Magnanimously the Spaniards allowed the Marquesans to stand around and watch while mass was celebrated.... When [the islanders] attempted to take two canoe loads of ... coconuts to the ships half the unarmed natives were killed and three of the bodies hung in the rigging in grim warning. The Spaniards were not only killing under orders, they were killing for target practice.
>
> —*Islands and Empires; Western Impact on the Pacific and East Asia*

All over the Pacific, islanders were "pacified" through violence or deception inflicted on them by the conquering nations of France, England, Spain, and others. Rivalries among the European nations were often acted out in the Pacific. For example, the Cook Islands, inhabited by a mixture of Polynesian and Maori peoples, were partly controlled by the Protestant Mission of the London Missionary Society, until the threat of incursions by French Catholics from Tahiti persuaded the British to declare the islands a protectorate of Britain. New Zealand eventually annexed the islands, and it controlled them until 1965.

Business interests frequently took precedence over islanders' sovereignty. In Hawaii, for instance, when Queen Liliuokalani proposed to limit the influence of the business community in island governance, a few dozen American business leaders—without the knowledge of the U.S. president or Congress, yet with the unauthorized help of 160 U.S. Marines—overthrew the Hawaiian monarch, installed Sanford Dole (of Dole Pineapple fame) as president, and petitioned Congress for Hawaii's annexation as a U.S. territory.

Whatever the method of acquisition, once the islands were under European or American control, the colonizing nations insisted that the islanders learn Western languages, wear Western clothing, convert to Christianity, and pay homage to faraway rulers whom they had never even seen.

This blatant Eurocentrism ignored the obvious—that the islanders already had rich cultural traditions that both predated European culture and constituted substantial accomplishments in technology, the arts, and social structure. Islanders were skilled in the construction of boats suitable for navigation on the high seas and of homes and religious buildings of varied architecture; they had perfected the arts of weaving and cloth-making, tattooing (the word itself is Tahitian), and dancing. Some cultures organized their political affairs much as did early New Englanders, with village meetings to decide issues by consensus, while others had developed strong chieftainships and kingships with an elaborate variety of rituals and taboos (a Tongan word) associated with the ruling elite. Island trade involving vast distances brought otherwise disparate people together. And although reading and writing were not known on most islands, some evidence of an ancient writing system has been found.

Despite these cultural attributes and a long history of skill in interisland or intertribal warfare, the islanders could not withstand the superior force of European firearms. Within just a few generations, the entire Pacific had been conquered and colonized by Britain, France, the Netherlands, Germany, the United States, and other nations.

CONTEMPORARY GROUPINGS

The Pacific islands today are classified into three racial/cultural groupings. The first, Micronesia, with a population of approximately 414,000 people, contains seven political entities, four of which are politically independent and three of which are affiliated with the United States. Guam is perhaps the best known of

(UN photo by Nagata Jr.)

In the South Pacific area of Micronesia, some 2,000 islands are spread over an ocean area of 3 million square miles. There remain many relics of the diverse cultures found on these islands; these boys are walking among the highly prized stone discs that were used as money on the islands of the Yap District.

these islands. Micronesians share much in common genetically and culturally with Asians. The term *Micronesia* refers to the small size of the islands in this group.

The second grouping, Melanesia, with a population of some 5.5 million (if New Guinea is included), contains six political entities, four of which are independent and two of which are affiliated with colonial powers. The best known of these islands is probably Fiji. The term *Melanesia* refers to the dark skin of the inhabitants, who, despite appearances, apparently have no direct ties with Africa.

Polynesia, the third grouping, with a population of approximately 536,000, contains 12 political entities, three of which are independent, while the remaining are affiliated with colonial powers. *Polynesia* means "many islands," the most prominent grouping of which is probably Hawaii. Most of the cultures in Polynesia have some ancient connections with the Marquesas Islands or Tahiti.

Subtracting the atypically large population of the island of New Guinea leaves about 2.2 million people in the region that we generally think of as the Pacific islands. Although it is possible that some of the islands were peopled by or had contact with ancient civilizations of South America, the overwhelming

TYPES OF PACIFIC ISLAND GOVERNMENTS

The official names of some of the Pacific island nations indicate the diversity of government structures found there:

Republic of Fiji
Federated States of Micronesia
Independent State of Papua New Guinea
Kingdom of Tonga

weight of scholarship places the origins of the Pacific islanders in Southeast Asia, Indonesia, and Australia.

Geologically, the islands may be categorized into the tall, volcanic islands, which have abundant water, flora, and fauna, and are suitable for agriculture; and the dry, flat, coral islands, which have fewer resources (though some are rich in phosphate). It also appears that the farther away an island is from the Asian and Australian continental landmasses, the less varied and plentiful are the flora and fauna.

THE CASE OF THE DISAPPEARING ISLAND

It wasn't much to begin with, but the way things are going, it won't be *anything* very soon. Nauru, a tiny, 8 1/2-square-mile dot of phosphate dirt in the Pacific, is being gobbled up by the Nauru Phosphate Corporation. Made of bird droppings (guano) mixed with marine sediment, Nauru's high-quality phosphate has a ready market in Australia, New Zealand, Japan, and other Pacific Rim countries, where it is used in industry, medicine, and agriculture.

Until recently, most Pacific Islanders would have envied the 12,809 Nauruans. With lots of phosphate to sell to willing buyers, Nauru earned so much income that its people had to pay no taxes and were provided with free health and dental care, bus transportation, and even their daily newspaper. Schooling (including higher education in Australia) was also paid for. Rent for government-built homes, supplied with telephones and electricity, was only US$5 a month. Nor did Nauruans have to work particularly hard for their living, since most phosphate pit laborers, about 3000 of them, were imported from other islands, while most managers and other professionals came from Australia (which once controlled the island), New Zealand, and Great Britain.

It sounds too good to be true, and it is, for Nauru's phosphate has been its only export, and now, the pits are nearly empty. Already there is only just a little fringe of green left along the shore, where everyone lives, and the government (Nauru is considered to be the smallest independent republic in the world) is debating what will happen when even the ground under people's homes is mined and shipped away. Some think that topsoil should be brought in to see if the moon-like surface of the excavated areas could be revitalized. Others think that moving away makes sense; that is, the government has been encouraged (by Australia and others) to just buy another island somewhere and move everyone there.

But all these dreams are now on hold while the tiny island deals with an even more urgent problem: near bankruptcy. As it turns out, Nauruans knew the phosphate was running out, so they put large amounts of their earnings into investments and trusts. They purchased several Boeing 737s, a number of hotels on other islands, and the tallest skyscraper in Melbourne, Australia. But many of their investments went sour, their former properties were taken over by GE Capital Corporation (which now says Nauru owes it US$176 million), and they began to drain their trust funds. Today, on the verge of bankruptcy, the government has frozen wages, eliminated many of the benefits people once enjoyed, and has lobbied Australia for more and more assistance. So now, along with having little land left, Nauru also has little of its economy left. Whether there will even be a Nauru in the near future is a very serious question.

THE PACIFIC COMMUNITY

During the early years of Western contact and colonization, maltreatment of the indigenous peoples and diseases such as measles and influenza greatly reduced their numbers and their cultural strength. Moreover, the carving up of the Pacific by different Western powers superimposed a cultural fragmentation on the region that added to the separateness resulting naturally from distance. Today, however, improved medical care is allowing the populations of the islands to rebound, and the withdrawal or realignment of European and American political power under the post–World War II United Nations policy of decolonization has permitted the growth of regional organizations.

First among the postwar regional groups was the South Pacific Commission. Established in 1947, when Western powers were still largely in control, many of its functions have since been augmented or superseded by indigenously created organizations such as the South Pacific Forum, which was organized in 1971 and has since spawned numerous other associations, including the South Pacific Regional Trade and Economic Agency and the South Pacific Islands Fisheries Development Agency. Through an executive body (the South Pacific Bureau for Economic Cooperation), these associations handle such issues as relief funding, the environment, fisheries, trade, and regional shipping. The organizations have produced a variety of duty-free agreements among countries, and have yielded joint decisions about regional transportation and cultural exchanges. As a result, regional art festivals and sports competitions are now a regular feature of island life, and a regional university in New Zealand attracts several thousand island students a year, as do schools in Hawaii. Sixteen Pacific nations (including East Timor, French Polynesia, and New Caledonia as observers) met in 2004 in Samoa and voted to improve regional transportation and create an AIDS policy.

Some regional associations have been able to deal forcefully with much more powerful countries. For instance, when the regional fisheries association set higher licensing fees for foreign fishing fleets (most fleets are foreign-owned, because island fishermen usually cannot provide capital for such large enterprises), the Japanese protested vehemently. Nevertheless, the association held firm, and many islands terminated their contracts with the Japanese rather than lower their fees. In 1994, the Cook Islands, the Federated States of Micronesia, Fiji, Kiribati, the Marshall Islands, Nauru, Niue, Papua New Guinea, the Solomon Islands, Tonga, Tuvalu, Vanuatu, and Western Samoa signed an agreement with the United States to establish a joint commercial commission to foster private-sector businesses and to open opportunities for trade, investment, and training. Through this agreement, the people of the islands hoped to increase the attractiveness of their products to the U.S. market.

Increasingly important issues in the Pacific are the testing of nuclear weapons and the disposal of toxic waste. Island leaders, with the occasional support of Australia and the strong support of New Zealand, have spoken out vehemently against the continuation of nuclear testing in the Pacific by the French government (Great Britain and the United States tested hydrogen bombs on coral atolls for years, but have now stopped) and against the burning of nerve-gas stockpiles by the United States on Johnston Atoll. In 1985, the 13 independent or self-governing countries of the South Pacific adopted their first collective agreement on regional security, the South Pacific Nuclear Free Zone Treaty. Encouraged by New Zealand and Australia, the group declared the Pacific a nuclear-free zone and issued a communique criticizing the dumping of nuclear waste in the re-

GUAM: THIS IS LIBERATION?

In 1994, the people of the U.S. Territory of Guam celebrated the 50th anniversary of their liberation by U.S. Marines and Army Infantry from the occupying troops of the Japanese Army. During the three years that they controlled the tiny, 30-mile-long island, the Japanese massacred some of the Guamanians and subjected many others to forced labor and internment in concentration camps.

Their liberation, therefore, was indeed a cause for celebration. But the United States quickly transformed the island into its military headquarters for the continuing battle against the Japanese. The entire northern part of the island was turned into a base for B-29 bombers, and the Pacific submarine fleet took up residence in the harbor. Admiral Chester W. Nimitz, commander-in-chief of the Pacific, made Guam his headquarters. By 1946, the U.S. military government in Guam had laid claim to nearly 80 percent of the island, displacing entire villages and hundreds of individual property owners.

Since then, some of the land has been returned, and large acreages have been handed over to the local civilian government—which was to have distributed most of it, but has not yet done so. The local government still controls about one third of the land, and the U.S. military

controls another third, meaning that only one third of the island is available to the residents for private ownership. Litigation to recover the land has been bitter and costly, involving tens of millions of dollars in legal expenses since 1975. The controversy has prompted some local residents to demand a different kind of relationship with the United States, one that would allow for more autonomy. It has also spurred the growth of nativist organizations such as the Chamorru Nation, which promotes the Chamorru language of the original Malayo–Polynesian inhabitants (spelled Chamorro by the Spanish) and organizes acts of civil disobedience against both civilian and military authorities.

Guam was first overtaken by Spain in 1565. It has been controlled by the United States since 1898, except for the brief Japanese interlude. Whether the local islanders, who now constitute a fascinating mix of Chamorro, Spanish, Japanese, and American cultures, will be able to gain a larger measure of autonomy after hundreds of years of colonization by outsiders is difficult to predict, but the ever-present island motto, *Tano Y Chamorro* ("Land of the Chamorros"), certainly spells out the objective of many of those who call Guam home.

gion. Some island leaders, however, see the storage of nuclear waste as a way of earning income to compensate those who were affected by the nuclear testing on Bikini and Enewetak Islands. The Marshall Islands, for example, are interested in storing nuclear waste on already-contaminated islands; however, the nearby Federated States of Micronesia, which were observers at the Nuclear Free Zone Treaty talks, oppose the idea and have asked the Marshalls not to proceed.

World leaders met in Jamaica in 1982 to sign into international law the Law of the Sea. This law, developed under the auspices of the United Nations, gave added power to the tiny Pacific island nations because it extended the territory under their exclusive economic control to 12 miles beyond their shores or 200 miles of undersea continental shelf. This put many islands in undisputed control of large deposits of nickel, copper, magnesium, and other valuable metals. The seabed areas away from continents and islands were declared the world's common heritage, to be mined by an international company, with profits channeled to developing countries. The United States has negotiated for years to increase the role of industrialized nations in mining the seabed areas; if modifications are made to the treaty, the United States will likely sign the document.

COMING OF AGE?

If the peoples of the Pacific islands are finding more reasons to cooperate economically and politically, they are still individually limited by the heritage of cultural fragmentation left them by their colonial pasts. Western Samoa, for example, was first annexed by Germany in 1900, only to be given to New Zealand after Germany's defeat in World War I. Today, the tiny nation of mostly Christian Polynesians, independent since 1962, uses both English and Samoan as official languages and embraces a formal governmental structure copied from Western parliamentary practice. Yet the structure of its hundreds of small villages

remains decidedly traditional, with clan chiefs ruling over large extended families, who make their not particularly profitable livings by farming breadfruit, taro, yams, bananas, and copra.

Political independence has not been easy for those islands that have embraced it, nor for those colonial powers that continue to deny it. Anticolonial unrest continues on many islands, especially the French ones. However, concern over economic viability has led most islands to remain in some sort of loose association with their former colonial overseers. After the defeat of Japan in World War II, the Marshall Islands, the Marianas, and the Carolines (now Federated States of Micronesia) were assigned by the United Nations to the United States as trust territories. The French Polynesian islands have remained overseas "departments" (similar to U.S. states) of France. In such places as New Caledonia, however, there has been a growing desire for autonomy, which France has attempted to meet in various ways while still retaining sovereignty. The UN decolonization policy has made it possible for most Pacific islands to achieve independence if they wish, but many are so small that true economic independence in the modern world will never be possible. In French Polynesia, the independence issue continues to flare up. Some 15,000 people marched against the removal of pro-independence president Oscar Temaru after parliament voted to censure him in 2004. A more conservative former president who does not advocate separation from France replaced Temaru.

In addition to relations with their former colonial masters, the Pacific islands have found themselves in the middle of many serious domestic political problems. For example, Fiji, a former British colony, weathered two military coups in 1987 and a bizarre coup in 2000. In the 2000 episode, native Fijian businessman George Speight captured the prime minister and other government leaders and held them hostage until they resigned their posts. The prime minister, Mahendra Chaudhry, was not a native Fijian but, rather, an Indian. So many immigrants from

India had settled in Fiji over the past few decades that they now constituted some 44 percent of the population, enough to wield political power on the island. This situation had frightened the native Fijians into stripping Indians of various civil rights (restored in 1997), but the election of the first Indian prime minister seemed to push some of the native Fijians past their tolerance limit. While not all native Fijians feel threatened by the rise of the Indian population, when a Suva television station criticized the latest coup and called for racial tolerance, the station was ransacked by Fijian rioters, who also destroyed some 20 other buildings in the capital city. Tensions continued in 2001 when the new prime minister, Laisenia Qarase, refused to seat Indo-Fijians in his cabinet until a court ruling ordered him to do so. In 2002, George Speight was sentenced to life in prison for his takeover of the Parliament, a development that produced a foiled plot by his supporters to kidnap government leaders and hold them hostage until Speight was released. Finally, in 2004, all coup leaders were convicted and sent to jail.

Similarly, tribal tensions have wracked the Solomon Islands. In the past few years, residents of the island of Malaita have migrated in large numbers to the larger, main island of Guadalcanal. The locals have complained that the newcomers have taken over their land and the government. Tribal violence related to this issue has killed some 50 islanders in the recent past; but in 2000, Malaita's prime minister, a Malaitan, was kidnapped in Honiara, the capital city, and forced to resign. Violence forced 20,000 residents to flee their homes, and foreign nationals were evacuated to avoid the fierce fighting by opposing paramilitary groups. It is clear that in Fiji, the Solomon Islands, and other places, Pacific islanders will have to find creative ways to allow all members of society to participate in the political process. Without political stability, other pressing issues will be neglected.

No amount of political realignment can overcome the economic dilemma of most of the islands. Japan, the single largest purchaser of island products, as well as the United States and others, are good markets for the Pacific economies, but exports are primarily of mineral and agricultural products (coffee, tea, cocoa, sugar, tapioca, coconuts, mother-of-pearl) rather than of the more profitable manufactured or "value-added" items produced by industrial nations. In addition, there will always be the cost of moving products from the vastness of the Pacific to the various mainland markets.

Another problem is that many of the profits from the island's resources do not redound to the benefit of the islanders. Tuna, for example, is an important and profitable fish catch, but most of the profits go to the Taiwanese, Korean, Japanese, and American fleets that ply the Pacific. Similarly, tourism profits largely end up in the hands of the multinational hotel owners. About 80 percent of visitors to the island of Guam since 1982 have been Japanese (more than half a million people annually)—seemingly a gold mine for local Guamanians, since each traveler typically spends more than $2,000. However, since the tourists tend to purchase their tickets on Japanese airlines and book rooms in Japanese-owned or -managed hotels, much of the money that they spend for their vacations in Guam never reaches the Guamanians. Some enterprising Fijians, together with Chinese, Malaysian, and Hong Kong gangs, have been attempting to make money their own way: in Fiji's capital of

Suva, they created the largest methamphetamine lab in the Southern Hemisphere. Located in a three-building complex stocked with drums of chemicals, the lab was capable of producing US$540 million worth of "meth." Police shut down the lab in 2004, but the case illustrates how desperate some islanders are for income.

The poor economies, especially in the outer islands, have prompted many islanders to move to larger cities (about 1 million islanders now live in the Pacific's larger cities) to find work. Indeed, there is currently a tremendous mixing of all of the islands' peoples. Hawaii, for example, is peopled now with Samoans, Filipinos, and many other islanders; pure Hawaiians are a minority, and despite efforts to preserve the Hawaiian language, it is used less and less. Similarly, Fiji, as we have seen, is now populated by nearly as many immigrants from India as by native Fijians. New Caledonians are outnumbered by Indonesians, Vietnamese, French, and others. And, of course, whites have long outnumbered the Maoris on New Zealand. Guam is peopled with islanders from all of Micronesia as well as from Samoa and other islands.

In addition to interisland migration, many islanders emigrate to Australia, New Zealand, the United States, or other countries and then send money back home to sustain their families. Those remittances are important to the economies of the islands, but the absence of parents or adult children for long periods of time does considerable damage to the social fabric. In a few cases, such as in the Cook Islands and American Samoa, more islanders live abroad than remain on the islands. For example, whereas American Samoa has about 65,000 residents, more than 130,000 Samoans live on the U.S. mainland, in such places as Los Angeles and Salt Lake City. Those who leave often find life abroad quite a shock, for the island culture, influenced over the decades by the missionary efforts of Mormons, Methodists, Seventh-day Adventists, and especially the London Missionary Society, is conservative, cautious, and personal. Metropolitan life, by contrast, is considered by some islanders to be wild and impersonal. Some young emigrants respond to the "cold" environment and marginality of big-city life by engaging in deviant behavior themselves, such as selling drugs and joining gangs.

Island society itself, moreover, is not immune from the social problems that plague larger societies. Many islands report an increasing number of crimes and suicides. Young Samoans, for example, are afflicted with many of the same problems—gangs, drugs, and unemployment—as are their U.S. inner-city counterparts. Samoan authorities have reported increases in incidences of rape, robbery, and other socially dysfunctional behaviors. In addition, the South Pacific Commission and the World Health Organization are now reporting an alarming increase in HIV/AIDS and other sexually transmitted diseases.

Pacific islanders are trying to cope with a variety of other problems as well. Tuvalu, for example, has a very threatening problem: its nine coral atolls are sinking into the sea. With sea levels rising due to global warming, the country is already subjected to monthly flooding. The 11,000 inhabitants are pleading with both Australia and New Zealand to allow or assist with relocation to those countries or to the island of Niue. The highest point on the islands is only 16 feet above sea level, and drinking water is already becoming brackish as the ocean penetrates groundwater reserves. In 50 years or less, the land will be entirely submerged. New Zealand has already accepted some islanders, but Australia has denied both access and relocation assistance. Other problems caused by nature also make life difficult for islanders. Low-lying islands often suffer serious damage from cyclones, such as the category 5 cyclone that hit Tokelau in 2004 and completely leveled the capital city.

For decades, and notwithstanding the imposition of foreign ways, islanders have shared a common culture; most people know how to raise bananas, coconuts, and yams, how to roast pigs and fish, and how to make breadfruit, tapioca, and poi. But much of island culture has depended on an identity shaped and preserved by isolation from the rest of the world. Whether the essence of island life—and especially the identity of the people—can be maintained in the face of increasing integration into a much larger world remains to be seen.

Japan

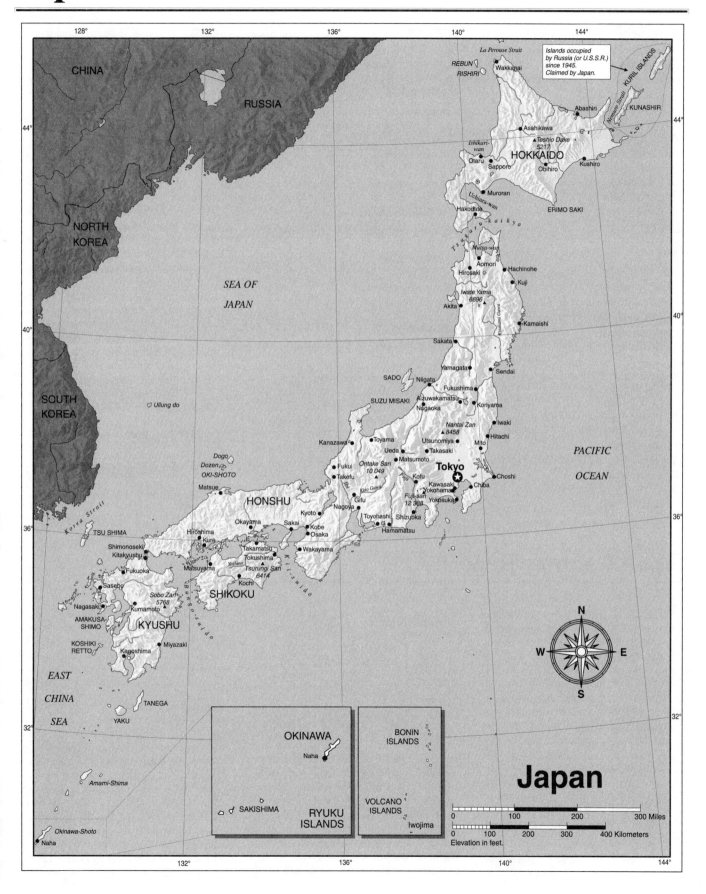

CHINA

RUSSIA

La Perouse Strait
REBUN
RISHIRI
Wakkanai

Abashiri

KURIL ISLANDS

Islands occupied
by Russia (or U.S.S.R.)
since 1945.
Claimed by Japan.

KUNASHIR

Asahikawa

Ishikari-wan
▲ *Teshio Dake*
5217

HOKKAIDO

NORTH
KOREA

Otaru
Sapporo
Obihiro

Kushiro

Muroran

Uchiura-wan

ERIMO SAKI

Hakodate

T s u g a r u - k a i k y o

Mutsu-wan

Aomori
Hirosaki
Hachinohe
Kuji

Iwate Yama
6696

SEA OF
JAPAN

Akita

Kamaishi

Sakata

SOUTH
KOREA

Ullung do

Yamagata
Sendai

SADO
Niigata
Fukushima

SUZU MISAKI
Aizuwakamatsu
Koriyama
Nagaoka

Nantai Zan
▲ 8458
Iwaki
Hitachi

PACIFIC

Dogo
Dozen
OKI-SHOTO

Kanazawa
Toyama
Ueda
Matsumoto

Utsunomiya
Takasaki
Mito

OCEAN

Fukui
Ontake San
10,049

Kofu
Choshi

Takefu

★ **Tokyo**

Matsue

Kawasaki
Yokohama
Yokosuka

Chiba

HONSHU

Gifu
Nagoya

Fuji-san
12,388

Kiso Gawa

Okayama
Sakai
Kyoto
Kobe
Osaka

Toyohashi
Shizuoka

Hamamatsu

Hiroshima
Kure

Takamatsu
Tokushima

Wakayama

Shimonoseki
Kitakyushu

Matsuyama

Yoshino
Tsurungi San
6414

TSU SHIMA

Fukuoka

Kochi

SHIKOKU

Sasebo

Sobo Zan
5768

Nagasaki
Kumamoto

AMAKUSA-
SHIMO

KYUSHU

KOSHIKI
RETTO
Kagoshima
Miyazaki

EAST

CHINA

TANEGA

SEA

YAKU

Amami-Shima

N
W   E
S

OKINAWA

Naha

BONIN
ISLANDS

Japan

Okinawa-Shoto
Naha

SAKISHIMA

RYUKU
ISLANDS

VOLCANO
ISLANDS

Iwojima

0   100   200   300 Miles

0   100   200   300   400 Kilometers

Elevation in feet.

Korea Strait

B u n g o - s u i d o

K i i - s u i d o

19

Japan Statistics

GEOGRAPHY

Area in Square Miles (Kilometers): 145,882 (377,835) (about the size of California)
Capital (Population): Tokyo (Greater Tokyo: 35.3 million)
Environmental Concerns: air and water pollution; acidification; depletion of global resources due to Japanese demand
Geographical Features: mostly rugged and mountainous
Climate: tropical in the south to cool temperate in the north

PEOPLE

Population

Total: 127,333,002
Annual Growth Rate: 0.08%
Rural/Urban Population Ratio: 21/79
Ethnic Makeup: 99% Japanese; 1% others (mostly Korean)
Major Language: Japanese
Religions: primarily Shinto and Buddhist; only 15% claims any formal religious affiliation

Health

Life Expectancy at Birth: 78 years (male); 84 years (female)
Infant Mortality: 3.28/1,000 live births
Physicians Available: 1/546
HIV/AIDS Rate in Adults: 0.1%

Education

Adult Literacy Rate: 99%
Compulsory (Ages): 6–15; free

COMMUNICATION

Telephones: 71.149 million main lines
Mobile phones: 86.66 million
Daily Newspaper Circulation: 578/1,000 people
Televisions: 619/1,000 people
Internet Users: 47,080,000 (2001)

TRANSPORTATION

Highways in Miles (Kilometers): 686,616 (1,144,360)
Railroads in Miles (Kilometers): 14,202 (23,671)
Usable Airfields: 174
Motor Vehicles in Use: 68,030,000

GOVERNMENT

Type: constitutional monarchy
Independence Date: traditional founding 660 B.C.; constitutional monarchy established 1947
Head of State/Government: Emperor Akihito; Prime Minister Junichiro Koizumi
Political Parties: Liberal Democratic Party; Social Democratic Party of Japan; New Komeito; Democratic Party of Japan; Japan Communist Party; Liberal Party; New Conservative Party

Suffrage: universal at 20

MILITARY

Military Expenditures (% of GDP): 1%
Current Disputes: various territorial disputes with Russia, others

ECONOMY

Currency ($ U.S. Equivalent): 102 yen = $1
Per Capita Income/GDP: $28,200/$3.582 trillion
GDP Growth Rate: 2.7%
Inflation Rate: -0.6%
Unemployment Rate: 5.3%
Labor Force by Occupation: 65% services; 30% industry; 5% agriculture
Natural Resources: fish
Agriculture: rice; sugar beets; vegetables; fruit; pork; poultry; dairy and eggs; fish
Industry: metallurgy; engineering; electrical and electronics; textiles; chemicals; automobiles; food processing
Exports: $447 billion (primary partners United States, China, South Korea)
Imports: $346.6 billion (primary partners China, United States, South Korea)

SUGGESTED WEB SITES

http://www.odci.gov/cia/
  publications/factbook/geos/
  ja.html
http://jin.jcic.or.jp

Japan Country Report

The Japanese nation is thought to have begun about 250 B.C., when ancestors of today's Japanese people began cultivating rice, casting objects in bronze, and putting together the rudiments of the Shinto religion. However, humans are thought to have inhabited the Japanese islands as early as 20,000 B.C. Some speculate that remnants of these or other early peoples may be the non-Oriental Ainu people (now largely Japanized) who still occupy parts of the northern island of Hokkaido. Asiatic migrants from China and Korea and islanders from the South Pacific occupied the islands between 250 B.C. and A.D. 400, contributing to the population base of modern Japan.

Between A.D. 300 and 710, military aristocrats from some of the powerful clans

DEVELOPMENT

Japan has entered a post smokestack era in which primary industries are being moved abroad, producing a hollowing effect inside Japan and increasing the likelihood of rising unemployment. Nevertheless, prospects for growth are excellent, despite the current economic woes.

into which Japanese society was divided established their rule over large parts of the country. Eventually, the Yamato clan leaders, claiming divine approval, became the most powerful. Under Yamato rule, the Japanese began to import ideas and technology from nearby China, including the Buddhist religion and the Chinese method of writing—which the elite somewhat awk-

wardly adapted to spoken Japanese, an entirely unrelated language. The Chinese bureaucratic style of government and architecture was also introduced; Japan's first permanent capital was constructed at the city of Nara between the years 710 and 794.

As Chinese influence waned in the period 794–1185, the capital was relocated to Kyoto, with the Fujiwara family wielding real power under the largely symbolic figurehead of the emperor. A warrior class controlled by *shoguns,* or generals, held power at Kamakura between 1185 and 1333 and successfully defended the country from invasion by the Mongols. Buddhism became the religion of the masses, although Shintoism was often embraced simultaneously. Between 1333 and 1568, a very rigid class structure developed, along

(Japan National Tourist Organization)

The Japanese emperor has long been a figurehead in Japan. In 1926, Hirohito (left) became emperor and ushered in the era named *Showa*. He died on January 7, 1989, having seen Japan through World War II and witnessed its rise to the economic world power it is today. He was succeeded by his son, Akihito, who named his reign *Heisei*, meaning "Achieving Peace."

with a feudalistic economy controlled by *daimyos,* feudal lords who reigned over their own mini-kingdoms.

In 1543, Portuguese sailors landed in Japan, followed a few years later by the Jesuit missionary Francis Xavier. An active trade with Portugal began, and many Japanese (perhaps half a million), including some feudal lords, converted to Christianity. The Portuguese introduced firearms to the Japanese and perhaps taught them Western-style techniques of building castles with moats and stone walls. Wealthier feudal lords were able to utilize these innovations to defeat weaker rivals; by 1600, the country was unified under a military bureaucracy, although feudal lords still retained substantial sovereignty over their fiefs. During this time, the general Hideyoshi attempted an unsuccessful invasion of nearby Korea.

The Tokugawa Era

In the period 1600 to 1868, called the Tokugawa Era, the social, political, and economic foundations of modern Japan were put in place. The capital was moved to Tokyo, cities began to grow in size, and a merchant class arose that was powerful enough to challenge the hegemony of the

FREEDOM

Japanese citizens enjoy full civil liberties, and opposition parties and ideologies are seen as natural and useful components of democracy. Certain people, however, such as those of Korean ancestry, have been subject to both social and official discrimination—an issue that is gaining the attention of the Japanese.

centuries-old warrior class. Strict rules of dress and behavior for each of the four social classes (samurai, farmer, craftsman, and merchant) were imposed, and the Japanese people learned to discipline themselves to these codes. Western ideas came to be seen as a threat to the established ruling class. The military elite expelled foreigners and put the nation into 2 1/2 centuries of extreme isolation from the rest of the world. Christianity was banned, as was most trade with the West. Even Japanese living abroad were forbidden from returning, for fear that they might have been contaminated with foreign ideas.

During the Tokugawa Era, indigenous culture expanded rapidly. Puppet plays and a new form of drama called *kabuki* became

popular, as did *haiku* poetry and Japanese pottery and painting. The samurai code, called *bushido,* along with the concept of *giri,* or obligation to one's superiors, suffused Japanese society. Literacy among males rose to about 40 percent, higher than most European countries of the day. Samurai busied themselves with the education of the young, using teaching methods that included strict discipline, hard work, and self-denial.

During the decades of isolation, Japan grew culturally strong but militarily weak. In 1853, a U.S. naval squadron appeared in Tokyo Bay to insist that Japan open up its ports to foreign vessels needing supplies and desiring to trade. Similar requests had been denied in the past, but the sophistication of the U.S. ships and their advanced weaponry convinced the Japanese military rulers that they no longer could keep Japan isolated from the outside.

The Era of Modernization:
The Meiji Restoration

Treaties with the United States and other Western nations followed, and the dislocations associated with the opening of the country to the world soon brought discredit to the ruling shoguns. Provincial samurai took control of the government. The em-

HEALTH/WELFARE

The Japanese live longer on average than any other people on Earth. Every citizen is provided with inexpensive medical care under a national health-care system, but many people still prefer to save substantial portions of their income for health emergencies and old age.

peror, long a figurehead in Kyoto, away from the center of power, was moved to Tokyo in 1868, beginning the period known as the Meiji Restoration.

Although the Meiji leaders came to power with the intention of ousting all the foreigners and returning Japan to its former state of domestic tranquillity, they quickly realized that the nations of the West were determined to defend their newly won access to the ports of Japan. To defeat the foreigners, they reasoned, Japan must first acquire Western knowledge and technology.

Thus, beginning in 1868, the Japanese leaders launched a major campaign to modernize the nation. Ambassadors and scholars were sent abroad to learn about Western-style government, education, and warfare. Implementing these ideas resulted in the abolition of the feudal system and the division of Japan into 43 prefectures, or states, and other administrative districts under the direct control of the Tokyo government. Legal codes that established the formal separation of society into social classes were abolished; and Western-style dress, music, and education were embraced. The old samurai class turned its attention from warfare to leadership in the government, in schools, and in business. Factories and railroads were constructed, and public education was expanded. By 1900, Japan's literacy rate was 90 percent, the highest in all of Asia. Parliamentary rule was established along the lines of the government in Prussia, agricultural techniques were imported from the United States, and banking methods were adopted from Great Britain.

Japan's rapid modernization soon convinced its leaders that the nation was strong enough to begin doing what other advanced nations were doing: acquiring empires. Japan went to war with China, acquiring the Chinese island of Taiwan in 1895. In 1904, Japan attacked Russia and successfully acquired Korea and access to Manchuria (both areas having been in the sphere of influence of Russia). Siding against Germany in World War I, Japan was able to acquire Germany's Pacific empire—the Marshall, Caroline, and Mariana Islands. Western nations were surprised at Japan's rapid empire-building but did little to stop it. Indeed, some Westerners

viewed Japan's aggression against others as a sign of a progressive nation.

The Great Depression of the 1930s caused serious hardships in Japan because, being resource-poor yet heavily populated, the country had come to rely on international trade to supply its people's basic needs. Many Japanese advocated the forced annexation of Manchuria as a way of providing needed resources. This was accomplished easily, albeit with much brutality, in 1931. With militarism on the rise, the Japanese nation began moving away from democracy and toward a military dictatorship. Political parties were eventually banned, and opposition leaders were jailed and tortured.

WORLD WAR II AND THE JAPANESE EMPIRE

The battles of World War II in Europe, initially won by Germany, promised to substantially re-align the colonial empires of France and other European powers in Asia. The military elite of Japan declared its intention of creating a Greater East Asia Co-Prosperity Sphere—in effect, a Japanese empire created out of the ashes of the European empires in Asia that were then dissolving. In 1941, under the guidance of General Hideki Tojo and with the tacit approval of the emperor, Japan captured the former French colony of Indochina (Vietnam, Laos, and Cambodia), bombed Pearl Harbor in Hawaii, and captured oil-rich Indonesia. These victories were followed by others: Japan captured all of Southeast Asia, including Burma (now called Myanmar), Thailand, Malaya, the Philippines, and parts of New Guinea; and expanded its hold in China and in the islands of the South Pacific. Many of these conquered peoples, lured by the Japanese slogan of "Asia for the Asians," were initially supportive of the Japanese, believing that Japan would rid their countries of European colonial rule. It soon became apparent, however, that Japan had no intention of relinquishing control of these territories and would go to brutal lengths to subjugate the local peoples. Japan soon dominated a vast empire, the constituents of which were virtually the same as those making up what we call the Pacific Rim today.

In 1941, the United States launched a counteroffensive against the powerful Japanese military. (American history books refer to this offensive as the Pacific Theater of World War II, but the Japanese call it the *Pacific War*. We use the term *World War II* in this text, for reasons of clarity and consistency.) By 1944, the U.S. and other Allied troops, at the cost of tens of thousands of lives, had ousted the Japanese from most of their conquered lands and were begin-

ning to attack the home islands themselves. Massive firebombing of Tokyo and other cities, combined with the dropping of two atomic bombs on Hiroshima and Nagasaki, convinced the Japanese military rulers that they had no choice but to surrender.

This was the first time in Japanese history that Japan had been conquered, and the Japanese were shocked to hear their emperor, Hirohito—whose voice had never been heard on radio—announce on August 14, 1945, that Japan was defeated. The emperor cited the suffering of the people—almost 2 million Japanese had been killed—as well as the devastation of Japan's cities brought about by the use of a "new and most cruel bomb," and the possibility that, without surrender, Japan as a nation might be completely "obliterated." Emperor Hirohito then encouraged his people to look to the future, to keep pace with progress, and to help build world peace by accepting the surrender ("enduring the unendurable and suffering what is insufferable").

This attitude smoothed the way for the American Occupation of Japan, led by General Douglas MacArthur. Defeat seemed to inspire the Japanese people to adopt the ways of their more powerful conquerors and to eschew militarism. Under the Occupation forces, the Japanese Constitution was rewritten in a form that mimicked that of the United States. Industry was restructured, labor unions encouraged, land reform accomplished, and the nation as a whole demilitarized. Economic aid from the United States, as well as the prosperity in Japan that was occasioned by the Korean War in 1953, allowed Japanese industry to begin to recover from the devastation of war. The United States returned the governance of Japan back to the Japanese people by treaty in 1951 (although some 47,000 troops still remain in Japan as part of an agreement to defend Japan from foreign attack).

By the late 1960s, the Japanese economy, stronger already than some European economies, was more than self-sustaining, and the United States was Japan's primary trading partner. Today, about 16 percent of the items Japan imports come from the United States, and the US buys about 25 percent of Japan's exports. For many years after World War II, Japan's trade with its former Asian empire, however, was minimal, because of lingering resentment against Japan for its wartime brutalities. As recently as the late 1970s, for example, anti-Japanese riots and demonstrations occurred in Indonesia upon the visit of the Japanese prime minister to that country, and the Chinese government raises the alarm each time Japan effects a modernization of its military.

On December 7, 1941, Japan entered World War II as a result of its bombing of Pearl Harbor in Hawaii. This photograph, taken from an attacking Japanese plane, shows Pearl Harbor and a line of American battleships.

Nevertheless, between the 1960s and early 1990s, Japan experienced an era of unprecedented economic prosperity. Annual economic growth was three times as much as in other industrialized nations. Japanese couples voluntarily limited their family size so that each child born could enjoy the best of medical care and social and educational opportunities. The fascination with the West continued, but eventually, rather than "modernization" or "Americanization," the Japanese began to speak of "internationalization," reflecting both their capacity for and their actual membership in the world community, politically, culturally, and economically. Militarily, however, Japan was forbidden by its U.S.–drafted Constitution to develop a war capacity. In recent years, Japanese troops have engaged in UN peacekeeping operations and have participated in non-combat duties in the Iraq War, but even that level of military involvement has caused tremendous domestic and international comment.

The Japanese government as well as private industry began to accelerate the drive for diversified markets and resources in the mid-1980s. This was partly in response to protectionist trends in countries in North America and Europe with which Japan had accumulated huge trade surpluses, but it was also due to changes in Japan's own internal social and economic conditions. Japan's recent resurgence of interest in its neighboring countries and the resulting rise of a bloc of rapidly developing countries we are calling the Pacific Rim can be explained by both external protectionism and internal changes. This time, however, Japanese influence—no longer linked with militarism—is being welcomed by virtually all nations in the region.

DOMESTIC CHANGE

What internal conditions are causing Japan's renewed interest in Asia and the Pacific? One change involves wage structure. For several decades, Japanese exports were less expensive than competitors' because Japanese workers were not paid as well as workers in North America and Europe. Today, however, the situation is reversed: Average manufacturing wages in Japan are now higher than those paid to workers in the United States. Schoolteachers, college professors, and many white-collar workers are also better off in Japan. These wage differentials are the result of successful union activity and demographic changes (although the high cost of living reduces the actual domestic buying power of Japanese wages to about 70 percent that of the United States).

Whereas prewar Japanese families—especially those in the rural areas—were large, today's modern household typically consists of a couple and only one or two children. As Japan's low birth rate began to affect the supply of labor, companies were forced to entice workers with higher wages. The cost of land, homes, food— even Japanese-grown rice—is so much higher in Japan than in most of its neighbor countries that employees in Japan expect high wages (household income in Japan is higher even than in the United States).

Given conditions like these, many Japanese companies have found that they cannot be competitive in world markets unless they move their operations to countries like the Philippines, Singapore, or China, where an abundance of laborers keeps wage costs 75 to 95 percent lower than in Japan. Abundant, cheap labor (as well as a desire to avoid import tariffs) is also the reason why so many Japanese companies have been constructed in the economically depressed areas of the U.S. Midwest and South.

Another internal condition that is spurring Japanese interest in the Pacific Rim is a growing public concern for the domestic environment. Beginning in the 1970s, the Japanese courts handed down several landmark decisions in which Japanese companies were held liable for damages to people caused by chemical and other industrial wastes. Japanese industry, realizing that it no longer had a carte blanche to make profits at the expense of the environment, began moving some of its smokestack industries to new locations in developing-world countries, just as other industrialized nations had done. This has turned out to be a wise move economically (although obviously not environmentally) for many companies, as it has put their operations closer to their raw materials. This, in combination with cheaper labor costs, has allowed them to remain globally competitive. It also has been a tremendous benefit to the host countries, although environmental groups in many Rim countries are also now becoming active, and industry in the future may be forced to effect actual improvements in their operations rather than move polluting technologies to "safe" areas.

Attitudes toward work are also changing in Japan. Although the average Japanese worker still works about six hours more per week than the typical North American, the new generation of workers—those born and raised since World War II—are not so eager to sacrifice as much for their companies as were their parents. Recent policies have eliminated weekend work in virtually all government offices and many industries, and sports and other recreational activities are becoming increasingly popular. Given these conditions, Japanese corporate leaders are finding it more cost effective to move operations abroad to countries like South Korea, where labor legislation is weaker and long work hours remain the norm.

MYTH AND REALITY OF THE ECONOMIC MIRACLE

The Japanese economy, like any other economy, must respond to market as well as social and political changes to stay vibrant. It just so happened that for several decades, Japan's attempt to keep its economic boom alive created the conditions that, in turn, furthered the economies of all the countries in the Asia/Pacific region. That a regional "Yen Bloc" (so called because of the dominance of the Japanese currency, the yen) had been created was revealed in the late 1990s when sluggishness in the Japanese economy contributed to dramatic downturns in the economies of surrounding countries.

For many years, world business leaders were of the impression that whatever Japan did—whether targeting a certain market or reorienting its economy toward regional trade—turned to gold, as if the Japanese possessed some secret that no one else understood. But when other countries in Asia began to copy the Japanese model, their economies also improved—until, that is, lack of moderation and an inflexible application of the model produced a major correction in the late 1990s, which slowed economic growth throughout the entire region. Rather than possessing a magical management style or other great business secret, the Japanese achieved their phenomenal success thanks to hard work, advanced planning, persistence, and, initially, large doses of outside financial help, in business, education, and other fields has been the result of, among other things, hard work, advance planning, persistence, and outside financial help.

However, even with those ingredients in place, Japanese enterprises often fall short. In many industries, for example, Japanese workers are less efficient than are workers in other countries. Japan's national railway system was once found to have 277,000 more employees on its payroll than it needed. At one point, investigators revealed that the system had been so poorly managed for so many years that it had accumulated a public debt of $257 billion. Multimillion-dollar train stations had been built in out-of-the-way towns, for no other reason than that a member of the *Diet* (the Japanese Parliament) happened to live there and had pork-barreled the project. Both government and industry have been plagued by bribery and corruption, as occurred in the Recruit Scandal of the late 1980s, which caused many implicated government leaders, including the prime minister, to resign.

Nor is the Japanese economy impervious to global market conditions. Values of stocks traded on the Tokyo Stock Exchange took a serious drop in 1992; investors lost millions of dollars, and many had to declare bankruptcy. Moreover, the tenacious recession that hit Japan in the early 1990s and that is continuing into the 2000s has seriously damaged many corporations. Large corporations such as Yamaichi Securities and Sogo (a department-store chain) have had to declare bankruptcy, as have many banks. Indeed, at one time, the majority of the top 10 banks in the world were Japanese; but by 2000, only two Japanese banks remained on that prestigious list. Still, the rise of Japan from utter devastation in 1945 to the second-largest economy in the world has been nothing short of phenomenal. It will be helpful to review in detail some of the reasons for that success. We might call these the 10 commandments of Japan's economic success.

THE 10 COMMANDMENTS OF JAPAN'S ECONOMIC SUCCESS

1. Some of Japan's entrenched business conglomerates, called *zaibatsu*, were broken up by order of the U.S. Occupation commander after World War II; this allowed competing businesses to get a start. Similarly, the physical infrastructure— roads, factories—was destroyed during the war. This was a blessing in disguise, for it paved the way for newer equipment and technologies to be put in place quickly.

2. The United States, seeing the need for an economically strong Japan in order to offset the growing attraction of Communist ideology in Asia, provided substantial reconstruction aid. For instance, Sony Corporation got started with help from the Agency for International Development (AID)—an organization to which the United States is a major contributor. Mazda Motors got its start by making Jeeps for U.S. forces during the Korean War. (Other Rim countries that are now doing well can also thank U.S. generosity: Taiwan received $5.6 billion and South Korea received $13 billion in aid during the period 1945–1978.)

3. Japanese industry looked upon government as a facilitator and received useful economic advice as well as political and financial assistance from government planners. (In this regard, it is important to note that many of Japan's civil servants are the best graduates of Japan's colleges and universities—government service always having pride of place among career choices.) Also, the advice and help coming from the government were fairly consistent over time, because the same political party, the Liberal Democratic Party, remained in power for almost the entire postwar period.

4. Japanese businesses selected an export-oriented strategy that stressed building market share over immediate profit.

5. Except in certain professions, such as teaching, labor unions in Japan were not as powerful as in Europe and the United States. This is not to suggest that unions were not effective in gaining benefits for workers, but the structure of the union movement—individual company unions rather than industry-wide unions—moderated the demands for improved wages and benefits.

6. Company managers stressed employee teamwork and group spirit and implemented policies such as "lifetime employment" and quality-control circles, which contributed to group morale. In this they were aided by the tendency of Japanese workers to grant to the company some of the same level of loyalty traditionally reserved for families. In certain ways, the gap between workers and management was minimized. (Because of changing internal and external conditions, many Japanese firms are now abandoning the lifetime employment model in favor of "employment by performance" models).

7. Companies benefited from the Japanese ethic of working hard and saving much. For most of Japan's postwar history, workers labored six days a week, arriving early and leaving late. The paychecks were carefully managed to include a substantial savings component—generally between 15 and 25 percent. This guaranteed that there were always enough cash reserves for banks to offer company expansion loans at low interest.

8. The government spent relatively little of its tax revenues on social-welfare programs or military defense, preferring instead to invest public funds in private industry.

9. A relatively stable family structure (i.e., few divorces and substantial family support for young people, many of whom remained at home until marriage at about age 27), produced employees who were reliable and psychologically stable.

10. The government as well as private individuals invested enormous amounts of money and energy into education, on the assumption that, in a resource-poor country, the mental energies of the people would need to be exploited to their fullest.

Some of these conditions for success are now part of immutable history; but others, such as the emphasis on education, are open to change as the conditions of Japanese life change. A relevant example is the practice of lifetime employment. Useful as a management tool when companies were small and *skilled* laborers were difficult to find, it is now giving way to a freer labor-market system. In some Japanese industries, as many as 30 percent of new hires quit after two years on the job. In other words, the aforementioned conditions for success were relevant to one particular era of Japanese and world history and may not be as effective in other countries or other times. Selecting the right strategy for the right era has perhaps been the single most important condition for Japanese economic success.

CULTURAL CHARACTERISTICS

All these conditions notwithstanding, Japan would never have achieved economic success without its people possessing certain social and psychological characteristics, many of which can be traced to the various religious/ethical philosophies that have suffused Japan's 2,000-year history. Shintoism, Buddhism, Confucianism, Christianity, and other philosophies of living have shaped the modern Japanese mind. This is not to suggest that Japanese are tradition-bound; nothing could be further from the truth. Even though many Westerners think "tradition" when they think Japan, it is more accurate to think of Japanese people as imitative, preventive, pragmatic, obligative, and inquisitive rather than traditional. These characteristics are discussed in this section.

Imitative

The capacity to imitate one's superiors is a strength of the Japanese people; rather than representing an inability to think creatively, it constitutes one reason for Japan's legendary success. It makes sense to the Japanese to copy success, whether it is a successful boss, a company in the West, or an educational curriculum in Europe. It is true that imitation can produce conformity; but, in Japan's case, it is often conformity based on respect for the superior qualities of someone or something rather than simple, blind mimicry.

Once Japanese people have mastered the skills of their superiors, they believe that they have the moral right to a style of their own. Misunderstandings on this point arise often when East meets West. One American schoolteacher, for example, was sent to Japan to teach Western art to elementary-school children. Considering her an expert, the children did their best to copy her work to the smallest detail. Misunderstanding that this was at once a compliment and the first step toward creativity, the teacher removed all of her art samples from the classroom in order to force the students to paint something from their own imaginations. Because the students found this to be a violation of their approach to creativity, they did not perform well, and the teacher left Japan believing that Japanese education teaches conformity and compliance rather than creativity and spontaneity.

This episode is instructive about predicting the future role of Japan vis-à-vis the West. After decades of imitating the West, Japanese people are now beginning to feel that they have the skills and the moral right to create styles of their own. We can expect to see, therefore, an explosion of Japanese creativity in the near future. Some observers have noted, for example, that the global fashion industry seems to be gaining more inspiration from designers in Tokyo than from those in Milan, Paris, or New York. And the Japanese have often registered more new patents with the U.S. Patent Office than any other nation except the United States. The Japanese are also now winning more Nobel prizes than in the past, including the prize for literature in 1994, the prizes for chemistry in 2000, 2001, and 2002, and the prize for physics in 2002.

Preventive

Japanese individuals, families, companies, and the government generally prefer long-range over short-range planning, and they greatly prefer foreknowledge over postmortem analysis. Assembly-line workers test and retest every product to prevent customers from receiving defective products. Some store clerks plug in and check electronic devices in front of a customer in order to prevent bad merchandise from sullying the good reputation of the store, and commuter trains in Japan have three times as many "Watch your step" and similar notices as do trains in the United States and Europe. Insurance companies do a brisk business in Japan; even though all Japanese citizens are covered by the government's national health plan, many people buy additional coverage—for example, cancer insurance—just to be safe.

This concern with prevention trickles down to the smallest details. At train stations, multiple recorded warnings are given of an approaching train to commuters standing on the platform. Parent–teacher associations send teams of mothers around the neighborhood to determine which streets are the safest for the children. They then post signs designating certain roads as "school passage roads" and instruct children to take those routes even if it takes longer to walk to school. The Japanese think that it is better to avoid an accident than to have an emergency team ready when a child is hurt. Whereas Americans say, "If it ain't broke, don't fix it," the Japanese say, "Why wait 'til it breaks to fix it?"

Pragmatic

Rather than pursue a plan because it ideologically fits some preordained philosophy, the Japanese try to be pragmatic on most points. Take drugs as an example. Many nations say that drug abuse is an insurmountable problem that will, at best, be contained but probably never eradicated, because to do so would violate civil liberties. But, as a headline in the *Asahi Evening News* proclaimed a few years ago, "Japan Doesn't Have a Drug Problem and Means to Keep It That Way." Reliable statistics support this claim, but that is not the whole story. In 1954, Japan had a serious drug problem, with 53,000 drug arrests in one year. At the time, the authorities concluded that they had a big problem on their hands and must do whatever was required to solve it. The government passed a series of tough laws restricting the production, use, exchange, and possession of all manner of drugs, and it gave the police the power to arrest all violators. Users were arrested as well as dealers: It was reasoned that if the addicts were not left to buy the drugs, the dealers would be out of business. Their goal at the time was to arrest all addicts, even if it meant that certain liberties were briefly circumscribed. The plan, based on a do-what-it-takes pragmatism, worked; today, Japan is the only industrialized country without a widespread drug problem. In this case, to pragmatism was added the Japanese tendency to work for the common rather than the individual good.

This approach to life is so much a part of the Japanese mind-set that many Japanese cannot understand why the United States and other industrialized nations have so many unresolved social and economic problems. For instance, when it comes to availability of money for loans to start up businesses or purchase homes, it is clear that one of the West's most serious problems is a low personal savings rate (about three cents saved for every dollar in the United States, compared to about 15 cents in Japan). This makes money scarce and interest rates on borrowed money relatively high. Knowing that this is a problem in the United States, the Japanese wonder why Americans simply do not start saving more. They think, "We did it in the past

with a poorer economy than yours; why can't you?"

Obligative

The Japanese have a great sense of duty toward those around them. Thousands of Japanese workers work late without pay to improve their job skills so that they will not let their fellow workers down. Good deeds done by one generation are remembered and repaid by the next, and lifelong friendships are maintained by exchanging appropriate gifts and letters. North Americans and Europeans are often considered untrustworthy friends because they do not keep up the level of close, personal communications that the Japanese expect of their own friends; nor do the Westerners have as strong a sense of place, station, or position.

Duty to the group is closely linked to respect for superior authority. Every group—indeed, every relationship—is seen as a mixture of people with inferior and superior resources. These differences must be acknowledged, and no one is disparaged for bringing less to the group than someone else. However, equality is assumed when it comes to basic commitment to or effort expended for a task. Slackers are not welcome. Obligation to the group along with respect for superiors motivated Japanese pilots to fly suicide missions during World War II, and it now causes workers to go the extra mile for the good of the company.

That said, it is also true that changes in the intensity of commitment are becoming increasingly apparent. More Japanese than ever before are beginning to feel that their own personal goals are more important than those of their companies or extended families. This is no doubt a result of the Westernization of the culture since the Meiji Restoration, in the late 1800s, and especially of the experiences of the growing number of Japanese—approximately half a million in a given year—who live abroad and then take their newly acquired values back to Japan. (About half of these "away Japanese" live in North America and Western Europe.)

There is no doubt that the pace of "individualization" of the Japanese psyche is increasing and that, more and more, the Japanese attitude toward work is approaching that of the West. Many Japanese companies are now allowing employees to set their own "flex-time" work schedules, and some companies have even asked employees to stop addressing superiors with their hierarchical titles and instead refer to everyone as *san,* or Mr. or Ms.

Inquisitive

The image of dozens of Japanese business-people struggling to read a book or news-

paper while standing inside a packed commuter train is one not easily forgotten, symbolizing as it does the intense desire among the Japanese for knowledge, especially knowledge of foreign cultures. Nearly 6 million Japanese travel abroad each year (many to pursue higher education), and for those who do not, the government and private radio and television stations provide a continuous stream of programming about everything from Caribbean cuisine to French ballet. The Japanese have a yen for foreign styles of dress, foreign cooking, and foreign languages. The Japanese study languages with great intensity. Every student is required to study English; many also study Chinese, Greek, Latin, Russian, Arabic, and other languages, with French being the most popular after English—although it is evident that, on the whole, the Japanese do not have a gift for languages and struggle in vain to gain fluency.

ACHIEVEMENTS

Japan has achieved virtually complete literacy. Although there are poor areas, there are no slums inhabited by a permanent underclass. The gaps between the social classes appear to be less pronounced than in many other societies. The country seems to be entering an era of remarkable educational accomplishment, and Japanese scientists have been awarded 5 Nobel Prizes in the past decade.

Observers inside and outside of Japan are beginning to comment that the Japanese are recklessly discarding Japanese culture in favor of foreign ideas and habits, even when they make no sense in the Japanese context. A tremendous intellectual debate, called *Nihonjin-ron,* is now taking place in Japan over the meaning of being Japanese and the Japanese role in the world. There is certainly value in these concerns, but, as was noted previously, the secret about Japanese traditions is that they are not traditional. That is, the Japanese seem to know that, in order to succeed, they must learn what they need to know for the era in which they live, even if it means modifying or eliminating the past. This is probably the reason why the Japanese nation has endured for more than 2,000 years while many other empires have fallen. In this sense, the Japanese are very forward-looking people and, in their thirst for new modes of thinking and acting, they are, perhaps, revealing their most basic and useful national personality characteristic: inquisitiveness. Given this attitude toward learning, it should come as no surprise that formal schooling in Japan is a very serious business to the government and to families. It is to that topic that we now turn.

SCHOOLING

Probably most of the things that the West has heard about Japanese schools are distortions or outright falsehoods. We hear that Japanese children are highly disciplined, for example; yet in reality, Japanese schools at the elementary and junior high levels are rather noisy, unstructured places, with children racing around the halls during breaks and getting into fights with classmates on the way home. Japan actually has a far lower percentage of its college-age population enrolled in higher education than is the case in the United States—35 percent as compared to 50 percent. Moreover, the Japanese government does not require young people to attend high school (they must attend only until age 15), although 94 percent do anyway. Given these and other realities of school life in Japan, how can we explain the consistently high scores of Japanese on international tests and the general agreement that Japanese high school graduates know almost as much as college graduates in North America?

Structurally, schools in Japan are similar to those in many other countries: There are kindergartens, elementary schools, junior high schools, and high schools. Passage into elementary and junior high is automatic, regardless of student performance level. But admission to high school and college is based on test scores from entrance examinations. Preparing for these examinations occupies the full attention of students in their final year of both junior high and high school, respectively. Both parents and school authorities insist that studying for the tests be the primary focus of a student's life at those times. For instance, members of a junior high soccer team may be allowed to play on the team only for their first two years; during their last year, they are expected to be studying for their high school entrance examinations. School policy reminds students that they are in school to learn and to graduate to the next level, not to play sports. Many students even attend after-hours "cram schools" (juku) several nights a week to prepare for the exams.

Time for recreational and other non-school activities is restricted, because Japanese students attend school more days per year than students in North America. When mandatory school attendance on Saturdays was eliminated in 2002 as part of a major school reform program, many parents protested vigorously. They worried that their children would fall behind. School leaders agreed to start voluntary Saturday schools, and when they did, as many as 75 percent of their students attended. Summer vacation is only about six weeks long, and students often attend school activities during most of that period. Japanese youths are expected to treat schooling as their top priority over part-time jobs (usually prohibited by school policy during the school year, except for the needy), sports, dating, and even family time.

Children who do well in school are generally thought to be fulfilling their obligations to the family, even if they do not keep their rooms clean or help with the dishes. The reason for this focus is that parents realize that only through education can Japanese youths find their place in society. Joining Japan's relatively small military is generally not an option, opportunities for farming are limited because of land scarcity, and most major companies will not hire a new employee who has not graduated from college or a respectable high school. Thus, the Japanese find it important to focus on education—to do one thing and do it well.

Teachers are held in high regard in Japan, partly because when mass education was introduced, many of the high-status samurai took up teaching to replace their martial activities. In addition, in modern times, the Japan Teacher's Union has been active in agitating for higher pay for teachers. As a group, teachers are the highest-paid civil servants in Japan. They take their jobs very seriously. Public-school teachers, for example, visit the home of each student each year to merge the authority of the home with that of the school, and they insist that parents (usually mothers) play active supporting roles in the school.

Some Japanese youths dislike the system, and discussions are currently under way among Japanese educators on how to improve the quality of life for students. Occasionally the pressure of taking examinations (called "exam hell") produces such stress that a desperate student will commit suicide rather than try and fail. Stress also appears to be the cause of *ijime,* or bullying of weaker students by stronger peers. In recent years, the Ministry of Education has worked hard to help students deal with school stress, with the result that Japan's youth suicide rate has dropped dramatically, far lower than the youth rate in the United States (although suicide among adults, especially those affected by the downturn in the economy, has risen to its highest level since 1947). Despite these and other problems, most Japanese youths enjoy school and value the time they have to be with their friends, whether in class, walking home, or attending cram school.

Some of those who fail their college entrance exams continue to study privately, some for many years, and take the exam each year until they pass. Others travel abroad and enroll in foreign universities that do not have such rigid entrance requirements. Still others enroll in vocational training schools. But everyone in Japan realizes that education—not money, name, or luck—is the key to success.

Parents whose children are admitted to the prestigious national universities—such as Tokyo and Kyoto Universities—consider that they have much to brag about. Other parents are willing to pay as much as $35,000 on average for four years of college at the private (but usually not as prestigious) universities. Once admitted, students find that life slows down a bit. For one thing, parents typically pay more than 65 percent of the costs, and approximately 3 percent is covered by scholarships. This leaves only about 30 percent to be earned by the students; this usually comes from tutoring high school students who are studying for the entrance exams. Contemporary parents are also willing to pay the cost of a son's or daughter's traveling to and spending a few months in North America or Europe either before college begins or during summer breaks—a practice that is becoming de rigueur for Japanese students, much as taking a "grand tour" of Europe was once expected of young, upper-class Americans and Canadians.

College students may take 15 or 16 courses at a time, but classes usually meet only once or twice a week, and sporadic attendance is the norm. Straight lecturing rather than class discussion is the typical learning format, and there is very little homework beyond studying for the final exam. Students generally do not challenge the professors' statements in class, but some students develop rather close, avuncular-type relationships with their professors outside of class. Hobbies, sports, and club activities (things the students did not have time to do while in public school) occupy the center of life for many college students. Japanese professors visiting universities in North America and Europe are often surprised at how diligently students in those places study during their college years. By contrast, Japanese students spend a lot of time making friendships that will last a lifetime and be useful in one's career and private life.

THE JAPANESE BUSINESS WORLD

Successful college graduates begin their work careers in April, when most large companies do their hiring (although this practice is slowly giving way to individual hiring throughout the year). They may have to take an examination to determine how much they know about electronics or stocks and bonds, and they may have to complete a detailed personality profile. Finally, they will have to submit to a very serious interview with company management. During interviews, the managers will watch their every move; the applicants will be careful to avoid saying anything that will give them "minus points."

Once hired, individuals attend training sessions in which they learn the company song and other rituals as well as company policy on numerous matters. They may be housed in company apartments (or may continue to live at home), permitted to use a company car or van, and advised to shop at company grocery stores. Almost never are employees married at this time, and so they are expected to live a rather spartan life for the first few years.

Employees are expected to show considerable deference to their section bosses, even though, on the surface, bosses do not appear to be very different from other employees. Bosses' desks are out in the open, near the employees; they wear the same uniform; they socialize with the employees after work; even in a factory, they are often on the shop floor rather than sequestered away in private offices. Long-term employees often come to see the section leader as an uncle figure (bosses are usually male) who will give them advice about life, be the best man at their weddings, and provide informal marital and family counseling as needed.

Although there are cases of abuse or unfair treatment of employees, Japanese company life can generally be described as somewhat like a large family rather than a military squad; employees (sometimes called associates) often obey their superiors out of genuine respect rather than forced compliance. Moreover, competition between workers is reduced because everyone hired at the same time receives more or less the same pay and most workers receive promotions at about the same time. Only later in one's career are individualistic promotions given. That said, it is important to note that, under pressure to develop a more inventive workforce to compete against new ideas and products from the West, many Japanese companies are now experimenting with new pay and promotion systems that reward performance, not longevity.

Employees are expected to work hard, for not only are Japanese companies in competition with foreign businesses, but they also must survive the fiercely competitive business climate at home. Indeed, the Japanese skill in international business was developed at home. There are, for example,

hundreds of electronics companies and thousands of textile enterprises competing for customers in Japan. And whereas the United States has only four automobile-manufacturing companies, Japan has nine. All these companies entice customers with deep price cuts or unusual services, hoping to edge out unprepared or weak competitors. Many companies fail. There were once, for instance, almost 40 companies in Japan that manufactured calculators, but today only half a dozen remain, the rest victims of tough internal Japanese competition.

At about age 30, after several years of working and saving money for an apartment, a car, and a honeymoon (wedding and reception costs of approximately US$27,000 on average are shared by the bride and groom's parents), the typical Japanese male worker marries (although there is an increasing number of Japanese males and females who opt to never marry). The average bride, about age 27, may have taken private lessons in flower arranging, the tea ceremony, sewing, cooking, and perhaps a musical instrument like the *koto*, the Japanese harp. She probably will not have graduated from college, although she may have attended a specialty college for a while. If she is working, she likely is paid much less than her husband, even if she has an identical position (despite equal-pay laws). She may spend her time in the company preparing and serving tea for clients and employees, dusting the office, running errands, and answering telephones. When she has a baby, she will be expected to quit—although more women today are choosing to remain on the job, and some are advancing into management or are leaving to start their own companies.

Because the wife is expected to serve as the primary caregiver for the children, the husband is expected always to make his time available for the company. He may be asked to work weekends, to stay out late most of the week (about four out of seven nights), or even to be transferred to another branch in Japan or abroad without his family. This loyalty is rewarded in numerous ways: Unless the company goes bankrupt or the employee is unusually inept, he may be permitted to work for the company until he retires, usually at about age 55 or 60, even if the company no longer really needs his services; he and his wife will be taken on company sightseeing trips; the company will pay most of his health-insurance costs (the government pays the rest); and he will have the peace of mind that comes from being surrounded by lifelong friends and workmates. His association with company employees will be his main social outlet, even after retirement; upon his death, it will be his former workmates who organize

and direct his funeral services. There are many younger employees who fear the strictures of the traditional company and prefer to work in small and medium-sized businesses with a more individualistic work environment.

THE FAMILY

The loyalty once given to the traditional Japanese extended family, called the *ie*, has been transferred to the modern company. This is logical from a historical perspective, since the modern company once began as a family business and was gradually expanded to include more workers, or "siblings." Thus, whereas the family is seen as the backbone of most societies, it might be more accurate to argue that the *kaisha*, or company, is the basis of modern Japanese society. As one Japanese commentator explained, "In the West, the home is the cornerstone of people's lives. In Tokyo, home is just a place to sleep at night.… Each family member— husband, wife, and children—has his own community centered outside the home."

Thus, the common image that Westerners hold of the centrality of the family to Japanese culture may be inaccurate. For instance, father absence is epidemic in Japan. It is an unusual father who eats more than one meal a day with his family. He may go shopping or to a park with his family when he has free time from work, but he is more likely to go golfing with a workmate. Schooling occupies the bulk of the children's time, even on weekends. And with fewer children than in earlier generations and with appliance-equipped apartments, many Japanese women rejoin the workforce after their children are self-maintaining.

Japan's divorce rate, while rising, is still considerably lower than in other industrialized nations (2.3 per thousand in Japan compared to 4.19 in the U.S.), a fact that may seem incongruent with the conditions described above. Yet, as explained by one Japanese sociologist, Japanese couples "do not expect much emotional closeness; there is less pressure on us to meet each other's emotional needs. If we become close, that is a nice dividend, but if we do not, it is not a problem because we did not expect it in the first place."

Despite these modifications to the common Western image of the Japanese family, Japanese families have significant roles to play in society. Support for education is one of the most important. Families, especially mothers, support the schools by being actively involved in the parent-teacher association, by insisting that children be given plenty of homework, and by saving for college so that the money for tu-

(UN photo by Jan Corash)

In Japan, not unlike in many other parts of the world, economic well-being often requires two incomes. Still, there is strong social pressure on women to stop working once they have a baby. All generations of family members take part in childrearing.

ition is available without the college student having to work.

Another important function of the family is mate selection. Somewhat fewer than half of current Japanese marriages are arranged by the family or have occurred as a result of far more family involvement than in North America. Families sometimes ask a go-between (an uncle, a boss, or another trusted person) to compile a list of marriageable candidates. Criteria such as social class, blood type, and occupation are considered. Photos of prospective candidates are presented to the unmarried son or daughter, who has the option to veto any of them or to date those he or she finds acceptable. Young people, however, increasingly select their mates with little or no input from parents.

Finally, families in Japan, even those in which the children are married and living away from home, continue to gather for the purpose of honoring the memory of deceased family members or to enjoy one another's company for New Year's Day, Children's Day, and other celebrations.

WOMEN IN JAPAN

Ancient Confucian values held that women were legally and socially inferior to men. This produced a culture in feudal Japan in which the woman was expected to walk several steps behind her husband when in public, to eat meals only after the husband had eaten, to forgo formal education, and to serve the husband and male members of the family whenever possible. A "good woman" was said to be one who would endure these conditions without complaint. This pronounced gender difference (though minimized substantially over the centuries since Confucius) can still be seen today in myriad ways, including in the preponderance of males in positions of leadership in business and politics, in the smaller percentage of women college graduates, and in the pay differential between women and men.

Given the Confucian values noted above, one would expect that all top leaders would be males. However, women's roles are also subject to the complexity of both ancient and modern cultures. Between A.D. 592 and 770, for instance, of the 12 reigning emperors, half were women. The debate on whether or not to allow a woman to become empress has been engaged again, and this because the heir-apparent's recently born only child is a girl. In rural areas today, women take an active decision-making role in farm associations. In the urban workplace, some women occupy typically pink-collar positions (nurses,

clerks, and so on), but many women are also doctors and business executives; 28,000 are company presidents.

Thus, it is clear that within the general framework of gender inequality imposed by Confucian values, Japanese culture, especially at certain times, has been rather lenient in its application of those values. There is still considerable social pressure on women to stop working once they marry, and particularly after they have a baby, but it is clear that many women are resisting that pressure: one out of every three employees in Japan is female, and nearly 60 percent of the female workforce are married. An equal-pay law was enacted in 1989 that makes it illegal to pay women less for doing comparable work (although it may take years for companies to comply fully). And the Ministry of Education has mandated that home economics and shop classes now be required for both boys and girls; that is, both girls and boys will learn to cook and sew as well as construct things out of wood and metal.

In certain respects, and contrary to the West's image of Japanese gender roles, some Japanese women seem more assertive than women in the West. For example, in a recent national election, a wife challenged her husband for his seat in the House of Representatives (something that

29

Timeline: PAST

20,000–4,500 B.C.
Prepottery, paleolithic culture

4,500–250 B.C.
Jomon culture with distinctive pottery

250 B.C.–A.D. 300
Yayoi culture with rice agriculture, Shinto religion, and Japanese language

A.D. 300–700
The Yamato period; warrior clans import Chinese culture

710–794
The Nara period; Chinese-style bureaucratic government at the capital at Nara

794–1185
The Heian period; the capital is at Kyoto

1185–1333
The Kamakura period; feudalism and shoguns; Buddhism is popularized

1333–1568
The Muromachi period; Western missionaries and traders arrive; feudal lords control their own domains

1568–1600
The Momoyama period; feudal lords become subject to one central leader; attempted invasion of Korea

1600–1868
The Tokugawa Era; self-imposed isolation from the West

1868–1912
The Meiji Restoration; modernization; Taiwan and Korea are under Japanese control

1912–1945
The Taisho and Showa periods; militarization leads to war and Japan's defeat

1945
Japan surrenders; the U.S. Occupation imposes major changes in the organization of society

1951
Sovereignty is returned to the Japanese people by treaty

1980s
The ruling party is hit by scandals but retains control of the government; Emperor Hirohito dies; Emperor Akihito succeeds

1990s
After years of a slow economy, Japan officially admits it is in a recession; a devastating earthquake in Kobe kills more than 6,000 people

PRESENT

2000s
A U.S. Navy submarine accidentally rams into and sinks a Japanese trawler; nine Japanese are killed; relations with the United States are strained

Experts predict that the worker/retiree ratio will drop from 6:1 to 2:1 by 2020, as the population ages

Japanese business undergoes a shift toward more openness and flexibility

Weaknesses in the Japanese economy are a concern all over the world

Japan's economy begins a slow rebound from a decade of stagnation

---

has not been done in the United States, where male candidates usually expect their wives to stump for them). Significantly, too, the former head of the Japan Socialist Party was an unmarried woman, and in 1999 Osaka voters elected a woman as mayor for the first time. Women have been elected to the powerful Tokyo Metropolitan Council and awarded professorships at prestigious universities such as Tokyo University. And, while women continue to be used as sexual objects in pornography and prostitution, certain kinds of misogynistic behavior, such as rape and serial killing, are less frequent in Japan than in Western societies. New laws against child pornography may reduce the abuse of young girls, but a spate of bizarre killings by youths and mafia in recent years is causing the sense of personal safety in Japan to dissipate. Signs in train stations warn of pickpockets, and signs on infrequently traveled paths warn of molesters.

Recent studies show that many Japanese women believe that their lives are easier than those of most Westerners. With their husbands working long hours and their one or two children in school all day, Japanese women find they have more leisure time than Western women. Gender-based social divisions remain apparent throughout Japanese culture, but modern Japanese women have learned to blend these divisions with the realities and opportunities of the contemporary workplace and home.

RELIGION/ETHICS

There are many holidays in Japan, most of which have a religious origin. This fact, as well as the existence of numerous shrines and temples, may leave the impression that Japan is a rather religious country. This is not true, however. Most Japanese people do not claim any active religious affiliation, but many will stop by a shrine occasionally to ask for divine help in passing an exam, finding a mate, or recovering from an illness. A recent increase in temple or shrine visits has been attributed to families asking for divine help for those unemployed due to the recession.

Nevertheless, modern Japanese culture sprang from a rich religious heritage. The first influence on Japanese culture came from the animistic Shinto religion, from whence modern Japanese acquired their respect for the beauty of nature. Confucianism brought a respect for hierarchy and education. Taoism stressed introspection, and Buddhism taught the need for good behavior now in order to acquire a better life in the future.

Shinto was selected in the 1930s as the state religion and was used as a divine jus-

tification for Japan's military exploits of that era, but most Japanese today will say that Japan is, culturally, a Buddhist nation. Some new Buddhist denominations have attracted thousands of followers. The rudiments of Christianity are also a part of the modern Japanese consciousness, but few Japanese have actually joined Christian churches. Sociologically, Japan, with its social divisions and hierarchy, is probably more of a Confucian society than it is Buddhist or any other philosophy, although Confucianism is so deeply woven into the fabric of Japanese life that few would recognize it as a distinct philosophy.

Most Japanese regard morality as springing from within the group rather than pronounced from above. That is, a Japanese person may refrain from stealing so as not to offend the owner of an object or bring shame upon the family, rather than because of a divine prohibition against stealing. Thus we find in Japan a relatively small rate of violent—that is, public—crimes, and a much larger rate of white-collar crimes such as embezzlement, in which offenders believe that they can get away with something without creating a public scandal for their families.

THE GOVERNMENT

The Constitution of postwar Japan became effective in 1947 and firmly established the Japanese people as the ultimate source of sovereignty, with the emperor as the symbol of the nation. The national Parliament, or *Diet,* is empowered to pass legislation. The Diet is divided into two houses: the House of Representatives, with 480 members elected for four-year terms; and the House of Councillors, with 252 members elected for six-year terms from each of the 47 prefectures (states) of Japan as well as nationally. The prime minister, assisted by a cabinet, is also the leader of the party with the most seats in the Diet. Prefectures are governed by an elected governor and an assembly, and cities and towns are governed by elected mayors and town councils. The Supreme Court, consisting of a chief judge and 14 other judges, is independent of the legislative branch of government.

Japan's Constitution forbids Japan from engaging in war or from having military capability that would allow it to attack another country. Japan does maintain a well-equipped self-defense force, but it relies on a security treaty with the United States in case of serious aggression against it. In recent years, the United States has been encouraging Japan to assume more of the burden of the military security of the Asian region, and Japan has increased its expenditures in absolute terms. But until the

Constitution is amended, Japan is not likely to initiate any major upgrading of its military capability. This is in line with the general wishes of the Japanese people, who, since the devastation of Hiroshima and Nagasaki, have become firmly committed to a pacifist foreign policy. Traditionally, Japanese politicians have feared that any dramatic increase in Japan's military posture would trigger a similar response from other Asian nations. However, in recent years, it has become clear that nearby countries, particularly China, are set on a course of rapid militarization, regardless of Japan's posture. Consequently, Prime Minister Koizumi and others have started becoming more vocal about Japan's need to build a stronger military and to consider the re-writing or elimination of Article 9 of the Constitution that prohibits offensive weapons. In 2003, the legislature passed historic bills that would allow Japan to deploy troops defensively, whether or not the country had been attacked. Many opposed the bills, while others argued that Japan should even develop its own nuclear capability. Thus, the debate is engaged. The urgency of this debate became clear in 1998, when North Korea fired a test missile over the Japanese archipelago. As China develops its military into superpower status, Japan may feel that it has no choice but to acquire offensive military capabilities. But it is a debate that the Japanese would prefer not to have at all.

This tendency toward not wanting to get involved militarily is reflected in one of Japan's most recent performances on the world stage. The Japanese were slow to play any significant part in supporting military expenditures for the Persian Gulf War, even when the outcome had a direct potential effect on their economy. The Iraqi invasion of Kuwait in August 1990 brought on the wrath—against Japan—of a coalition of countries led by the United States in January 1991, but it generated an initial commitment from Japan of only $2 billion (later increased to $9 billion, still a small fraction of the cost) and no personnel of any kind. This meager support was criticized by some foreign observers, who pointed out that Japan relies heavily on Gulf oil. When the United States invaded Iraq, Japan sent 1,000 air and naval forces to support US troops, and the prime minister has consistently supported the Bush administration's Iraq policy.

In 1992, the Japanese government announced its intention of building its own F-16–type jet-fighter planes; and subsequently, amid protests from the public, the Diet voted to send as many as 1,800 Japanese soldiers—the first to go abroad since World War II—to Cambodia to assist in the UN–supervised peacekeeping effort. Countries that had experienced the full force of Japanese domination in the past, such as China and Korea, expressed dismay at these evidences of Japan's modern military capability, but the United States welcomed the moves as an indication of Japan's willingness to share the costs of providing military security to Asia.

The Japanese have formed numerous political parties to represent their views in government. Among these have been the Japan Communist Party, the Social Democratic Party, and the New Frontier Party. For nearly 40 years, however, the most powerful party was the Liberal Democratic Party (LDP). Formed in 1955, it guided Japan to its current position of economic strength, but a series of sex and bribery scandals caused it to lose control of the government in 1993. A shaky coalition of eight parties took control for about a year but was replaced by an even more unlikely coalition of the LDP and the Japan Socialists—historic enemies who were unable to agree on most policies. Eventually, the LDP was able to regain some of its lost political clout; but with some half a dozen changes in the prime ministership in the 1990s and continuous party realignments, it would be an understatement to say that Japan's government has been in flux. Under the current LDP Prime Minister, Junichiro Koizumi, politics seem to have returned to normal (although normal also includes occasional brawls on the floor of the Diet, as in 2004, when members of the upper house physically assaulted each other after an all-night debate on sending troops to Iraq). Currently, the LDP rules in an unlikely coalition with the Buddhist-influenced New Komeito Party, but in upper house elections in 2004, the Democratic Party of Japan and other opposition parties reduced the LDP's majority on the strength of public dissatisfaction with Koizumi's pension reform plans and his unwavering support of the Bush Administration's Iraq War. Koizumi, who has relied more on his personal popularity than on playing "factional" politics, has nevertheless had to reshuffle his cabinet several times since coming to power in 2001 in order to please various constituencies within his ruling coalition. With his popularity dropping in opinion polls, it is likely he will step down in 2006, after about 5 years in office.

Part of the reason for Japan's political instability can be explained by Japan's party faction system. Party politics in Japan has always been a mixture of Western-style democratic practice and feudalistic personal relationships. Japanese parties are really several parties rolled into one. That is, parties are divided into several factions, each comprised of a group of loyal younger members headed by a powerful member of the Diet. The senior member has a duty to pave the way for the younger members politically, but they, in turn, are obligated to support the senior member in votes and in other ways. The faction leader's role in gathering financial support for faction members is particularly important, because Diet members are expected by the electorate to be patrons of numerous causes, from charity drives to the opening of a constituent's fast-food business. Because parliamentary salaries are inadequate to the task, outside funds, and thus the faction, are crucial. The size and power of the various factions are often the critical elements in deciding who will assume the office of prime minister and who will occupy which cabinet seats. The role of these intraparty factions is so central to Japanese politics that attempts to ban them have never been successful.

The factional nature of Japanese party politics means that cabinet and other political positions are frequently rotated. This would yield considerable instability in governance were it not for the stabilizing influence of the Japanese bureaucracy. Large and powerful, the career bureaucracy is responsible for drafting more than 80 percent of the bills submitted to the Diet. Many of the bureaucrats are graduates from the finest universities in Japan, particularly Tokyo University, which provides some 80 percent of the senior officials in the more than 20 national ministries. Many of them consider their role in long-range forecasting, drafting legislation, and implementing policies to be superior to that of the elected officials under whom they work. They reason that, whereas the politicians are bound to the whims of the people they represent, bureaucrats are committed to the nation of Japan—to, as it were, the *idea* of Japan. So superior did bureaucrats feel to their elected officials (and bosses) that until recently, elected officials were not questioned directly in the Diet; career bureaucrats represented their bosses to the people. Generally speaking, government service is considered a higher calling than are careers in private business, law, and other fields. However, in recent years, a number of scandals involving both corrupt and incompetent bureaucrats in the Ministry of Finance and other ministries have tarnished the image of this once-unassailable category of professionals.

In addition to the bureaucracy, Japanese politicians have leaned heavily on big business to support their policies of postwar reconstruction, economic growth, and social reform. Business has accepted heavy taxation so that social-welfare programs such as the national health plan are feasible, and they have provided political candidates with substantial financial help. In turn, the government has seen its role as that of facilitating the growth of private industry (some critics claim that the relationship between government and business is so close that Japan is best described not as a nation but as "Japan, Inc."). Consider, for example, the powerful Ministry of Economy, Trade, and Industry (METI, formerly MITI). Over the years, it has worked closely with business, particularly the Federation of Economic Organizations (Keidanren) to forecast potential market shifts, develop strategies for market control, and generally pave the way for Japanese businesses to succeed in the international marketplace. The close working relationship between big business and the national government is an established fact of life in Japan, and, despite criticism from countries with a more laissez-faire approach to business, it will undoubtedly continue into the future, because it has served Japan well.

The prolonged recession in Japan has uncovered inconsistencies in the structure of Japanese society. For instance, we have seen that the Japanese are an inquisitive, change-oriented people, yet the administrative structure of the government and of the business community seems impervious to change. Decisions take an agonizingly long time to make, and outsiders as well as the Japanese public find the government's mode of operation to be far from open and transparent. For instance, in late 1999, one of Japan's 51 nuclear reactors leaked radiation and endangered the lives of many people. At first the government decided not to announce the leak at all. Then, when it finally did, it claimed that only a few people had been affected. In fact, more than 400 people had been affected, and the government seemed incapable of telling the whole truth about the matter. Many blamed bureaucratic rigidity. Similarly, in 1999–2000, the government injected more than $62 billion into the economy (the ninth such stimulus package since the recession started in 1989), yet the economy remained sluggish. Again, many blamed the snail-paced decision-making process as a factor that makes it difficult for companies to quickly take advantage of new opportunities in the marketplace. Whether Japan can make the structural changes necessary to remain competitive in the new global economy remains to be seen.

There are, of course, many other factors besides Japan's decision-making process that slow or disrupt economic growth. For instance, like other Pacific Rim countries,

Japan is regularly afflicted by serious natural disasters. Japan is located in such a seismically active area that the country experiences some 300 small earthquakes (many of them unfelt by the average person) per day. Tsunami warnings were posted in early 2005 when an offshore earthquake raised wave action along Japan's lengthy coastline. Gale-force winds occasionally hit eastern Japan, causing blackouts and stranding ships. Fourteen people were injured in a 7.3 magnitude earthquake in western Japan in 2004, and the strongest typhoon in several decades hit the island of Okinawa. In the same year, tropical storm Meari caused the deaths of 22 people and triggered landslides that drove 10,000 from their homes. Another storm killed 77 people. Also in 2004, one of Japan's most active volcanoes, Mount Asama, erupted, spewing molten rocks and smoke into the air and covering Tokyo—the world's largest city with over 35 million inhabitants—for several days with volcanic ash. Several powerful earthquakes also killed 39 people in northern Japan, derailing a high-speed train and destroying many homes and roads. Like the tsunami that hit South Asia in late 2004, a large tsunami hit southern Japan in 1993 killing hundreds and obliterating entire towns, and in 1995, a large earthquake hit the heavily populated Kobe area killing more than 6,000 people.

Sometimes, Japan's disasters are manmade. Mercury poisoning in the 1980s, for example, killed some 1,700 people and caused mothers to give birth to severely deformed babies. The pollutant, dumped by an irresponsible Japanese company, affected drinking water for an entire town. It took two decades to resolve the matter in the Japanese courts. Accidents at some of Japan's many nuclear power plants have also killed workers and forced many residents from their homes. So angry are the citizens over incidents like this that it is now virtually impossible for new power plants to be approved by the government (some 11 new plants have been proposed), yet the old plants are aging and will likely produced even more serious problems. Finally, despite Japan's strong advocacy of the Kyoto Protocol on global warming, experts warn that Japan, itself, is not likely to meet greenhouse emission targets, as it continues to send tons of pollutants into the air each day.

THE FUTURE

In the postwar years of political stability, the Japanese have accomplished more than anyone, including themselves, thought possible. Japan's literacy rate is 99 percent, 99 percent of Japanese households have

telephones, 99 percent have color televisions, and 75 percent own automobiles. Nationalized health care covers every Japanese citizen, and the Japanese have the longest life expectancy in the world. In other ways too, Japan has dramatically improved life for its people. For instance, the average size of Japanese homes, once criticized by Westerners as too small to live in, has been increasing every year until it is now just about the same as the average size of European homes. Home ownership also compares favorably or even exceeds other industrialized nations: 60 percent of homes are owned in Japan compared to 66 percent in the United States, 53 percent in France and 38 percent in Germany. Japan's success has been noted and emulated by people all over the world. For instance, as of 2003, 2.35 million foreigners were studying Japanese and the number of schools and colleges teaching the language increased 10-fold in the last two decades. Millions of tourists, businesspeople, educators, and others (5.8 million people from 190 countries in 2002) travel to Japan each year to observe Japan's success for themselves, and in 2002 over 35,000 marriages were made in which one spouse was Japanese and the other a foreigner.

With only half the population of the United States, a land area about the size of Great Britain, and extremely limited natural resources (it has to import 99.6 percent of its oil, 99.8 percent of its iron, and 86.7 percent of its coal), Japan has nevertheless created the second-largest economy in the world (or third-largest, depending on which way one counts). Japan is about to create the world's largest bank, the economy is back on investors' A-Lists despite a long recession, and Tokyo consistently ranks as one of the world's most sophisticated cities (and the most expensive). Where does Japan go from here?

When the Spanish were establishing hegemony over large parts of the globe, they were driven in part by the desire to bring Christianity to the "heathen." The British, for their part, believed that they were taking "civilization" to the "savages" of the world. China and the former Soviet Union were once strongly committed to the ideals of communism, while the United States has felt that its mission is that of expanding democracy and capitalism.

What about Japan? For what reason do Japanese businesses buy up hotels in New Zealand and skyscrapers in New York? What role does Japan have to play in the world in addition to spawning economic development? What values will guide and perhaps temper Japan's drive for economic dominance?

These are questions that the Japanese people themselves are attempting to answer; but, finding no ready answers, they are beginning to encounter more and more difficulties with the world around them and within their own society. Animosity over the persistent trade imbalance in Japan's favor continues to simmer in Europe and North America as well as in some countries of the Pacific Rim. To deflect these criticisms, Japan has substantially increased its gift-giving to foreign governments, including allocating money for the stabilization or growth of democracy in Central/Eastern Europe and for easing the foreign debt burden of Mexico and other countries.

What Japan has been loathe to do, however, is remove the "structural impediments" that make it difficult for foreign companies to do business in Japan. For example, 50 percent of the automobiles sold in Iceland are Japanese, which means less profit for the American and European manufacturers who used to dominate car sales there. Yet because of high tariffs and other regulations, very few American and European cars have been sold in Japan. Beginning in the mid-1980s, Japan reluctantly began to dismantle many of these trade barriers, and the process has been so successful that Japan now has a lower overall average tariff on nonagricultural products than the United States—its severest critic in this arena.

But Japanese people worry that further opening of their markets may destroy some fundamentals of Japanese life. Rice, for instance, costs much more in Japan than it should, because the Japanese government protects rice farmers with subsidies and limits most rice imports from abroad. The Japanese would prefer to pay less for rice at the supermarket, but they also argue that foreign competition would prove the undoing of many small rice farmers, whose land would then be sold to housing developers. This, in turn, would destroy more of Japan's scarce arable land and weaken the already shaky traditions of the Japanese countryside—the heart of traditional Japanese culture and values.

Today, thousands of foreign firms do business in Japan; some of them, like Polaroid and Schick, control the Japanese market in their products. Foreign investment in Japan has grown about 16 percent annually since 1980. In the case of the United States, the profit made by American firms doing business in Japan (nearly 800 of them) in a single year is just about equal to the amount of the trade imbalance between Japan and the United States. Japanese supermarkets are filled with foreign foodstuffs, and the radio and television airwaves are filled with the sounds and sights of Western music and dress. Japanese youths are as likely to eat at McDonald's or

Kentucky Fried Chicken outlets as at traditional Japanese restaurants, and many Japanese have never worn a kimono nor learned to play a Japanese musical instrument. It is clear to many observers that, culturally, Japan already imports much more from the West than the West does from Japan.

Given this overwhelming Westernization of Japan as well as Japan's current capacity to continue imbibing Western culture, even the change-oriented Japanese are beginning to ask where they, as a nation, are going. Will national wealth, as it slowly trickles down to individuals, produce a generation of hedonistic youths who do not appreciate the sacrifices of those before them? Will there ever be a time when, strapped for resources, the Japanese will once again seek hegemony over other nations? What future role should Japan assume in the international arena, apart from economic development? If these questions remain to be answered, circumstances of international trade have at least provided an answer to the question of Japan's role in the Pacific Rim countries: It is clear that, for the next several decades, Japan, albeit increasingly looking over its shoulder at the new China, will continue to shape the pace and nature of economic development, and thus the political environment, of the entire Pacific Rim.

Australia (Commonwealth of Australia)

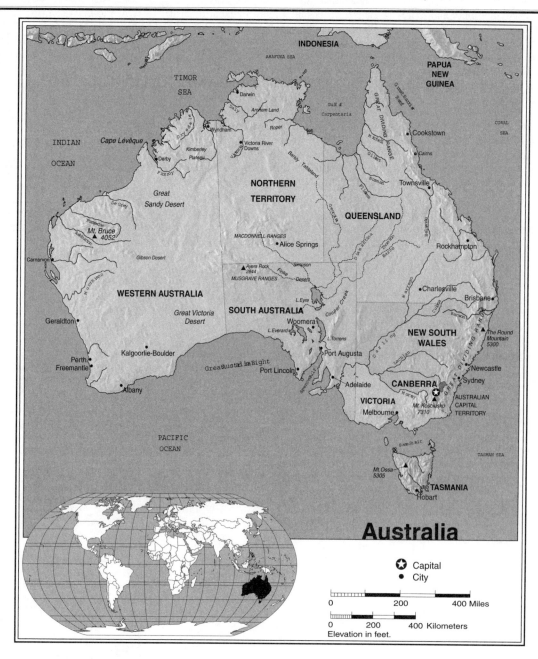

Australia

⭐ Capital
• City

0   200   400 Miles

0   200   400 Kilometers
Elevation in feet.

Australia Statistics

GEOGRAPHY

Area in Square Miles (Kilometers): 2,867,896 (7,686,850) (slightly smaller than the United States)

Capital (Population): Canberra (387,000)

Environmental Concerns: depletion of the ozone layer; pollution; soil erosion and excessive salinity; desertification; wildlife habitat loss; degradation of Great Barrier Reef; limited freshwater resources

Geographical Features: mostly low plateau with deserts; fertile plain in southeast

Climate: generally arid to semiarid; temperate to tropical

PEOPLE

Population

Total: 19,913,144

Annual Growth Rate: 0.9%

Rural/Urban Population (Ratio): 15/85

Major Languages: English; indigenous languages

Ethnic Makeup: 92% Caucasian; 7% Asian; 1% Aboriginal and others

Religions: 26% Anglican; 26% Roman Catholic; 24% other Christian; 24% unaffiliated or other

Health

Life Expectancy at Birth: 77 years (male); 83 years (female)
Infant Mortality: 4.9/1,000 live births
Physicians Available: 1/400 people
HIV/AIDS Rate in Adults: 0.15%

Education

Adult Literacy Rate: 100%
Compulsory (Ages): 6–15; free

COMMUNICATION

Telephones: 10.815 million main lines
Daily Newspaper Circulation: 297/1,000 people
Televisions: 641/1,000 people
Internet Users: 10,630,000 (2002)

TRANSPORTATION

Highways in Miles (Kilometers): 566,060 (913,000)
Railroads in Miles (Kilometers): 20,968 (33,819)

Usable Airfields: 444
Motor Vehicles in Use: 10,705,000

GOVERNMENT

Type: federal parliamentary state
Independence Date: January 1, 1901 (federation of U.K. colonies)
Head of State/Government: Queen Elizabeth II; Prime Minister John Howard
Political Parties: Liberal Party; National Party; Australian Labour Party; Australian Democratic Party; Green Party; One Nation Party; others
Suffrage: universal and compulsory at 18

MILITARY

Military Expenditures (% of GDP): 2.8%
Current Disputes: territorial claim in Antarctica; boundary issues with East Timor and Indonesia

ECONOMY

Currency ($ U.S. Equivalent): 1.54 Australian dollars = $1
Per Capita Income/GDP: $29,000/$571 billion
GDP Growth Rate: 3%
Inflation Rate: 2.8%
Unemployment Rate: 6%
Labor Force by Occupation: 73% services; 22% industry; 5% agriculture
Natural Resources: bauxite; diamonds; coal; copper; iron ore; oil; natural gas; other minerals
Agriculture: wheat; barley; sugarcane; fruit; livestock
Industry: mining; industrial and transportation equipment; food processing; chemicals; steel; tourism
Exports: $68.7 billion (primary partners developing countries, Japan, ASEAN)
Imports: $82.9 billion (primary partners developing countries, Europe, United States, Japan)

Australia Country Report

Despite its out-of-the-way location, far south of the main trading routes between Europe and Asia, seafarers from England, Spain, and the Netherlands began exploring parts of the continent of Australia in the seventeenth century. The French later made some forays along the coast, but it was the British who first found something to do with a land that others had disparaged as useless: They decided to send their prisoners there. The British had long believed that the easiest solution to prison overcrowding was expulsion from Britain. Convicts had been sent to the American colonies for many years, but after American independence was declared in 1776, Britain began to send prisoners to Australia.

DEVELOPMENT

Mining of nickel, iron ore, and other metals continues to supply a substantial part of Australia's gross domestic product. In recent years, Japan has become Australia's primary trading partner rather than Great Britain. Seven out of 10 of Australia's largest export markets are Asian countries.

Australia seemed like the ideal spot for a penal colony: It was isolated from the centers of civilization; it had some good harbors; and, although much of the continent was a flat, dry, riverless desert with only sparse vegetation, the coastal fringes were well suited to human habitation. Indeed, although the British did not know it in the 1700s, they had come across a huge continent endowed with abundant natural resources. Along the northern coast (just south of present-day Indonesia and Papua New Guinea) was a tropical zone with heavy rainfall and tropical forests. The eastern coast was wooded with valuable pine trees, while the western coast was dotted with eucalyptus and acacia trees. Minerals, especially coal, gold, nickel, petroleum, iron, and bauxite, were plentiful, as were the many species of unique animals: kangaroos, platypus, and koalas, to name a few. (As it happens, these unique animals are lucky to be alive, for it was just off the northern coast of Australia that a meteorite six miles across smashed into the earth 250 million years ago, leaving a 125-mile wide crater and wiping out some 70 percent of the world's land species).

Today, grazing and agricultural activities generally take place in the central basin of the country, which consists of thousands of miles of rolling plains. The bulk of the population resides along the coast, where the relatively few good harbors are found. Many of Australia's lakes are saltwater lakes left over from an ancient inland sea, much like the Great Salt Lake in North America. Much of the interior is watered by deep artesian wells that are drilled by lo-

cal ranchers, since Australia's mountain ranges are generally not high enough to provide substantial water supplies.

FREEDOM

Australia is a parliamentary democracy adhering to the ideals incorporated in English common law. Constitutional guarantees of human rights apply to all of Australia's citizens. However, social discrimination continues, and, despite improvements since the 1960s, the Aborigines remain a neglected part of society.

The British chose to build their first penal colony alongside a good harbor that they called Sydney. By the 1850s, when the practice of transporting convicts stopped, more than 150,000 prisoners, including hundreds of women, had been sent there and to other colonies. Most of them were illiterate English and Irish from the lower socioeconomic classes. Once they completed their sentences, they were set free to settle on the continent. These released prisoners, their guards, gold prospectors, and of course, the Aborigines, constituted the foundation of today's Australian population. The population of 19.9 million people is still quite small compared to the size of the continent.

(Australian Information Service photo)

Most Aborigines eventually adapted to the Europeans' customs, but some continue to live in their traditional ways on tribal reservations.

RACE RELATIONS

Convicts certainly did not constitute the beginning of human habitation on the continent. Tens of thousands of Aborigines (literally, "first inhabitants") inhabited Australia and the nearby island of Tasmania when Europeans first made contact. Living in scattered tribes and speaking hundreds of entirely unrelated languages, the Aborigines, whose origin is unknown (various scholars see connections to Africa, the Indian subcontinent, or the Melanesian Islands), survived by fishing and nomadic hunting. Succumbing to European diseases, violence, forced removal from their lands, and, finally, neglect, thousands of Aborigines died during the first centuries of contact. Indeed, the entire Tasmanian grouping of people (originally

numbering 5,000) is now extinct. Today's Aborigines continue to suffer discrimination. A 1997 bill that would have liberalized land rights for Aborigines failed in Parliament. In 2002, Australia's highest court dismissed a land claim by the Yorta Yorta tribe that would have guaranteed their access to 800 square miles of ancestral lands. Even after 10 years of effort, the Council on Aboriginal Reconciliation was not able to persuade the government to issue a formal apology for past abuses (some 62 percent of the population were opposed to a formal apology). In 2004, the conservative coalition government of John Howard announced plans to abolish Australia's 14-year old Aboriginal and Torres Strait Islander Commission. Some see the Commission has having failed to improve life for the Aborigines, but Aborigine leaders warned of a violent backlash if the Commission were dissolved. With only one Aboriginal member in Parliament to represent the views of the 400,000 Aboriginals (2 percent of the total population), it has always been difficult for the Aboriginal voice to be heard in the halls of government. Occasionally, with Aborigines living in high crime areas and having to deal with alcoholism, drug abuse and high unemployment, frustrations boil over. In 2004, rioting continued for 9 hours in a poor Aboriginal neighborhood in Sydney after a teenager was killed in a chase with police. Rioters threw gas bombs and set fire to a train station, injuring 40 police officers. Despite these problems, Howard, age 65, was elected to an unprecedented 4th term in office, gaining 52 percent of the vote for his coalition government compared to 48 percent for the Labour Party in the 2004 elections. If he completes this term, he will have been prime minister for twelve years. His success in building a robust economy (unemployment at nearly all-time lows and inflation at just 2 percent) seems to be the secret behind his frequent wins.

After Europeanization of the continent, most Aborigines adopted European ways, including converting to Christianity. Today, many live in the cities or work for cattle and sheep ranchers. Others reside on reserves (tribal reservations) in the central and northern parts of Australia. Yet modernization has affected even the reservation Aborigines—some have telephones, and some dispersed tribes in the Northern Territories communicate with one another by satellite-linked video conferencing—but in the main, they continue to live as they have always done, organizing their religion around plant or animal sacred symbols, or totems, and initiating youth into adulthood through lengthy and sometimes painful rituals.

Whereas the United States began with 13 founding colonies, Australia started with six, none of which felt a compelling need to unite into a single British nation until the 1880s, when other European powers began taking an interest in settling the continent. It was not until 1901 that Australians formally separated from Britain (while remaining within the British Commonwealth, with the Queen of England as head of state). Populated almost entirely by whites from Britain or Europe (people of European descent still constitute about 95 percent of Australia's population despite a recent influx of Asians and Middle Easterners), Australia has maintained close cultural and diplomatic links with Britain and the West, at the expense of ties with the geographically closer nations of Asia.

HEALTH/WELFARE

Australia has developed a complex and comprehensive system of social welfare. Education is the province of the several states. Public education is compulsory. Australia boasts several world-renowned universities. The world's first voluntary-euthanasia law passed in Northern Territory in 1996, but legal challenges have thus far prevented its use.

Reaction against Polynesians, Chinese, and other Asian immigrants in the late 1800s produced an official "White Australia" policy, which remained intact until the 1960s and effectively excluded nonwhites from settling in Australia. During the 1960s, however, the government made an effort to relax these restrictions and to restore land and some measure of self-determination to Aborigines. In the 1990s, Aborigines successfully persuaded the federal government to block a dam project on Aboriginal land that would have destroyed sacred sites. In this rare case, the federal government sided with the Aborigines against white developers and local government officials. In 1993, despite some public resistance, the government passed laws protecting the land claims of Aborigines and set up a fund to assist Aborigines with land purchases. Evidence of continued racism can be found, however, in such graffiti painted on walls of high-rise buildings as "Go home Japs!" (in this case, the term *Jap,* or, alternatively, *wog,* refers to any Asian, regardless of nationality). The unemployment rate of Aborigines is four times that of the nation as a whole, and there are substantially higher rates of chronic health problems and death by infectious diseases among this population.

ECONOMIC PRESSURES

Despite lingering discriminatory attitudes against nonwhites, events since World War II have forced Australians to reconsider their position, at least economically, vis-à-vis Asia and Southeast Asia. Despite its historic links to the developed economies of Europe, and despite never having directly suffered the destruction of war as have most of its Asian neighbors, the Australians are worried that their economy may not be able to withstand the competitive pressures coming from Japan, China, and other Asian nations. Both Japan (despite a decade of economic stagnation), and Hong Kong (despite its return to mainland China), have per capita gross domestic products only slightly below Australia's. Like New Zealand, Australia historically relied on the export of primary goods (minerals, wheat, beef, and wool) rather than more lucrative manufactured goods to support its economy. In more recent years, the country has developed a strong manufacturing sector. But if the Asian economies continue to grow at rates much faster than Australia's, the day may come when Australia will find itself lagging behind its neighbors, and that will be a new a difficult experience for Australians, whose standard of living has been the highest in the Pacific Rim for decades. Building on a foundation of sheep (imported in the 1830s and now supplying more than a quarter of the world's supply of wool), mining (gold was discovered in 1851), and agriculture (Australia is nearly self-sufficient in food), the country has now developed its manufacturing sector such that Australians are able to enjoy a standard of living equal in most respects to that of North Americans.

But Australians are wary of the growing global tendency to create mammoth regional trading blocs, such as the North American Free Trade Association, consisting of the United States, Canada, Mexico, and others; the European Union (formerly the European Community), now including many countries of Eastern Europe; the ASEAN nations of Southeast Asia; and an informal "yen bloc" in Asia, headed by Japan. These blocs might exclude Australian products from preferential trade treatment or eliminate them from certain markets altogether. Beginning in 1983, the Labour government of then–prime minister Robert Hawke began to establish collaborative trade agreements with Asian countries, a plan that seemed to have the support of the electorate, even though it meant reorienting Australia's foreign policy away from its traditional posture Westward.

In the early 1990s, under Labour prime minister Paul Keating, the Asianization plan intensified. The Japanese prime minister and the governor of Hong Kong visited Australia, while Australian leaders made calls on the leaders of South Korea, China, Thailand, Vietnam, Malaysia, and Laos. Trade and security agreements were signed with Singapore and Indonesia, and a national-curriculum plan was implemented whereby 60 percent of Australian schoolchildren will be studying Japanese and other Asian languages by the year 2010. The Liberal Party prime minister, John Howard, reelected in 2002, has also moderated his views on Asian immigration and now advocates a nondiscriminatory immigration policy rather than the restrictive policy that he championed in the 1980s.

ACHIEVEMENTS

The vastness and challenge of Australia's interior lands, called the "outback," have inspired a number of Australian writers to create outstanding poetry and fictional novels. In 1973, Patrick White became the first Australian to win a Nobel Prize in Literature. Jill Ker Conway, Thomas Keneally, and Colleen McCullough are other well-known Australian authors.

But if immigration is being reluctantly tolerated, illegal immigration is not. Strict new laws have been passed against illegals, and the government is taking proactive measures to curtail the number of refugees from Afghanistan, Iraq, and Iran who come by way of nearby Indonesia and are desperate to settle in the country. The government's "Pacific Solution" allows the navy to capture refugee ships at sea and send the occupants to other Pacific islands. Then the Australian government pays the smaller islands for handling the refugees. Nauru, for example, has received more than $15 million for taking in 1,200 refugees. The government resorted to these measures after a hunger strike by 500 of the 2,000 refugees in a detention camp drew negative international publicity. Some detainees sewed their mouths shut, while others tried suicide in order to pressure the government to grant them asylum. Handling these matters is made more difficult by the anti-immigrant, ultraright One Nation Party, which advocates shutting the door to immigrants and returning to Australia's former "whites only" policy.

Despite government efforts to bring Australia within the orbit of Asian trade, the economic threat to Australia remains. Even in the islands of the Pacific, an area that Australia and New Zealand generally have considered their own domain for economic investment and foreign aid, investments by Asian countries are beginning to winnow Australia's sphere of influence. U.S. president Bill Clinton, in a 1996 visit, promised Australian leaders that they would not be left out of the emerging economic structures of the region. Indeed, economic conditions have improved; recent budgets have shown surpluses and even allowed for tax cuts. A free-trade agreement with the United States that will eliminate tariffs on 99 percent of U.S. goods imported into Australia and which will take effect in 2005 should help the economy, as will a similar agreement with Thailand. Still, years of recession, an unemployment rate sometimes nearing 9 percent (6 percent in 2004), and 2 million people living in poverty leave many Australians concerned about their economic future.

Labor tension erupted in 1998 when dockworkers found themselves locked out of work by employers who claimed they were inefficient workers. Eventually the courts found in favor of the workers, but not until police and workers clashed and national attention was drawn to the protracted sluggish economy.

THE AMERICAN CONNECTION

By any standard, Australia is a democracy solidly embedded in the traditions of the West. Political power is shared back and forth between the Labour Party and the Liberal–National Country Party coalition, and the Constitution is based on both British parliamentary tradition and the U.S. model. Thus, it has followed that Australia and the United States have built a warm friendship as both political and military allies. A military mutual-assistance agreement, ANZUS (for Australia, New Zealand, and the United States), was concluded after World War II (New Zealand withdrew in 1986). And just as it had sent troops to fight Germany during World Wars I and II, Australia sent troops to fight in the Korean War in 1950, the Vietnam War in the 1960s, the anti-Taliban war in Afghanistan in 2002, and the U.S.-led war in Iraq in 2003. Regarding Iraq, Prime Minister John Howard has, like Britain's Tony Blair, strongly supported U.S. President George Bush and has defiantly refused to even consider withdrawing Australian troops, despite some Australians having been taken hostage and executed by militants in Iraq. Australia also joined the United States and other countries in 1954 in establishing the Southeast Asia Treaty Organization, an Asian counterpart to the North Atlantic Treaty Organization designed to contain the spread of communism.

In 1991, when the Philippines refused to renew leases on U.S. military bases there, there was much discussion about transferring U.S. operations to the Cockburn Sound Naval Base in Australia. Singapore was eventually chosen for some of the op-

Timeline: PAST

1600s
European exploration of the Australian coastline begins

1688
British explorers first land in Australia

1788
The first shipment of English convicts arrives

1851
A gold rush lures thousands of immigrants

1901
Australia becomes independent within the British Commonwealth

1940s
Australia is threatened by Japan during World War II

1947
Australia proposes the South Pacific Commission

1951
Australia joins New Zealand and the United States in the ANZUS military security agreement

1954
Australia joins the South East Asian Treaty Organization

Relations with the United States are strained over the Vietnam War **1960s** I

1972
The Australian Labour Party wins for the first time in 23 years; Gough Whitlam is prime minister

1975
After a constitutional crisis, Whitlam is replaced by opposition leader J. M. Fraser

1980s
Australia begins to strengthen its economic ties with Asian countries

1990s
After 13 years in power, the Labour Party is defeated by Liberal Party leader John Howard

Australia condemns nuclear testing in the Pacific

PRESENT

2000s
Aboriginal rights are increasingly part of the public debate in Australia

Sydney hosts the 2000 Summer Olympic Games

Summer wildfires near Sydney destroy 1.6 million acres of eucalyptus forests and 170 buildings

A widespread sex abuse scandal involving Anglican priests and others causes the resignation of the Anglican archbishop of Adelaide and the arrest of some 150 people involved with child pornography

The government introduces legislation to ban same-sex marriages and to prevent gays from adopting foreign children

Australia signs a sweeping free-trade agreement with the United States

erations, but the incident reveals the close relationship of the two nations. U.S. military aircraft already land in Australia, and submarines and other naval craft call at Australian ports. The Americans also use Australian territory for surveillance facilities. There is historical precedence for this level of close cooperation: Before the U.S. invasion of the Japanese-controlled Philippines in the 1940s, the United States based its Pacific-theater military headquarters in Australia; moreover, Britain's inability to lead the fight against Japan forced Australia to look to the United States.

A few Australians resent the violation of sovereignty represented by the U.S. bases, but most regard the United States as a solid ally. Indeed, many Australians regard their country as the Southern Hemisphere's version of the United States: Both countries have immense space and vast resources, both were founded as disparate colonies that eventually united and obtained independence from Britain, and both share a common language and a Western cultural heritage.

There is yet another way that some Australians would like to be similar to the United States: They want to be a republic. Polls in advance of a 1999 referendum to decide whether or not Australia should remain a constitutional monarchy, headed by the king or queen of England, showed that more than 60 percent of the population favored severing ties with England. However, the actual vote found 55 percent in favor of the status quo—just enough to keep Queen Elizabeth II as head of state. The queen visited Australia right after the vote to show her thanks to the residents of working-class and rural areas, where support for the monarchy was strongest.

Unlike New Zealand, which has distanced itself from the United States by refusing to allow nuclear-armed ships to enter its ports and has withdrawn from the ANZUS security treaty with the United States, Australia has sided with the U.S. in attempting to dissuade South Pacific states from declaring the region a nuclear-free zone. Yet it has also maintained good ties with the small and vulnerable societies of the Pacific through its leadership in such regional associations as the South Pacific Commission, the South Pacific Forum, and the ever-more-influential Asia-Pacific Economic Cooperation group (APEC). It has also condemned nuclear-bomb testing programs in French-controlled territories.

AUSTRALIA AND THE PACIFIC

Australia was not always possessed of good intentions toward the islands around it. For one thing, white Australians thought of themselves as superior to the brown-skinned islanders; and for another, Australia preferred to use the islands' resources for its own economic gain, with little regard for the islanders themselves. At the end of World War I, for example, the phosphate-rich island of Nauru, formerly under German control, was assigned to Australia as a trust territory. Until phosphate mining was turned over to the islanders in 1967, Australian farmers consumed large quantities of the island's phosphates but paid just half the market price. Worse, only a tiny fraction of the proceeds went to the people of Nauru. Similarly, in Papua New Guinea, Australia controlled the island without taking significant steps toward its domestic development until the 1960s, when, under the guidance of the United Nations, it did an about-face and facilitated changes that advanced the successful achievement of independence in 1975.

In addition to forgoing access to cheap resources, Australia was reluctant to relinquish control of these islands because it saw them as a shield against possible military attack. It learned this lesson well in World War II. In 1941, Japan, taking advantage of the Western powers' preoccupation with Adolf Hitler, moved quickly to expand its imperial designs in Asia and the Pacific. The Japanese first disabled the U.S. Navy by attacking its warships docked in Pearl Harbor, Hawaii. They then moved on to oust the British in Hong Kong and the Gilbert Islands, and the Americans in Guam and Wake Island. Within a few months, the Japanese had taken control of Burma, Malaya, Borneo, the Philippines, Singapore, and hundreds of tiny Pacific islands, which they used to create an immense defensive perimeter around the home islands of Japan. They also had captured part of New Guinea and were keeping a large force there, which greatly concerned the Australians. Yet fighting was kept away from Australia proper when the Japanese were successfully engaged by Australian and American troops in New Guinea. Other Pacific islands were regained from the Japanese at a tremendous cost in lives and military hardware. Japan's defeat came only when islands close enough to Japan to be attacked by U.S. bomber aircraft were finally captured. Japan surrendered in 1945, but the colonial powers had learned that possession of small islands could have strategic importance. This experience is part of the reason for colonial powers' reluctance to grant independence to the vast array of islands over which they have exercised control. Australia is now faced with the question of whether or not to grant independence to the 4,000 inhabitants of Christmas Island who recently voted to become a self-ruling territory within Australia.

There is no doubt that stressful historical periods and events such as World War II drew the English-speaking countries of the South Pacific closer together and closer to the United States. But recent realignments in the global economic system are creating strains. When the United States insists that Japan take steps to ease the U.S.–Japan trade imbalance, Australia sometimes comes out the loser. For instance, both Australia and the United States are producers of coal. Given the nearly equal distance between those two countries and Japan, it would be logical to expect that Japan would buy coal at about the same price from both countries. In fact, however, Japan pays about $7 a ton more for American coal than for Australian coal, a discrepancy directly attributable to Japan's attempt to reduce the trade imbalance with the United States. Resentment against the United States over such matters is likely to grow, and managing such international tensions will no doubt challenge the skills of the leadership of Australia in the coming years.

Brunei (State of Brunei Darussalam)

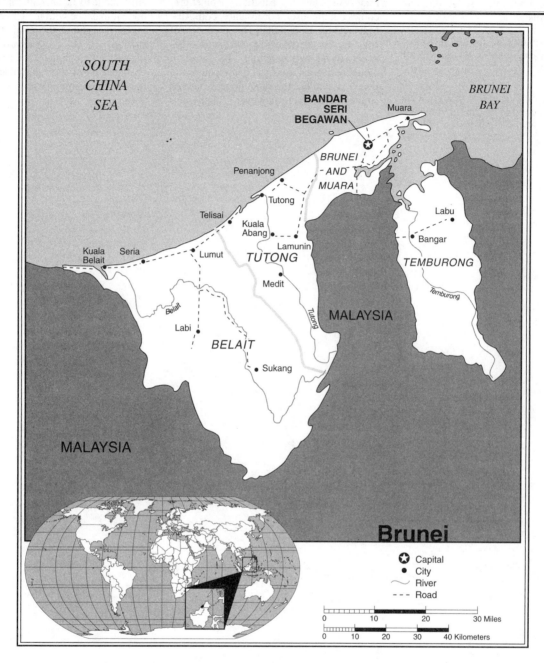

Brunei Statistics

GEOGRAPHY

Area in Square Miles (Kilometers): 2,228
(5,770) (about the size of Delaware)

Capital (Population): Bandar Seri
Begawan (46,000)

Environmental Concerns: water pollution;
seasonal smoke/haze resulting from
forest fires in Indonesia

Geographical Features: flat coastal plain
rises to mountains in east; hilly lowlands
in west

Climate: tropical; hot, humid, rainy

PEOPLE

Population

Total: 365,251

Annual Growth Rate: 1.95%

Rural/Urban Population (Ratio): 28/72

Major Languages: Malay; English;
Chinese; Iban; native dialects

Ethnic Makeup: 67% Malay; 15%
Chinese; 18% others

Religions: 67% Muslim; 13% Buddhist;
10% Christian; 10% indigenous beliefs
and others

Health

Life Expectancy at Birth: 72 years (male); 77 years (female)
Infant Mortality: 13.05/1,000 live births
Physicians Available: 1/1,398 people
HIV/AIDS Rate in Adults: 0.2%

Education

Adult Literacy Rate: 88%
Compulsory (Ages): 5–17; free

COMMUNICATION

Telephones: 79,000 main lines
Daily Newspaper Circulation: 70/1,000 people
Televisions: 308/1,000 people
Internet Users: 28,000 (2001)

TRANSPORTATION

Highways in Miles (Kilometers): 1521 (2,525)
Railroads in Miles (Kilometers): 8 (13)
Usable Airfields: 2
Motor Vehicles in Use: 165,000

GOVERNMENT

Type: constitutional sultanate (monarchy)
Independence Date: January 1, 1984 (from the United Kingdom)
Head of State/Government: Sultan and Prime Minister Sir Hassanal Bolkiah is both head of state and head of government
Political Parties: Brunei Solidarity National Party (the only legal party); Brunei People's Party (banned); Brunei National Democratic Party (deregistered)
Suffrage: none

MILITARY

Military Expenditures (% of GDP): 5.9%
Current Disputes: dispute over the Spratly Islands

ECONOMY

Currency ($ U.S. Equivalent): 1.73 Brunei dollars = $1

Per Capita Income/GDP: $18,600/$6.5 billion
GDP Growth Rate: 3%
Inflation Rate: 1%
Unemployment Rate: 10%
Labor Force by Occupation: 48% government; 42% industry and services; 10% agriculture
Natural Resources: petroleum; natural gas; timber
Agriculture: rice; cassava (tapioca); bananas; water buffalo
Industry: petroleum; natural gas; contruction
Exports: $3.43 billion (primary partners Japan, South Korea, Thailand, Australia)
Imports: $1.4 billion (primary partners Singapore, United Kingdom, Malaysia)

SUGGESTED WEBSITES

```
http://www.odci.gov/cia/
    publications/factbook/geos/
    bx.html
http://www.brunet.bn
http://www.brunei.gov.bn/
    index.htm
```

Brunei Country Report

Home to only 351,000 people, Brunei rarely captures the headlines. But perhaps it should, for despite its tiny size (about the size of Delaware or Prince Edward Island), the country boasts one of the highest living standards in the world. Moreover, the sultan of Brunei, with assets of $37 billion, is considered the richest person in the world. The secret? Oil. First exploited in Brunei in the late 1920s, today petroleum and natural gas almost entirely support the sultanate's economy. The government's annual income is nearly twice its expenditures, despite the provision of free education and medical care, subsidized food and housing, and the absence of income taxes. Brunei is currently in the middle of a five-year plan designed to diversify its economy, but 98 percent of the nation's revenues continue to derive from the sale of oil and natural gas.

DEVELOPMENT

Brunei's economy is a mixture of the modern and the ancient: foreign and domestic entrepreneurship, government regulation and welfare statism, and village tradition. Chronic labor shortages are managed by the importation of thousands of foreign workers.

Muslim sultans ruled over the entire island of Borneo and other nearby islands during the sixteenth century. Advantageously located on the northwest coast of the island of Borneo, along the sea lanes of the South China Sea, Brunei was a popular resting spot for traders. During the 1700s, it became known as a haven for pirates. Tropical rain forests and swamps occupy much of the country—conditions that are maintained by heavy monsoon rains for about five months each year. Oil and natural-gas deposits are found both on- and offshore.

FREEDOM

Although Islam is the official state religion, the government practices religious tolerance. The Constitution provides the sultan with supreme executive authority, which he has used to suppress opposition groups and political parties.

In the 1800s, the sultan then in power agreed to the kingdom becoming a protectorate of Britain, in order to safeguard his domain from being further whittled away by aggressors bent on empire-building. The Japanese easily overtook Brunei in 1941, when they launched their Southeast Asian offensive in search of oil and gas for

their war machine. Today, the Japanese Mitsubishi Corporation has a one-third interest in the Brunei gas company.

HEALTH/WELFARE

The country's massive oil and natural-gas revenues support wide-ranging benefits to the population, such as subsidized food, fuel, and housing, and free medical care and education. This distribution of wealth is reflected in Brunei's generally favorable quality-of-life indicators.

In the 1960s, it was expected that Brunei, which is cut in two and surrounded on three sides by Malaysia, would join the newly proposed Federation of Malaysia; but it refused to do so, preferring to remain under British control. The decision to remain a colony was made by Sultan Sir Omar Ali Saifuddin. Educated in British Malaya, the sultan retained a strong affection for British culture and frequently visited the British Isles. (Brunei's 1959 Constitution, promulgated during Sir Omar's reign, reflected this attachment: It declared Brunei a self-governing state, with its foreign affairs and defense remaining the responsibility of Great Britain.)

In 1967, Sir Omar abdicated in favor of his son, who became the 29th ruler in succession. Sultan (and Prime Minister) Sir Hassanal Bolkiah Mu'izzaddin Waddaulah (a shortened version of his name!) oversaw Brunei's gaining of independence, in 1984. When the present Sultan leaves the throne, his son, Crown Prince Al-Muhtadee Billah Bolkiah, age 30, will assume the office. The Crown Prince made news in 2004 when he married a 17-year old half-Swiss commoner. The traditional Malay Muslim ceremony was held at the enormous main palace and was attended by 2,000 royalty and other dignitaries from all over the world.

ACHIEVEMENTS

An important project has been the construction of a modern university accommodating 1,500 to 2,000 students. Since independence, the government has tried to strengthen and improve the economic, social, and cultural life of its people.

Not all Bruneians have been pleased with the present sultan's control over the political process. Despite a constitutional provision calling for regular elections, the last general elections were held over four decades ago, in 1962. Voters who were dissatisfied with the outcome at that time launched an armed revolt, which was put down with British assistance. The sultan declared a state of emergency, which has remained in effect ever since. The 21-member appointed Legislative Council has not met for 20 years, and there are, in effect, no operative political parties in Brunei. But, at last, in late 2004, the sultan convened the legislature to consider several constitutional amendments. These were subsequently signed into law and included a new provision for direct election of up to 15 members of a newly redesigned Legislative Council of 45 members.

Brunei's largest ethnic group is Malay, accounting for 64 percent of the population. Indians and Chinese constitute sizable minorities, as do indigenous peoples such as Ibans and Dyaks. Despite Brunei's historic ties with Britain, Europeans make up only a tiny fraction of the population.

Brunei is an Islamic nation with Hindu roots. Islam is the official state religion, and in recent years, the sultan has proposed bringing national laws more closely in line with Islamic ideology. Modern Brunei is officially a constitutional monarchy, headed by the sultan, a chief minister, and a Council; in reality, however, the sultan and his family control all aspects of state decision making. The extent of the sultan's control of the government is revealed by his multiple titles: in addition to sultan, he is Brunei's prime minister, minister of defense, and minister of finance. The Constitution provides the sultan with supreme executive authority in the state. The concentration of near-absolute power in the hands of one family has produced some unsavory results. In 2000, the sultan's brother agreed to return billions of dollars of assets that he had stolen from the state while heading Brunei's overseas investment company. Court documents in the $15 billion lawsuit against him claimed that he had personally consumed $2.7 billion (!) to support his lavish lifestyle.

Brunei, along with China and several other Southeast Asian nations, claims the Spratly Islands. Recently, the various claimants agreed to work amicably on a long-term solution to the problem. In other international matters, Brunei joined with other ASEAN nations in 1995 to declare its country a nuclear-free zone. And in 2001, at the ASEAN meetings held in Brunei, the country agreed with others to establish a free-trade zone in the region within 10 years.

In recent years, Brunei has been plagued by a chronic labor shortage. The government and Brunei Shell (a consortium owned jointly by the Brunei government and Shell Oil) are the largest employers in the country. They provide generous fringe benefits and high pay. Non-oil private-sector companies with fewer resources find it difficult to recruit within the country and have, therefore, employed many foreign workers. Indeed, one third of all workers today in Brunei are foreigners. This situation is of considerable concern to the government, which is worried that social tensions between foreigners and residents may flare up at any time.

Timeline: PAST

A.D. 1521
Brunei is first visited by Europeans

1700
Brunei is known as haven for pirates

1800s
Briton James Brooke is given Sarawak as reward for help in a civil war

1847
The island of Labuan is ceded to Britain

1849
Britain attacks and ends pirate activities in Brunei

1888
The remainder of Brunei becomes a British protectorate

1963
Brunei rejects confederation with Malaysia

1984
Brunei gains its independence

1990s
Foreign workers are "imported" to ease the labor shortage; Brunei joins the International Monetary Fund

PRESENT

2000s

The sultan's brother agrees to return billions of dollars in stolen state assets

Brunei declares itself a nuclear-free zone

The sultan signs a new Constitution allowing for direct elections of 15 members of a re-designed Legislative Council

The Crown Prince weds a half-Swiss commoner in a lavish ceremony

Cambodia (Kingdom of Cambodia)

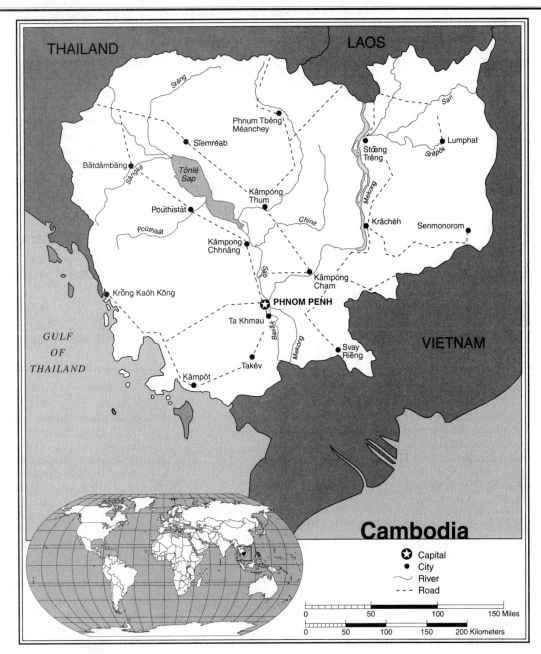

Cambodia Statistics

GEOGRAPHY

Area in Square Miles (Kilometers):
69,881 (181,040) (about the size of Oklahoma)

Capital (Population): Phnom Penh (1,109,000)

Environmental Concerns: habitat loss and declining biodiversity; soil erosion; deforestation; lack of access to potable water

Geographical Features: mostly low, flat plains; mountains in the southwest and north

Climate: tropical; rainy and dry seasons

PEOPLE

Population

Total: 13,363,421
Annual Growth Rate: 1.8%
Rural/Urban Population (Ratio): 84/16

Major Languages: Khmer; French; English

Ethnic Makeup: 90% Khmer (Cambodian); 5% Chinese; 5% others

Religions: 95% Theravada Buddhist; 5% others

Health

Life Expectancy at Birth: 55 years (male); 59 years (female)

Infant Mortality: 64/1,000 live births

Physicians Available: 1/7,900 people
HIV/AIDS Rate in Adults: 4.04%

Education
Adult Literacy Rate: 69%
Compulsory Ages: 6–12

COMMUNICATION
Telephones: 22,000 main lines
Televisions: 8/1,000 people
Internet Users: 6,000 (2001)

TRANSPORTATION
Highways in Miles (Kilometers): 21,461
(35,769)
Railroads in Miles (Kilometers): 365
(603)
Usable Airfields: 20
Motor Vehicles in Use: 30,000

GOVERNMENT
Type: multiparty democracy under a
constitutional monarchy

Independence Date: November 9, 1953
(from France)
Head of State/Government: King
Norodom Sihamoni; Prime Minister
(Premier) Hun Sen
Political Parties: National United Front…
(Funcinpec); Cambodian People's Party;
Buddhist Liberal Party; Democratic
Kampuchea (Khmer Rouge); Khmer
Citizen Party; Cambodian People's
Party; Sam Rangsi Party
Suffrage: universal at 18

MILITARY
Military Expenditures (% of GDP): 3%
Current Disputes: border disputes with
Thailand and Vietnam

ECONOMY
Currency ($ U.S. Equivalent): 3,835 riels
= $1
Per Capita Income/GDP: $1,900/$25.02
billion

GDP Growth Rate: 5%
Inflation Rate: 1.6%
Unemployment Rate: 2.5%
Labor Force by Occupation: 80%
agriculture
Population Below Poverty Line: 36%
Natural Resources: timber; gemstones;
iron ore; manganese; phosphates;
hydropower potential
Agriculture: rice; rubber; corn; vegetables
Industry: rice processing; fishing; wood
and wood products; rubber; cement; gem
mining; textiles
Exports: $1.05 billion (primary partners
United States, Vietnam, Germany)
Imports: $1.4 billion (primary partners
Singapore, Thailand, Hong Kong)

SUGGESTED WEBSITES
http://www.cambodia.org
http://www.odci.gov/cia/
publications/factbook/geos/
cb.html

Cambodia Country Report

Two hundred miles of steamy jungle from the Cambodian capital city of Phnom Penh is one of Southeast Asia's most impressive architectural complexes: the Hindu temples of Angkor. Built nearly a thousand years ago, when Cambodia ruled an empire stretching from the borders of China to the Bay of Bengal, the 145-acre Angkor complex became the seat of government of a long line of Khmer or Cambodian kings, as well as a pilgrimage destination for thousands of Hindu faithful. Today, the government is attempting to parlay the fame of Angkor into the basis for the country's fledgling tourist industry. The promotional plan seems to be working, with about 700,000 tourists visiting the site and the nearby provincial capital of Siem Reap in 2004. It is estimated that those tourists contributed US$525 million to the Cambodian economy—a massive infusion of money by recent Cambodian standards.

In Khmer (Cambodian), the word *Kampuchea,* which for a time during the 1980s was the official name of Cambodia, means "country where gold lies at the foothill." But Cambodia is certainly not a land of gold, nor of food, freedom, or stability. The average Cambodian is malnourished, illiterate, and will die before age 57, having experienced neither freedom nor peace. Despite a new Constitution, massive

United Nations aid, and a formal cease-fire, the horrific effects of Cambodia's bloody civil war continue.

Cambodia was not always a place to be pitied. In fact, at times it was the dominant power of Southeast Asia. Around the fourth century A.D., India, with its pacifist Hindu ideology, began to influence in earnest the original Chinese base of Cambodian civilization. The Indian script came to be used, the name of its capital city was an Indian word, its kings acquired Indian titles, and many of its Khmer people believed in the Hindu religion. The mile-square Hindu temple Angkor Wat, built in the twelfth century, still stands as a symbolic reminder of Indian influence, Khmer ingenuity, and the Khmer Empire's glory.

DEVELOPMENT

In the past, China, the United States, France, and others built roads and industries in Cambodia, but as a result of two decades of internecine warfare, the country remains an impoverished state whose economy rests on fishing, farming, and massive infusions of foreign aid. Beginning in 2001, the government launched an effort to attract tourists to some of the famous temples of ancient Cambodia. The plan seems to be working, although the infrastructure is not in place to support the burgeoning tourist population.

But the Khmer Empire, which at its height included parts of present-day Myanmar (Burma), Thailand, Laos, and Vietnam, was gradually reduced both in size and power until, in the 1800s, it was paying tribute to both Thailand and Vietnam. Continuing threats from these two countries as well as wars and domestic unrest at home led the king of Cambodia to appeal to France for help. France, eager to offset British power in the region, was all too willing to help. A protectorate was established in 1863, and French power grew apace until, in 1887, Cambodia became a part of French Indochina, a conglomerate consisting of the countries of Laos, Vietnam, and Cambodia.

The Japanese temporarily evicted the French in 1945 and, while under Japanese control, Cambodia declared its "independence" from France. Heading the country was the young King Norodom Sihanouk. Controlling rival ideological factions, some of which were pro-West while others were pro-Communist, was difficult for Sihanouk, but he built unity around the idea of permanently expelling the French, who finally left in 1955. King Sihanouk then abdicated his throne in favor of his father so that he could, as premier (prime minister), personally enmesh himself in political governance. He used the title Prince Sihanouk until 1993, when he declared himself, once again, King Sihanouk.

From the beginning, Sihanouk's government was bedeviled by border disputes with Thailand and Vietnam and by the incursion of Communist Vietnamese soldiers into Cambodia. Sihanouk's ideological allegiances were confusing at best; but, to his credit, he was able to keep Cambodia officially out of the Vietnam War, which raged for years (1950–1975) on its border. In 1962, Sihanouk announced that his country would remain neutral in the Cold War struggle.

Neutrality, however, was not seen as a virtue by the United States, whose people were becoming more and more eager either to win or to quit the war with North Vietnam. A particularly galling point for the U.S. military was the existence of the so-called Ho Chi Minh Trail, a supply route through the tropical mountain forests of Cambodia. For years, North Vietnam had been using the route to supply its military operations in South Vietnam, and Cambodia's neutrality prevented the United States, at least legally, from taking military action against the supply line.

All this changed in 1970, when Sihanouk, out of the country at the time, was evicted from office by his prime minister, General Lon Nol, who was supported by the United States and South Vietnam. Shortly thereafter, the United States, then at its peak of involvement in the Vietnam War, began extensive military action in Cambodia. The years of official neutrality came to a bloody end.

THE KILLING FIELDS

Most of these international political intrigues were lost on the bulk of the Cambodian population, only half of whom could read and write, and almost all of whom survived, as their forebears had before them, by cultivating rice along the lush Mekong River Valley. Always an agricultural economy, Cambodia's monsoon rains contribute to the numerous large rivers that sustain farming on the country's large central plain and produce tropical forests consisting of such useful products as coconuts, palms, bananas, and rubber. Despite these abundant resources, the country had almost always been poor, so villagers had long since learned that, even in the face of war, they could survive by hard work and reliance on extended-family networks. Most farmers probably thought that the war next door would not seriously alter their lives. But they were profoundly wrong, for, just as the United States had an interest in having a pro–U.S. government in Cambodia, the North Vietnamese desperately wanted Cambodia to be pro-Communist.

North Vietnam wanted the Cambodian government to be controlled by the Khmer Rouge, a Communist guerrilla army led by Pol Pot, one of a group of former students influenced by the left-wing ideology taught in French universities during the 1950s. After 5 years of bloody battles with government troops, the rebel Khmer Rouge took control of the country and launched, in 1975, 3 1/2-years of hellish extermination of the people, resulting in the deaths of between 1 million and 3 million fellow Cambodians—that is, between one fifth and one third of the entire Cambodian population. The official goal was to eliminate anyone who had been "polluted" by prerevolutionary thinking, but what actually happened was random violence, torture, and murder.

It is impossible to fully convey the mayhem and despair that engulfed Cambodia during those years. Cities were emptied of people. Teachers and doctors were killed or sent as slaves to work in the rice paddies. Despite the centrality of Buddhism to Cambodian culture (Hinduism having long since been displaced by Buddhist thought), thousands of Buddhist monks were killed or died of starvation as the Khmer Rouge carried out its program of eliminating religion. Some people were killed for no other reason than to terrorize others into submission. Explained Leo Kuper in *International Action Against Genocide*:

> Those who were dissatisfied with the new regime were… "eradicated," along with their families, by disembowelment, by beating to death with hoes, by hammering nails into the backs of their heads and by other cruel means of economizing on bullets. Persons associated with the previous regime were special targets for liquidation. In many cases, the executions included wives and children. There were summary executions too of intellectuals, such as doctors, engineers, professors, teachers and students, leaving the country denuded of professional skills.

The Khmer Rouge wanted to alter the society completely. Children were removed from their families, and private ownership of property was eliminated. Money was outlawed. Even the calendar was started over, at year 0. Vietnamese military leader Bui Tin explained just how totalitarian the rulers were:

> [In 1979] there was no small piece of soap or handkerchief anywhere. Any person who had tried to use a toothbrush was considered bourgeois and punished. Any person wearing glasses was considered an intellectual who must be punished.

It is estimated that before the Khmer Rouge came to power in 1975, Cambodia had 1,200 engineers, 21,000 teachers, and 500 doctors. After the purges, the country was left with only 20 engineers, 3,000 teachers, and 54 doctors.

A kind of bitter relief came in late 1978, when Vietnamese troops (traditionally Cambodia's enemy) invaded Cambodia, drove the Khmer Rouge to the borders of Thailand, and installed a puppet government headed by Hun Sen, a former Khmer Rouge soldier who defected and fled to Vietnam in the 1970s. Although almost everyone was relieved to see the Khmer Rouge pushed out of power, the Vietnamese intervention was almost universally condemned by other nations. This was because the Vietnamese were taking advantage of the chaos in Cambodia to further their aim of creating a federated state of Vietnam, Laos, and Cambodia. Its virtual annexation of Cambodia eliminated Cambodia as a buffer state between Vietnam and Thailand, destabilizing the relations of the region even more.

COALITION GOVERNANCE

The United States and others refused to recognize the Vietnam-installed regime, instead granting recognition to the Coalition Government of Democratic Kampuchea. This entity consisted of three groups: the Communist Khmer Rouge, led by Khieu Samphan and Pol Pot and backed by China; the anti-Communist Khmer People's National Liberation Front, led by former prime minister Son Sann; and the Armee Nationale Sihanoukiste, led by Sihanouk. Although it was doubtful that these former enemies could constitute a workable government for Cambodia, the United Nations granted its Cambodia seat to the coalition and withheld its support from the Hun Sen government.

Vietnam had hoped that its capture of Cambodia would be easy and painless. Instead, the Khmer Rouge and others resisted so much that Vietnam had to send in 200,000 troops, of which 25,000 died. Moreover, other countries, including the United States and Japan, strengthened their resolve to isolate Vietnam in terms of international trade and development financing. After 10 years, the costs to Vietnam of remaining in Cambodia were so great that Vietnam announced it would pull out its troops.

A 1992 diplomatic breakthrough allowed the United Nations to establish a peacekeeping force in the country of some 22,000 troops, including Japanese soldiers—the first Japanese military presence outside Japan since World War II. These troops were to keep the tenacious Khmer Rouge faction under control. The agreement, signed in Paris by 17 nations, called for the release of political prisoners; inspections of prisons; and voter registration for national elections, to be held in 1993. Most important, the warring factions, consisting of some 200,000 troops, including 25,000 Khmer Rouge troops, agreed to disarm under UN supervision.

Unfortunately, the Khmer Rouge, although a signatory to the agreement, refused to abide by its provisions. With revenues gained from illegal trading in lumber and gems with Thailand, it launched new attacks on villages, trains, and even the UN peacekeepers themselves, and it refused to participate in the elections of 1993, although it had been offered a role in the new government if it would cooperate.

Despite a violent campaign, 90 percent of those eligible voted in elections that, after some confusion, resulted in a new Constitution; the reenthronement of Sihanouk as king; and the appointment of Sihanouk's son, Prince Norodom Ranariddh of the Royalist Party, as first prime minister and Hun Sen of the Cambodian People's Party as second prime minister.

The new Parliament outlawed the Khmer Rouge, but relations between the two premiers was rocky at best. Both began negotiating separately with the Khmer Rouge to entice them to lay down arms in return for amnesty. Thousands accepted the offer, fatally weakening the rebel army. But Hun Sen soon claimed that Norodum Ranariddh was recruiting soldiers for a future coup attempt. Despite Norodom Ranariddh's standing as the son of King Sihanouk and one of the two premiers, Hun Sen deposed him in a bloody coup, and Ranariddh fled the country. Eventually, the king persuaded Hun Sen to allow his son to return to Cambodia, where he became head of the National Assembly.

International attempts to bring the aging leadership of the Khmer Rouge to justice have been frustrated by Premier Hun Sen, who apparently has persuaded many of the leaders to lay down arms in return for amnesty. Notorious leaders such as Khieu Samphan and Nuon Chea had surrendered by the late 1990s, and Ta Mok had been captured. The most infamous of all, Pol Pot, had died in 1998. The United Nations and the United States wanted the top 12 leaders to be tried for genocide by an international court, but Hun Sen refused to cooperate on the grounds that having outsiders as judges would violate Cambodia's sovereignty. Despite deep frustration at the United Nations, talks on this issue have resumed.

Cambodia's third national elections were held in 2003. Hun Sen's Cambodian People's Party (CPP) was pitted against two opposition party, the Sam Rainsy Party (SRP) and Ranariddh's royalist party or FUNCINPEC. The opposition campaigned for human rights and the elimination of corruption in government. As usual, the election was filled with violence; several members of the SRP were killed, as was the head of the 30,000-member Cambodian Free Trade Union of Workers. However, compared to earlier local elections in which some 20 candidates were killed, the 2003 election was relatively rational. However, the results were unexpected: Hun Sen's party lost its parliamentary majority and was required to negotiate with the opposition. After nearly a year of bitterness, an agreement was reached whereby Hun Sen's CCP would hold 60% of all cabinet seats, and Ranariddh's party would control the remaining 40%—some of which it would share with the SRP.

Barely had the dust settled on the election when King Norodom Sihanouk, 81, announced that, due to his failing health, he was abdicating the throne. A nine-member Throne Council of politicians and Buddhist monks was created to pick the royal successor. Although, under the constitution, the kingship is not hereditary, the Council nevertheless selected one of the king's sons, Norodom Sihamoni, age 51, who had served as Cambodian ambassador to UNESCO. His career had taken him to North Korean and Europe where he became a ballet dancer and professor of classical dance in Paris. He was fluent in French, Czech, English, and Russian, and had spent most of his life away from Cambodia, living in a modest Paris apartment as a bachelor with no children.

Timeline: PAST

A.D. 1863
France gains control of Cambodia

1940s
Japanese invasion; King Norodom Sihanouk is installed

1953
Sihanouk wins Cambodia's independence of France

1970
General Lon Nol takes power in a U.S.–supported coup

1975
The Khmer Rouge, under Pol Pot, overthrows the government and begins a reign of terror

1978
Vietnam invades Cambodia and installs a puppet government

1990s
Vietnam withdraws troops from Cambodia; a Paris cease-fire agreement is violated by the Khmer Rouge; Pol Pot dies and other Khmer Rouge leaders surrender

PRESENT

2000s
Hun Sen is reelected

Prince Ranariddh becomes president of the National Assembly

Local elections are held for the first time in more than 20 years

Hun Sen loses parliamentary majority in the 2003 elections and agrees to share power with opposition parties

King Sihanouk abdicates the throne. His son, Norodom Sihamoni, age 51, is selected to replace him

China (People's Republic of China)

China Statistics

GEOGRAPHY

Area in Square Miles (Kilometers):
3,705,386 (9,596,960) (about the same size as the United States)

Capital (Population): Beijing (10,836,000)

Environmental Concerns: air and water pollution; water shortages; desertification; trade in endangered species; acid rain; loss of agricultural land; deforestation

Geographical Features: mostly mountains, high plateaus, and deserts in the west; plains, deltas, and hills in the east

Climate: extremely diverse, from tropical to subarctic

PEOPLE

Population

Total: 1,300,000,000

Annual Growth Rate: 0.57%

Rural/Urban Population Ratio: 68/32

Major Languages: Standard Chinese (Putonghua) or Mandarin; Yue (Cantonese); Wu (Shanghainese); Minbei (Fuzhou); Minuan (Hokkien-Taiwanese); Xiang; Gan; Hahka

Ethnic Makeup: 92% Han Chinese; 8% minority groups (the largest being Chuang, Hui, Uighur, Yi, Miao, Mongolian, and Tibetan)

Religions: officially atheist; but Taoism, Buddhism, Islam, Christianity, ancestor worship, and animism exist

Health

Life Expectancy at Birth: 70 years (male); 74 years (female)

Infant Mortality: 25.2/1,000 live births

Physicians Available: 1/628 people

HIV/AIDS Rate in Adults: less than 0.1%

47

Education

Adult Literacy Rate: 91%
Compulsory (Ages): 7–17

COMMUNICATION

Telephones: 263,000,000 main lines
Cell Phones: 402 million
Daily Newspaper Circulation: 23/1,000 people
Televisions: 189/1,000 people
Internet Users: 45,800,000 (2002)

TRANSPORTATION

Highways in Miles (Kilometers): 840,000 (1,400,000)
Railroads in Miles (Kilometers): 39,390 (65,650)
Usable Airfields: 507
Motor Vehicles in Use: 11,450,000

GOVERNMENT

Type: one-party Communist state

Independence Date: unification in 221 B.C.; People's Republic established October 1, 1949
Head of State/Government: President Hu Jintao; Premier Wen Jiabao
Political Parties: Chinese Communist Party; eight registered small parties controlled by the CCP
Suffrage: universal at 18 in village and urban district elections

MILITARY

Military Expenditures (% of GDP): 5% (official figure)
Current Disputes: minor border disputes with a few countries, and potentially serious disputes over Spratly and Paracel Islands with several countries

ECONOMY

Currency ($ U.S. equivalent): 8.28 yuan = $1
Per Capita Income/GDP: $5,000/$6.45 trillion
GDP Growth Rate: 9.1%

Inflation Rate: 1.2%
Unemployment Rate: about 10% in urban areas
Labor Force by Occupation: 50% agriculture; 27% services; 23% industry
Natural Resources: coal; petroleum; iron ore; tin; tungsten; antimony; lead; zinc; vanadium; magnetite; uranium; hydropower
Agriculture: food grains; cotton; oilseed; pork; fish; tea; potatoes; peanuts
Industry: iron and steel; coal; machinery; light industry; textiles and apparel; food processing; consumer durables and electronics; telecommunications; armaments
Exports: $436.1 billion (primary partners United States, Hong Kong, Japan)
Imports: $397.4 billion (primary partners Japan, Taiwan, South Korea)

SUGGESTED WEBSITES

http://www.odci.gov/cia/
publications/factbook/geos/
ch.html

People's Republic of China Country Report

In early 2003, the world's first commercial magnetic levitation train or "maglev" floated silently out of Shanghai China's glitzy downtown and glided at 260 miles per hour to the ultramodern Pudong International Airport 19 miles away in just 14 minutes. Constructed by a German firm, the US$1.2 billion train, which floats above a single elevated monorail track, is not only the world's most modern form of land transportation, but it is also the face of the new China. Together with the US$1.24 billion Shanghai Formula 1 racetrack and the launching of satellites and other unmanned spacecraft, the train symbolizes a China that most people never thought would exist.

But to arrive at this point, the Chinese have had to endure many years of suffering throughout their long history. China is a very old culture. Human civilization appeared in China as early as 20,000 years ago, and the first documented Chinese dynasty, the Shang, began about 1523 B.C. Unproven legends suggest the existence of an even earlier Chinese dynasty (about 2000 B.C.), making China one of the oldest societies with a continuing cultural identity. Over the centuries of documented history, the Chinese people have been ruled by a dozen imperial dynasties; have enjoyed hundreds of years of stability and amazing cultural progress; and have en-

dured more centuries of chaos, military mayhem, and hunger. Yet China and the Chinese people remain intact—a strong testament to the tenacity of human culture.

DEVELOPMENT

In the early years of Communist control, authorities stressed the value of establishing heavy industry and collectivizing agriculture. More recently, China has become the recipient of millions of dollars of investment money from around the world—money that China is using to convert itself into a modern, capitalist economy. Labor being cheap in China, many companies outsource their manufacturing to China. The world's largest dam is currently under construction, despite the objections of environmentalists and engineers.

A second major characteristic is that the People's Republic of China (P.R.C.) is very big. It is the fourth-largest country in the world, accounting for 6.5 percent of the world's landmass. Much of China—about 40 percent—is mountainous; but large, fertile plains have been created by the country's numerous rivers, most of which flow toward the Pacific Ocean. China is blessed with substantial reserves of oil, minerals, and many other natural resources. Its large size and geopolitical location—it is bor-

dered by Russia, Kazakhstan, Pakistan, India, Nepal, Bhutan, Myanmar, Laos, Vietnam, North Korea, and Mongolia—have caused the Chinese people over the centuries to think of their land as the "Middle Kingdom": that is, the center of world civilization.
.

FREEDOM

Until the late 1970s, the Chinese people were controlled by Communist Party cadres who monitored both public and private behavior. Some economic and social liberalization occurred in the 1980s, when some villages were allowed to directly elect their leaders (although campaigning was not allowed). However, the 1989 Tiananmen Square massacre reminded Chinese and the world that despite some reforms, China is still very much a dictatorship.

However, its unwieldy size has been the undoing of numerous emperors who found it impossible to maintain its borders in the face of outside "barbarians" determined to possess the riches of Chinese civilization. During the Ch'in Dynasty (221–207 B.C.), a 1,500-mile-long, 25-foot-high wall—the so-called Great Wall—was erected along the northern border of China, in the futile hope that invasions from the north could be stopped. Although most of China's na-

(UN photo/John Issac)

In China today, urban couples are permitted to have only one child, and they can be severely penalized if they dare to have a second, or if they marry before the legal ages of 22 for men and 20 for women.

tional boundaries are now recognized by international law, recent Chinese governments have found it necessary to "pacify" border areas by settling as many Han Chinese there as possible (for example, in Tibet), to prevent secession by China's numerous ethnic minorities.

Another important characteristic of modern China is its huge population. With 1.3 billion people, China is home to about one fifth of all human beings alive today. About 92 percent of China's people are Han, or ethnic, Chinese; the remaining 8 percent are divided into more than 50 separate minority groups. Many of these ethnic groups speak mutually unintelligible languages, and although they often appear to be Chinese, they derive from entirely different cultural roots; some are Muslims, some are Buddhists, some are animists. As one moves away from the center of Chinese civilization and toward the western provinces, the influence of the minorities increases. The Chinese government has accepted the reality of ethnic influence and has granted a degree of limited autonomy

to some provinces with heavy populations of minorities.

One glaring exception to this is Tibet. A land of rugged beauty north of Nepal, India, and Bhutan, Tibet was forcefully annexed by China in 1959. Many Tibetans regard the spiritual leader, the Dalai Lama, not the Chinese government, as their true leader, but the Dalai Lama and thousands of other Tibetans live in exile in India, reducing the percentage of Tibetans in Tibet to only 44 percent. China suppresses any dissent against its rule in Tibet (even photos of the Dalai Lama, for example, cannot be displayed in public) and has diluted Tibetan culture by resettling thousands of Han Chinese in the region. Whereas the world community generally would act to protect Taiwan from an aggressive Chinese takeover, it has done very little to protest China's takeover of Tibet.

In the 1950s, Chinese Communist Party (CCP) chairman Mao Zedong encouraged couples to have many children, but this policy was reversed in the 1970s, when a formal birth-control program was inaugu-

rated in China. Urban couples today are permitted to have only one child. If they have more, penalties include expulsion from the Communist Party, dismissal from work, or a 10 percent reduction in pay for up to 14 years after the birth of the second child. The policy is strictly enforced in the cities, but it has had only a marginal impact on overall population growth because three quarters of China's people live in rural areas, where they are allowed more children in order to help with the farmwork. In the city of Shanghai, with a population of about 12.8 million people, authorities have recently removed second-child privileges for farmers living near the city and for such former exceptional cases as children of revolutionary martyrs and workers in the oil industry. Despite these and other restrictions, it is estimated that 15 million to 17 million people are now born each year in China.

Over the centuries, millions of people have found it necessary or prudent to leave China in search of food, political stability, or economic opportunity. Those who emigrated a thousand or more years ago are now fully assimilated into the cultures of Southeast Asia and elsewhere, and identify themselves accordingly. More recent émigrés (in the past 200 years or so), however, constitute visible, often wealthy, minorities in their new host countries, where they have become the backbone of the business community. Ethnic Chinese constitute the majority of the population in Singapore and a sizable minority in Malaysia. Important world figures such as Corazon Aquino, the former president of the Philippines, and Goh Chok Tong, the former prime minister of Singapore, are ethnically part or full Chinese. The Chinese constituted the first big wave of the 13 million Asian or part-Asians who call the United States home. Large numbers of Hong Kong Chinese immigrated to Canada in the mid-1990s, raising the number of Asians in Canada to nearly 12 million by 2004. Thus the influence of China continues to spread far beyond its borders, due to the influence of what are called "overseas Chinese."

Another crucial characteristic of China is its history of imperial and totalitarian rule. Except for a few years in the early 1900s, China has been controlled by imperial decree, military order, and patriarchal privilege. Confucius taught that a person must be as loyal to the government as a son should be to his father. Following Confucius by a generation or two was Shang Yang, of a school of governmental philosophy called Legalism, which advocated unbending force and punishment against wayward subjects. Compassion and pity were not considered qualities of good government.

HEALTH/WELFARE

The Communist government has overseen dramatic improvements in the provision of social services for the masses. Life expectancy has increased from 45 years in 1949 to 72 years (overall) today. Diverse forms of health care are available at low cost to the patient. The government has attempted to eradicate such diseases as malaria and tuberculosis. SARS remains a worrisome problem for China as does the avian flu.

Mao Zedong, building on this heritage as well as that of the Soviet Union's Joseph Stalin and Vladimir Lenin, exercised strict control over both the public and private lives of the Chinese people. Dissidents were summarily executed (generally people were considered guilty once they were arrested), the press was strictly controlled, and recalcitrants were forced to undergo "reeducation" to correct their thinking. Religion of any kind was suppressed, and churches were turned into warehouses. It is

estimated that, during the first three years of CCP rule, more than 1 million opponents of Mao's regime were executed. During the so-called Cultural Revolution (1966–1976), Mao, who apparently thought that a new mini-revolution in China might restore his eroding stature in the Chinese Communist Party, encouraged young people to report to the authorities anyone suspected of owning books from the West or having contact with Westerners. Even party functionaries were purged from the ranks if it were believed that their thinking had been corrupted by Western influences.

THE TEACHINGS OF CONFUCIUS

Confucius (550–478 B.C.) was a Chinese intellectual and minor political figure. He was not a religious leader, nor did he ever claim divinity for himself or divine inspiration for his ideas. As the feudalism of his era began to collapse, he proposed that society could best be governed by paternalistic kings who set good examples. Especially important to a stable society, he taught, were respect and reverence for one's elders. Within the five key relationships of society (ruler and subject, husband and wife, father and son, elder brother and younger brother, and friend and friend), people should always behave with integrity, propriety, and goodness.

The writings of Confucius—or, rather, the works written about him by his followers and entitled the *Analects*—eventually became required knowledge for anyone in China claiming to be an educated person. However, rival ideas such as Legalism were at times more popular with the elite; at one point, 460 scholars were buried alive for teaching Confucianism. Nevertheless, much of the hierarchical nature of Asian culture today can be traced to Confucian ideas.

ORIGINS OF THE MODERN STATE

Historically, authoritarian rule in China has been occasioned, in part, by China's mammoth size; by its unwieldy population; and by the ideology of some of its intellectuals. The modern Chinese state has arisen from those same pressures as well as some new ones. It is to these that we now turn.

The Chinese had traded with such non-Asian peoples as the Arabs and Persians for hundreds of years before European contact. For example, about the year 1400, almost a hundred years before Columbus reached the New World, Chinese explorer Admiral Zheng He made multiple sailings to both India and Africa. But in the 1700s and 1800s, the British and others extracted something new from China in exchange for merchandise from the West: the permission for foreign citizens to live in parts of

China without being subject to Chinese authority. Through this process of granting extraterritoriality to foreign powers, China slowly began to lose control of its sovereignty. The age of European expansion was not, of course, the first time in China's long history that its ability to rule itself was challenged; the armies of Kublai Khan successfully captured the Chinese throne in the 1200s, as did the Manchurians in the 1600s. But these outsiders, especially the Manchurians, were willing to rule China on-site and to imbibe as much Chinese culture as they could. Eventually they became indistinguishable from the Chinese.

ACHIEVEMENTS

Chinese culture has, for thousands of years, provided the world with classics in literature, art, pottery, ballet, and other arts. Under communism the arts have been marshaled in the service of ideology and have lost some of their dynamism. Since 1949, literacy has increased dramatically and now stands at 91 percent—the highest in Chinese history.

The European powers, on the other hand, preferred to rule China (or, rather, parts of it) from afar, as a vassal state, with the proceeds of conquest being drained away from China to enrich the coffers of the European monarchs. Aggression against Chinese sovereignty increased in 1843, when the British forced China to cede Hong Kong Island. Britain, France, and the United States all extracted unequal treaties from the Chinese that gave them privileged access to trade and ports along the eastern coast. By the late 1800s, Russia was in control of much of Manchuria, Germany and France had wrested special economic privileges from the ever-weakening Chinese government, and Portugal had long since controlled Macau. Further affecting the Chinese economy was the loss of many of its former tributary states in Southeast Asia. China lost Vietnam to France, Burma (today called Myanmar) to Britain, and Korea to Japan. During the violent Boxer Rebellion of 1900, the Chinese people showed how frustrated they were with the declining fortunes of their country.

Thus weakened internally and embarrassed internationally, the Manchu rulers of China began to initiate reforms that would strengthen their ability to compete with the Western and Japanese powers. A constitutional monarchy was proposed by the Manchu authorities but was preempted by the republican revolutionary movement of Western-trained Sun Yat-sen. Sun and his armies wanted an end to imperial rule; their dreams were realized in 1912, when

(UN/photo by A. Holcombe)

During Mao Zedong's "Great Leap Forward," huge agricultural communes were established, and farmers were denied the right to grow crops privately. The government's strict control of these communes met with chaotic results; there were dramatic drops in agricultural output.

Sun's Kuomintang (Nationalist Party, or KMT) took control of the new Republic of China, and the last emperor was forced to relinquish the throne.

Sun's Western-style approach to government was received with skepticism by many Chinese who distrusted the Western European model and preferred the thinking of Karl Marx and the philosophy of the Soviet Union. In 1921, Mao Zedong and others organized the Soviet-style Chinese Communist Party (CCP), which grew quickly and began to be seen as an alternative to the Kuomintang. After Sun's death, in 1925, Chiang Kai-shek assumed control of the Kuomintang and waged a campaign to rid the country of Communist influence. Although Mao and Chiang cooperated when necessary—for example, to resist Japanese incursions into Manchuria—they eventually came to be such bitter enemies that they brought a ruinous civil war to all of China.

Mao derived his support from the rural areas of China, while Chiang depended on the cities. In 1949, facing defeat, Chiang Kai-shek's Nationalists retreated to the island of Taiwan, where, under the name Republic of China (R.O.C.), they continued to insist on their right to rule all of China. The Communists, however, controlled the mainland and insisted that Taiwan was just a renegade province of the People's Republic of China. These two antagonists are officially (but not in actuality) still at war. Sometimes tensions between Taiwan and China reach dangerous levels. In the 1940s, the United States had to intervene to prevent an attack from the mainland. In 1996, U.S. warships once again patrolled the 150 miles of ocean named the Taiwan Strait to warn China not to turn its military exercises, including the firing of missiles in the direction of Taiwan, into an actual invasion. China used the blatantly aggressive actions as a warning to the newly elected Taiwanese president not to take any steps toward declaring Taiwan an independent nation. Each Chinese ruler since 1949 has re-stated the demand that Taiwan return to the fold of China; for their part, the people of Taiwan seem to prefer to live with de facto independence but without officially declaring it.

For many years after World War II, world opinion sided with Taiwan's claim to be the legitimate government of China. Taiwan was granted diplomatic recognition by many nations and given the China seat in the United Nations. In the 1970s, however, world leaders came to believe that it was dysfunctional to withhold recognition and standing from such a large and potentially powerful nation as the P.R.C. Because both sides insisted that there could not be two Chinas, nor one China and one Taiwan, the United Nations proceeded to give the China seat to mainland China. Dozens of countries subsequently broke off formal diplomatic relations with Taiwan in order to establish a relationship with China.

PROBLEMS OF GOVERNANCE

The China that Mao came to control was a nation with serious economic and social problems. Decades of civil war had disrupted families and wreaked havoc on the economy. Mao believed that the solution to China's ills was to wholeheartedly embrace socialism. Businesses were nationalized, and state planning replaced private initiative. Slowly, the economy improved. In 1958, however, Mao decided to enforce the tenets of socialism more vigorously so that China would be able to take an economic "Great Leap Forward." Workers were assigned to huge agricultural communes and were denied the right to grow their soybeans, cabbage, corn, rice, onions and other crops privately. All enterprises came under the strict control of the central government. The result was economic chaos and a dramatic drop in both industrial and agricultural output.

Exacerbating these problems was the growing rift between the P.R.C. and the Soviet Union. China insisted that its brand of communism was truer to the principles of Marx and Lenin and criticized the Soviets for selling out to the West. As relations with (and financial support from) the Soviet Union withered, China found itself increasingly isolated from the world community, a circumstance worsened by serious conflicts with India, Indonesia, and other nations. To gain friends, the P.R.C. provided substantial aid to Communist insurgencies in Vietnam and Laos, thus contributing to the eventual severity of the Vietnam War.

In 1966, Mao found that his power was waning in the face of Communist Party leaders who favored a more moderate approach to internal problems and external relations. To regain lost ground, Mao urged young students called Red Guards to fight against anyone who might have liberal, capitalist, or intellectual leanings. He called it the Great Proletarian Cultural Revolution, but it was an *anti*cultural purge: Books were burned, and educated people were arrested and persecuted. In fact, the entire country remained in a state of domestic chaos for more than a decade.

Soon after Mao died, in 1976, Deng Xiaoping, who had been in and out of Communist Party power several times before, came to occupy the senior position in the CCP. A pragmatist, he was willing to modify or forgo strict socialist ideology if he believed that some other approach would work better. Despite pressure by hard-liners to tighten governmental control, he nevertheless was successful in liberalizing the economy and permitting exchanges of scholars with the West. In 1979, he accepted formalization of relations with the United States—an act interpreted as a signal of China's opening up to the world.

China's opening has been dramatic, not only in terms of its international relations but also internally. During the 1980s, the P.R.C. joined the World Bank, the International Monetary Fund, the Asian Development Bank, and other multilateral organizations. It also began to welcome foreign investment of almost any kind and permitted foreign companies to sell their products within China itself (although many companies have pulled out of China, in frustration over unpredictable business policies or because they were unexpectedly shut down by the authorities). Trade between Taiwan and China—mostly carried on via Hong Kong, but now also permitted through several small Taiwanese islands adjacent to the mainland—was nearly $6 billion by the early 1990s, and has increased exponentially since then. And while Hong Kong was investing some $25 billion in China, China was investing $11 billion in Hong Kong. More Chinese firms were permitted to export directly and to keep more of the profits. Special Economic Zones (SEZs)—capitalist enclaves adjacent to Hong Kong and along the coast into which were sent the most educated of the Chinese population—were established to catalyze the internal economy. In coastal cities, especially in south China, construction of apartment complexes, new manufacturing plants, and roads and highways began in earnest. Indeed, the south China area, along with Hong Kong and Taiwan, seemed to be emerging as a mammoth trading bloc—"Greater China"—which economists predict will one day eclipse the economies of both Japan and the United States. Indeed, if one counts gross domestic product on the basis of purchasing-power parity, China already has moved to second place in the world, after the United States.

When capitalism was introduced in China, stock exchanges were opened in Shanghai and Shenzhen, and other dramatic changes were implemented, even in the western rural areas. The collectivized farm system imposed by Mao was replaced by a household contract system with hereditary contracts (that is, one step away from actual private land ownership), and free markets replaced most of the system of mandatory agricultural sales to the government. New industries were established in rural villages, and incomes improved such that many families were able to add new rooms onto their homes or to purchase two-story and even three-story homes. Predictions of China's economic dominance have had to be revised downward in the face of the Asian financial crisis, but the country's economy remains a force to be reckoned with. China's entry into the World Trade Association (WTO), though likely to cause some difficult internal adjustments, will, in the long run, strengthen China's position in the world economy.

TIANANMEN SQUARE

A strong spirit of entrepreneurship took hold throughout the country in the 1980s. Many people, especially the growing body of educated youth, interpreted economic liberalization as the overture to political democratization. College students, some of whom had studied abroad, pressed the government to crack down on corruption in the Communist Party and to permit greater freedom of speech and other civil liberties.

In 1989, tens of thousands of college students staged a prodemocracy demonstration in Beijing's Tiananmen Square. The call for democratization received wide international media coverage and soon became an embarrassment to the Chinese leadership, especially when, after several days of continual protest, the students constructed a large statue in the square similar in appearance to the Statue of Liberty in New York Harbor. Some party leaders seemed inclined at least to talk with the students, but hard-liners apparently insisted that the prodemocracy movement be crushed in order that the CCP remain in control of the government. The official policy seemed to be that it would be the Communist Party, and not some prodemocracy movement, that would lead China to capitalism.

The CCP leadership had much to fear; it was, of course, aware of the quickening pace of Communist party power dissolution in the Soviet Union and Central/Eastern Europe, but it was even more concerned about corruption and the breakdown of CCP authority in the rapidly capitalizing rural regions of China, the very areas that had spawned the Communist Party under Mao. Moreover, economic liberalization had spawned inflation, higher prices, and spot shortages, and the general public was disgruntled. Therefore, after several weeks of pained restraint, the authorities moved against the students in what has become known as the Tiananmen Square massacre. Soldiers injured thousands and killed hundreds of students; hundreds more were systematically hunted down and brought to trial for sedition and for spreading counter-revolutionary propaganda.

In the wake of the brutal crackdown, many nations reassessed their relationships with the People's Republic of China. The United States, Japan, and other nations halted or canceled foreign assistance, exchange programs, and special tariff privileges. The people of Hong Kong, anticipating the return of their British colony to P.R.C. control in 1997, staged massive demonstrations against the Chinese government's brutality, and they continue to do so annually, much to the frustration of the Chinese leadership, who, according to the agreement with Britain, must allow such demonstrations in Hong Kong, but brutally suppresses them within China proper. Foreign tourism all but ceased, and foreign investment declined abruptly.

The withdrawal of financial support and investment was particularly troublesome to the Chinese leadership, as it realized that China's economy was far behind other nations. Even Taiwan, with a similar heritage and a common history until the 1950s, but having far fewer resources and much less land, had long since eclipsed the mainland in terms of economic prosperity. The Chinese understood that they needed to modernize (although not, they hoped, to Westernize), and they knew that large capital investments from such countries as Japan, Hong Kong, and the United States were crucial to their economic reform program. Moreover, they knew that they could not tolerate a cessation of trade with their new economic partners. By the end of the 1980s, about 13 percent of China's imports came from the United States, 18 percent from Japan, and 25 percent from Hong Kong. Similarly, Japan received 16 percent of China's exports, and Hong Kong received 43 percent.

Once the worldwide furor over the 1989 Tiananmen massacre died down, and fortunately for the Chinese economy, the investment and loan-assistance programs from other countries were reinstated in most cases. China was even able to close a $1.2 billion contract with McDonnell Douglas Corporation to build 40 jetliners; and as a result of decisions to separate China's human-rights issues from trade issues, the United States repeatedly renewed China's "most favored nation" trade status. U.S. president Bill Clinton and China's leader, Jiang Zemin, engaged in an unprecedented public debate on Chinese television in June 1998; and Clinton was allowed to engage students and others in direct dialogue in which he urged religious freedom, free speech, and the protection of other human rights.

These and other events suggest that China is trying to address some of the major concerns voiced against it by the industrialized world—one of which is copyright violations by Chinese companies. Some have estimated that as much as 88 percent of China's exports of CDs consists of illegal copies. A 1995 copyright agreement is having some effect, but still, much of China's trade deficit with the United States comes from illegal products. This did not stop the United States, Japan, Australia, and other countries from pushing for China to join the World Trade Organization. In fact, many business leaders believe that WTO membership will force China to adhere more faithfully to international rules of fair production and trade.

Improved trade notwithstanding, the Tiananmen Square massacre and continuing instances of brutality against citizens have convinced many people, both inside and outside China, that the Communist Party has lost, not necessarily its legal, but certainly its moral, authority to govern. Amnesty International's 1996 report claimed that human-rights violations in China occur "on a massive scale" and noted that torture is used on political prisoners held in *laogai,* Chinese gulags similar to those in the former Soviet Union. Although some religions are permitted in China, others are suppressed, the most notable being the Falun Gong, a meditative religion that China banned in 2000 and whose adherents are systematically arrested and jailed. The cases of China's human-rights violations would fill volumes, from Tibetan leaders imprisoned for "revolutionary incitement," to U.S. prodemocracy activists deported for talking to striking workers, to the jamming of Radio Free Asia and the Voice of America, to the blocking of Internet search engines and Web sites. One Chinese dissident, Xu Wenli, was imprisoned for 16 years just for organizing an independent political party. He was released in 2002 after the United States pressured the Chinese government. In 2004 alone, numerous instances of suppression of free speech and other civil rights were reported: a Chinese Buddhist leader was dragged away for attempting to hold a religious ceremony not approved by the government; a Catholic bishop was arrested twice; a military surgeon was jailed for writing a letter in which he urged the government to take responsibility for the Tiananmen massacre; three Protestant activists were sent to prison for three years for "leaking state secrets"; some 54 people were jailed when they wrote their opinions on the internet about the SARS epidemic and other topics; and, after some of the 400 million mobile phone users started sending text messages that exposed the national cover-up of the SARS epidemic, the Chinese government shut down 20 internet service providers and began censoring text messages. There are so many mobile phone users in China (an average of 7,000 messages every second, or more than the rest of the world combined), that the government regards the technology as a direct threat to its ability to control the flow of information and thus to govern. When villagers in Wangying protested illegal taxes on their crops, the government countered by ransacking their homes, dousing people with boiling water, and beating and whipping them.

Then, of course, there is the issue of Tibet, where repression of religion has kept the Dalai Lama and many others in exile for more than 40 years and where police in a western province shot and killed the head of a Buddhist monastery in 2004 after he complained about being beaten in custody. One political prisoner, a Tibetan nun, was released in 2002 after 11 years in prison. She said that since the age of 13 she was tortured with beatings, solitary confinement, and electric shocks while in custody in the Drapchi prison. Despite all the economic liberalization, it is very clear to all observant China-watchers that China remains a dictatorship where basic human freedoms are regularly denied, and where any challenge to the government, however, mild, will be met with suppression.

Despite China's controlled press, reports of other forms of social unrest are occasionally heard. For example, in 1999, some 3,000 farmers in the southern Hunan Province demonstrated against excessive taxation. One protester was killed. Labor riots break out from time to time as workers protest the growing disparity between rich and poor. The government, despite its socialist rhetoric, admits that it is trying to create a middle-income group, but corruption at every level of society causes seething resentments. In western regions with large Islamic populations, anti-Beijing sentiment sometimes erupts in the form of bombings of government buildings, and through underground antigovernment meetings. In Langchenggang, a town in central China, deadly street fights broke out between Muslims and Han Chinese in 2004, and many other villages have erupted in violence, although news about the events are usually suppressed.

THE SOCIAL ATMOSPHERE

In 1997, the aged Deng Xiaoping died. He was replaced by a decidedly more forward-looking leader, Jiang Zemin. Under his charge, the country was able to avoid many of the financial problems that affected other Asian nations in the late 1990s (although many Chinese banks are dangerously overextended, and real-estate speculation in Shanghai and other major cities has left many high-rise office buildings severely underoccupied). Despite many problems yet to solve, including serious human-rights abuses, it is clear that the Chinese leadership has actively embraced capitalism and has effected a major change in Chinese society. Historically, the loyalty of the masses of the people was placed in their extended families and in feudal warlords, who, at times of weakened imperial rule, were nearly sovereign in their own provinces. Communist policy has been to encourage the masses to give their loyalty instead to the centrally controlled Communist Party. The size of families has been reduced to the extent that "family" as such

Timeline: PAST

1523–1027 B.C.
The Shang Dynasty is the first documented Chinese dynasty

1027–256 B.C.
The Chou Dynasty and the era of Confucius, Laotze, and Mencius

211–207 B.C.
The Ch'in Dynasty, from which the word *China* is derived

202 B.C.–A.D. 220
The Han Dynasty

A.D. 220–618
The Three Kingdoms period; the Tsin and Sui Dynasties

618–906
The T'ang Dynasty, during which Confucianism flourished

906–1279
The Five Dynasties and Sung Dynasty periods

1260–1368
The Yuan Dynasty is founded by Kublai Khan

1368–1644
The Ming Dynasty

1644–1912
The Manchu or Ch'ing Dynasty

1834
Trading rights and Hong Kong Island are granted to Britain

1894–1895
The Sino-Japanese War

1912
Sun Yat-sen's republican revolution ends centuries of imperial rule; the Republic of China is established

1921
The Chinese Communist Party is organized

1926
Chiang Kai-shek begins a long civil war with the Communists

1949
Mao Zedong's Communist Army defeats Chiang Kai-shek

1958
A disastrous economic reform, the Great Leap Forward, is launched by Mao

1966–1976
The Cultural Revolution; Mao dies

1980s
Economic and political liberalization begins under Deng Xiaoping; the P.R.C. and Britain agree to return Hong Kong to the Chinese

China expands its relationship with Taiwan; the Tiananmen Square massacre provokes international outrage

1990s
Crackdowns on dissidents and criminals result in hundreds of arrests and executions

Deng Xiaoping dies; Jiang Zemin becomes president

PRESENT

2000s
China bans the Falun Gong religion

The pace of China's modernization and political influence accelerates; military spending increases

Hu Jintao becomes president, in another peaceful transfer of power

The 2,900-member Legislature amends the Constitution, adding phrases about the preservation of human rights and the inviolability of private property rights

China joins Southeast Asian nations in a free trade accord that will take effect in 2010 and will become the world's largest free trade area with nearly 2 billion consumers.

---

has come to play a less important role in the lives of ordinary Chinese.

Historical China was a place of great social and economic inequality between the classes. The wealthy feudal lords and their families and those connected with the imperial court and bureaucracy had access to the finest in educational and cultural opportunities, while around them lived illiterate peasants who often could not feed themselves, let alone pay the often heavy taxes imposed on them by feudal and imperial elites. The masses often found life to be bitter, but they found solace in the teachings of the three main religions of China (often adhered to simultaneously): Confucianism, Taoism, and Buddhism. Islam, animism, and Christianity have also been significant to many people in China.

The Chinese Communist Party under Mao, by legal decree and by indoctrination, attempted to suppress people's reliance on religious values and to reverse the ranking of the classes; the values of hard, manual work and rural simplicity were elevated, while the refinement and education of the urban elites were denigrated. Homes of formerly wealthy capitalists were taken over by the government and turned into museums, and the opulent life of the capitalists was disparaged. During the Cultural Revolution, high school students who wanted to attend college had first to spend two years doing manual labor in factories

and on farms to help them learn to relate to the peasants and the working class. So much did revolutionary ideology and national fervor take precedence over education that schools and colleges were shut down for several years during the 1960s and 1970s and the length of compulsory education was reduced.

One would imagine that after 40 years of communism, the Chinese people would have discarded the values of old China. However, the reverse seems to be true. When the liberalization of the economy began in the late 1970s, many of the former values also came to the fore: the Confucian value of scholarly learning and cultural refinement, the desirability of money, and even Taoist and Buddhist religious values.

Thousands of Chinese are studying abroad with the goal of returning to China to establish or manage profitable businesses. Indeed, some Chinese, especially those with legitimate access to power, such as ranking Communist Party members, have become extremely wealthy. Along with the privatization of state enterprises has come the unemployment of hundreds of thousands of "redundant" workers (2 million workers lost their jobs in one province in a single year in the early 1990s). Many others have had to settle for lower pay or unsafe work conditions as businesses strive to enter the world of competitive production. Each year, numerous

accidents at various workplaces remind everyone that China's physical and managerial infrastructure is woefully inadequate: a gas explosion in a coal mine in Shaanxi province traps hundreds of miners; a chlorine gas leak and explosion at a chemical plant in Chongqing kills 233 and forces the evacuation of 150,000; an accident in a tin mine kills 81 people and the communist party secretary pays bribes to conceal the tragedy; and 2,000 fireworks factories are closed after a series of fatal explosions. Over 4,000 coal miners died in various accidents in 2004 alone. Demonstrations and more than 300 strikes by angry laborers exploded in early 1994; even those with good jobs were finding it difficult to keep up with inflation, which in recent years had been as high as 22 percent. Many of the most modern features of the new China are simply unattainable by the average Chinese person. For instance, the cost of riding the maglev train mentioned at the beginning of this chapter was so high that few were buying tickets on it; the government recently had to slash prices. Likewise, many modern high-rise office buildings in Shanghai and Beijing remain largely empty because business owners cannot afford the high rent. Nevertheless, those with an entrepreneurial spirit were finding ways to make more money than they had ever dreamed possible in an officially communist country.

Some former values may help revitalize Chinese life, while others, once suppressed by the Communists, may not be so desirable. For instance, despite being an unabashed womanizer himself, Mao attempted to eradicate prostitution, eliminate the sale of women as brides, and prevent child marriages. Some of those customs are returning, and gender-based divisions of labor are making their way into the workplace.

Particularly hard hit by the results of recent population policies are females. So many couples are having female fetuses aborted in favor of males that there are about 117 males for every 100 females. Some women would rather sell their female babies than have abortions, and some of these children end up being smuggled and sold by baby-trafficking gangs. Many other girls and women are exploited on hundreds of internet pornography sites (which the government is trying hard to eliminate by shutting down internet cafes and making nudity on the internet illegal), while prostitution has made a comeback all over China.

Predicting China's future is difficult, but in recent years, the Chinese government has accelerated the pace of China's modernization and its role as a major power. The results are evident at every turn, from the $8 billion upgrading of the railway system, to the purchase by Chinese companies of major gas and oil fields in Indonesia, to the projected hosting of the 2008 Summer Olympics in Beijing. China is now the world's largest recipient of foreign direct investment, overtaking the United States. With more than US$50 billion annually flowing into China from such companies as Goldman Sachs, General Motors, Volkswagen, Ford Motor, and Toyota Motor, to name a few, China's annual growth exceeds that of most countries of the world. China is also engaged in the world's largest public-works project, Three Gorges Dam. Costing $25 billion and employing some 20,000 laborers, the dam, to be completed by 2009, is expected to generate so much electricity that its cost will be recovered by the year 2014. The impact on the environment, however, has been and will continue to be severe: some 1 million people will have to be resettled, and whole towns and many historic sites will be flooded when the dam is completed. Along with other sectors, China, with the second-largest military budget in the world, is moving quickly to modernize its military, a fact that is causing consternation among all other Asian states, which worry about China's intentions with its new missiles and submarines. China has also been quietly flexing its muscles in the South China Sea, and it bristles at any suggestion of a breach of its sovereignty, the most recent example being the emergency landing (after a forced collision with a Chinese fighter jet) of a U.S. surveillance plane on Hainan Island. The Chinese government exploited the situation fully, detaining the crew for two weeks and refusing to allow the repaired plane to take off; rather, it had to be dismantled and removed piece by piece. Another example was the accidental bombing of the Chinese Embassy in Yugoslavia by NATO planes engaged in the Kosovo conflict. Upon news of the bombing, the Chinese government organized anti–U.S. and anti–NATO demonstrations all over China. The U.S. Embassy in Beijing and the consulate in Shanghai were damaged, and the United States, in particular, was vilified in the press.

Regardless of how China develops in the future, every country in the world now recognizes that it will have to find new ways of dealing with Asia's colossus.

East Timor (Democratic Republic of Timor-Leste)

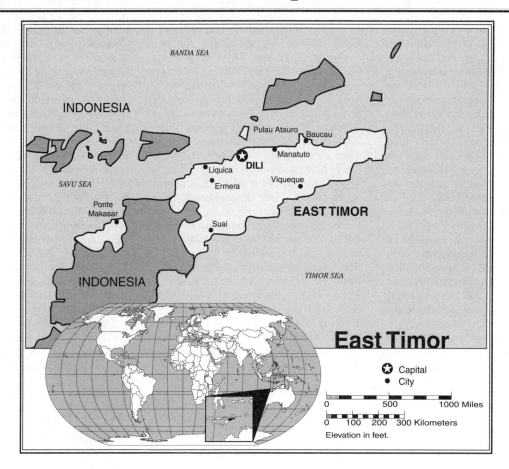

East Timor Statistics

GEOGRAPHY

Area in Square Miles (Kilometers): 5,793 (15,007) (about the size of Connecticut)
Capital (Population): Dili (140,000)
Environmental Concerns: deforestation; soil erosion
Geographical Features: mountainous; part of the Malay Archipelago
Climate: tropical

PEOPLE

Population

Total: 1,019,252
Annual Growth Rate: 2.11%
Rural/Urban Population Ratio: 76/24
Major Languages: Tetum; Portuguese; Indonesian
Ethnic Makeup: Malayo-Polynesian; Papuan; small Chinese minority

Religions: 90% Roman Catholic; 4% Muslim; animism

Health

Life Expectancy at Birth: 63 years (male); 67 years (female)
Infant Mortality: 49/1,000 live births

Education

Adult Literacy Rate: 48%

TRANSPORTATION

Highways in Miles (Kilometers): 2,280 (3,800)
Railroads in Miles (Kilometers): none
Usable Airfields: 8

GOVERNMENT

Type: republic

Independence Date: May 20, 2002 (from Indonesia)
Head of State/Government: President Kay Rala Xanana Gusmao: Prime Minister Mari Bin Amude Alkatiri
Political Parties: Revolutionary Front of Independent East Timor; Democratic Party; many others
Suffrage: universal at 17

MILITARY

Current Disputes: boundary disputes with Australia and Indonesia

ECONOMY

Currency ($ U.S. Equivalent): East Timor uses the U.S. dollar
Per Capita Income/GDP: $500/$440 million
GDP Growth Rate: -3%

56

Unemployment Rate: 50%

Population Below Poverty Line: 42%

Natural Resources: petroleum; gold;
 natural gas; timber; marble; manganese

Agriculture: rice; cassava; coffee;
 soybeans; fruits; vanilla; cabbage
Industry: printing; textiles; handicrafts; soap
Exports: $8 million
Imports: $237 million

East Timor Country Report

Among the athletes competing in Sydney in the 2002 Summer Olympics were four from East Timor, the Pacific Rim's newest nation. With the sovereignty of their country still in the hands of a transitional United Nations administration, the athletes had to march behind the flag of the International Olympic Committee. But they marched to thunderous applause from the spectators, who were obviously pleased that the painful struggle to bring peace and freedom to the troubled island had finally ended in success.

DEVELOPMENT

Under Indonesian occupation, new roads, bridges, schools, hospitals, and community health centers were constructed. Government offices were built in Dili and in smaller towns, but militias destroyed many of them in 1999. Australia signed an oil-development deal in 2002, but East Timor wants to clarify maritime boundaries before approving further oil development.

THE PORTUGUESE SETTLE IN

It was the fragrant sandalwood forests that first attracted the Portuguese to the tiny island of Timor. While other European explorers were busy trying to find their way to the fabled riches and spices of China by sailing west, the Portuguese and the Dutch tried sailing east around Africa. Along the way, they came across Timor. Located about 400 miles northwest of Australia, the mountainous Timor (280 miles long and 65 miles across, with some mountains nearly 10,000 feet high) is one of the 13,000 islands that comprise the Malay (Indonesian) Archipelago. The sandalwood trees, useful for woodcarvings, as well as teak, rosewood, bamboo, and eucalyptus, convinced the Portuguest to stay.

By 1520 the Portuguese had settled a colony on the island; they would, no doubt, have controlled the whole island had it not been for the Dutch, who claimed the island as well. Eventually, control of Timor was divided between the Netherlands in the west, and Portugal in the east.

Thus was born East Timor. When, after World War II, the rest of the Dutch East Indies gained independence as the new nation of Indonesia, East Timor remained a colony of Portugal.

The Portuguese never invested heavily in the development of East Timor. Portuguese-owned companies harvested the sandalwood trees, extracted marble from quarries, and grew coffee on the mountainsides, but Portugal regarded the distant island as somewhat peripheral to its interests. Yet when some of the Timorese people (many of whom were of mixed Portuguese-Timorese descent and most of whom had become Christians) declared their desire for independence, Portugal initially responded with force and then, in 1974, attempted to establish a provisional local government. Fighting broke out between those favoring independence and those preferring integration with Indonesia. Rather than try to solve the problem, Portugal withdrew. In 1975, the Revolutionary Front of Independent East Timor (Fretilin) declared victory and proclaimed the formation of the Democratic Republic of East Timor. After 450 years of outside control, East Timor was free. But the freedom lasted only nine days.

FREEDOM

Under UN supervision, peace has been restored to East Timor, but some 15% of the population are still held in camps across the border in West Timor. The Indonesian government released Xanana Gusmao after 20 years as a political prisoner, and he became the president of East Timor after independence from Indonesia in 2002.

INDONESIA INVADES

In December 1975, General Suharto of Indonesia invaded the island, with some 35,000 air, land, and naval troops. The attackers were brutal: They dropped napalm bombs on isolated mountain villages; they shot indiscriminately into crowds of unarmed civilians; they tortured, raped, and mutilated their victims. By the time the vi-

olence had stopped four years later, some 200,000 people (almost a third of the population) had been killed, including Nicolau Lobato, the East Timorese president. Many resisters had been placed in concentrations camps. Furthermore, nearly 80 percent of the farmers (most islanders are farmers) had been displaced from their lands, wreaking havoc on the economy. The world community condemned the invasion, and the United Nations refused to accept Indonesia's claim that East Timor had now become a province of Indonesia. But Indonesia held onto the island for 25 years, until 1999, when a new group of Indonesian leaders reluctantly agreed to a UN–brokered referendum on independence.

HEALTH/WELFARE

East Timor is the poorest nation on Earth. A very weak infrastructure limits access to medical care. Years of fighting have destroyed hospitals and most governmental functions.

When the votes were counted, 79 percent of the people (almost the exact percentage of the non-Indonesian part of the population) voted for independence. The result produced outrage among the mostly Muslim Indonesian troops living in East Timor. Refusing to accept the result, they launched a hellish blitz of random violence and destruction—much of it with anti-Christian overtones. Bands of marauding militias killed, raped, and destroyed everyone and everything in sight. They attacked United Nations officials as well as the Australian ambassador, and as they launched their scorched-earth campaign, they declared that "a free East Timor will eat stones." Despite a global outcry against the violence, Indonesian leaders seemed to do nothing to stop it. An international force led by Australia eventually drove out the Indonesians and brought an end to the brutality, but by that time hundreds of civilians had been killed, dozens of villages had been burned, and 70 percent of the buildings of the capital city of Dili had been intentionally destroyed. From these painful beginnings was born the Pacific Rim's newest nation.

INDEPENDENCE

On May 20, 2002, East Timor became a sovereign nation. Former U.S. president Bill Clinton attended the independence ceremony, as did the president of Indonesia, who, in a gesture of reconciliation, shared the platform with President Gusmao.

ACHIEVEMENTS

The Nobel Peace Prize of 1996 was awarded to two East Timorese: Jose Ramos-Horta and Bishop Carlos Filipe Ximenes Belo.

In an interesting development, East Timor adopted the U.S. dollar as its currency. UN troops from Australia, New Zealand, Japan, and Portugal remain in East Timor to help stabilize a still-restive population, and residents forcefully displaced to West Timor (where some of them became sex slaves to the militias) continue to slowly make their way back across the border.

With plentiful supplies of oil in the oceans surrounding East Timor, the country could become economically viable if the oil extraction is handled properly. However, a number of interests claim rights to the same oil fields as East Timor. For example, Australia claims it controls a large share of the oil, as does Indonesia, ConocoPhillips, and others. President Gusmao accused Australia of an "oil grab" when it started drilling for oil immediately upon the expiration of the former sea boundary agreement with Indonesia in 1999. Gusmao claims that his country's economy will be doomed if Australia continues to drill. It appears it will be many years before the maritime boundaries and oil rights issues are settled by the East Timor-Indonesia Boundary Committee, but if traditional maritime law is followed, most oil deposits will likely be located inside East Timor's waters.

THE PEOPLE OF EAST TIMOR

Full and accurate statistics on the population of East Timor are difficult to come by. In 1975, just before Suharto's invasion, the 680,000 East Timorese were divided as follows: 97 percent Timorese (including mestizos); 2 percent Chinese; 1 percent Portuguese. In 1999, the population was estimated at 800,000: 78 percent Timorese; 2 percent Chinese; 20 percent Indonesian.

Like many other Southeast Asian nations, East Timor is a social gumbo of religions, languages, and ethnicities. The first known inhabitants were the Atoni, a people of Melanesian descent. Most of the people today are of mixed Malay, Polynesian, and Papuan descent, with some Chinese and many Portuguese-Timorese mestizos added in to the mix. Official statistics show that about 90 percent of the population are Roman Catholic (the harbor city of Dili boasts the largest Catholic cathedral in Southeast Asia), but animism is still widely practiced in many of the hundreds of isolated villages.

Isolation has meant that not everyone in East Timor speaks the same language. Despite the small size of the country, East timor is comprised of 12 ethnic groups speaking some sixteen Austronesian and Papuan languages. The most frequently used language is Tetum, spoken by about 60 percent of the people, but Portuguese is also an official language.

Although international attention has focused on the capital city of Dili and the ravages that city-dwellers suffered at the hands of the Indonesian militias, the typical East Timorese is a country person, a farmer using slash-and-burn methods and growing only enough corn and vegetables to support a family. Many grow coffee, which was once the country's most profitable export. Handicraft enterprises (household utensils, clothing, and farm tools) also provide a livelihood for some. But years of warfare and a drought in the early 2000s, have destroyed some 70 percent of the economy. International aid has helped rebuild part of the country, but with an economy actually declining at about –3 percent a year, the $500 per capita income makes East Timor the poorest country in the world.

Solving these economic problems and restoring normalcy will tax the skills of East Timor's leaders, prominent among whom are: President Kay Rala Xanana Gusmao (formerly using the name Jose Alexander Xanana Gusmao)—once the head of the Timorese resistance movement and a political prisoner for 20 years; and Jose Ramos-Horta, also a member of the resistance leadership and co-winner (with Catholic bishop Carlos Filipe Ximenes Belo) of the Nobel Peace Prize.

Timeline: PAST

A.D. 1000
Traders from Java and China arrive to obtain sandalwood

1520
The Portuguese settle a colony on the island

1613
Dutch traders land on Timor

1859
The island is divided between the Netherlands in the west and Portugal in the east

1894–1912
The Portuguese put down an independence movement

1950s–1970s
East Timorese rebel against Portuguese rule

1975
Portugal withdraws; East Timor declares independence from Portugal; Indonesia invades

1975–1989
East Timor is closed to the outside world

1976
Indonesia proclaims East Timor its 27th province

1978
East Timor's president is killed by Indonesian soldiers

1990s
Indonesia agrees to independence referendum; 79% favor independence; Indonesian militias ravage the country

The UN takes control of the country and restores peace

PRESENT

2000s
"Xanana" Gusmao is elected president, in UN–sponsored elections

East Timor becomes a sovereign nation

A committee to settle maritime boundaries with Indonesia and others continues its work of mapping the actual border of the world's newest country.

Hong Kong (Hong Kong Special Administrative Region)

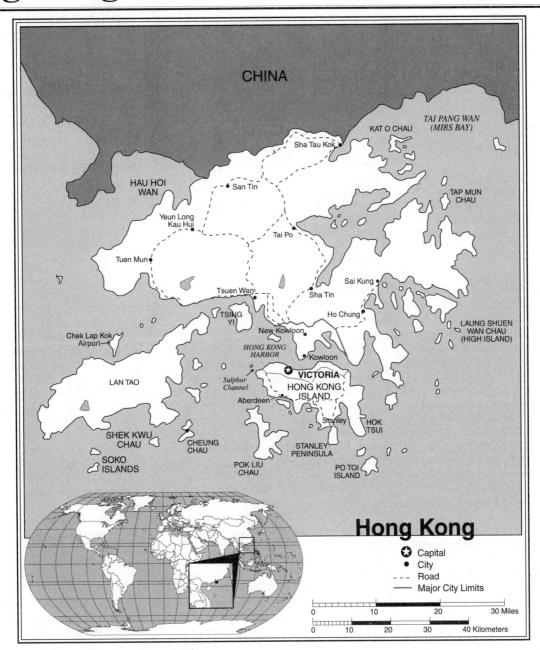

Hong Kong Statistics

GEOGRAPHY

Area in Square Miles (Kilometers): 671 (1,054) (about 6 times the size of Washington, D.C.)

Capital (Population): Victoria (na)

Environmental Concerns: air and water pollution

Geographical Features: hilly to mountainous, with steep slopes; lowlands in the north

Climate: tropical monsoon

PEOPLE

Population

Total: 6,855,125

Annual Growth Rate: 0.65%

Rural/Urban Population Ratio: 9/91

Major Languages: Chinese (Cantonese); English

Ethnic Makeup: 95% Chinese (mostly Cantonese); 5% others

Religions: 90% a combination of Buddhism and Taoism; 10% Christian

Health

Life Expectancy at Birth: 78 years (male); 84 years (female)

Infant Mortality: 5.7/1,000 live births

Physicians Available: 1/1,000 people

HIV/AIDS Rate in Adults: 0.06%

Education

Adult Literacy Rate: 93%

COMMUNICATION

Telephones: 3,839,000 main lines
Internet Users: 4,350,000 (2002)

TRANSPORTATION

Highways in Miles (Kilometers): 1,135
  (1,831)
Railroads in Miles (Kilometers): 22 (34)
Usable Airfields: 3

GOVERNMENT

Type: Special Administrative Region
  (SAR) of China
Head of State/Government: Hu Jintao;
  Chief Executive Tung Chee-hwa
Political Parties: Democratic Alliance for
  the Betterment of Hong Kong;
  Democratic Party; Association for

Democracy and People's Livelihood;
Hong Kong Progressive Alliance;
Citizens Party; Frontier Party; Liberal
Party
Suffrage: direct elections universal at 18
  for residents who have lived in Hong
  Kong for at least 7 years

MILITARY

Military Expenditures (% of GDP):
defense is the responsibility of China
Current Disputes: none

ECONOMY

Currency ($ U.S. Equivalent): 7.79 Hong
  Kong dollars = $1
Per Capita Income/GDP: $28,800/$213
  billion
GDP Growth Rate: 3.3%

Inflation Rate: -2.6%
Unemployment Rate: 7.9%
Natural Resources: outstanding deepwater
  harbor; feldspar
Agriculture: vegetables; poultry; fish; pork
Industry: textiles; clothing; tourism;
  electronics; plastics; toys; clocks;
  watches
Exports: $226 billion (primary partners
  China, United States, Japan)
Imports: $230 billion (primary partners
  China, Japan, Taiwan)

SUGGESTED WEBSITES

http://www.info.gov.hk/
  sitemap.htm
http://www.odci.gov/cia/
  publications/factbook/geos/
  hk.html

Hong Kong Country Report

Opium started it all for Hong Kong. The addictive drug from which such narcotics as morphine, heroin, and codeine are made, opium had become a major source of income for British merchants in the early 1800s. When the Chinese government declared the opium trade illegal and confiscated more than 20,000 large chests of opium that had been on their way for sale to the increasingly addicted residents of Canton, the merchants persuaded the British military to intervene and restore their trading privileges. The British Navy attacked and occupied part of Canton. Three days later, the British forced the Chinese to agree to their trading demands, including a demand that they be ceded the tiny island of Hong Kong (meaning "Fragrant Harbor"), where they could pursue their trading and military business without the scrutiny of the Chinese authorities.

DEVELOPMENT

Hong Kong is one of the preeminent financial and trading dynamos of the world. Annually, it exports billions of dollars' worth of products. Hong Kong's political future may be uncertain, but its fine harbor as well as its recently built $21 billion airport, a new Disney theme park, and information technology "cyberport," are sure to continue to fuel its economy.

Initially, the British government was not pleased with the acquisition of Hong Kong; the island, which consisted of nothing more than a small fishing village, had been an-

nexed without the foreknowledge of the authorities in London. Shortly, however, the government found the island's magnificent deepwater harbor a useful place to resupply ships and to anchor military vessels in the event of further hostilities with the Chinese. It turned out to be one of the finest natural harbors along the coast of China. On August 29, 1842, China reluctantly signed the Treaty of Nanking (or Nanjing), which ended the first Opium War and gave Britain ownership of Hong Kong Island "in perpetuity."

Twenty years later, a second Opium War caused China to lose more of its territory; Britain acquired permanent lease rights over Kowloon, a tiny part of the mainland facing Hong Kong Island. By 1898, Britain had realized that its miniscule Hong Kong naval base would be too small to defend itself against sustained attack by French or other European navies seeking privileged access to China's markets. The British were also concerned about the scarcity of agricultural land on Hong Kong Island and nearby Kowloon Peninsula. In 1898, they persuaded the Chinese to lease them more than 350 square miles of land adjacent to Kowloon. Thus, Hong Kong consists today of Hong Kong Island (as well as numerous small, uninhabited islands nearby), the Kowloon Peninsula, and the agricultural lands that came to be called the New Territories.

From its inauspicious beginnings, Hong Kong grew into a dynamic, modern society, wealthier than its promoters would have ever dreamed in their wildest imaginations. Hong Kong is now home to 6.9 million people. Most of the New Territo-

ries are mountainous or are needed for agriculture, so the bulk of the population is packed into about one tenth of the land space. This gives Hong Kong the dubious honor of being one of the most densely populated human spaces ever created. Millions of people live stacked on top of one another in 30-story-high buildings. Even Hong Kong's excellent harbor has not escaped the population crunch: Approximately 10 square miles of harbor have been filled in and now constitute some of the most expensive real estate on Earth.

Why are there so many people in Hong Kong? One reason is that, after occupation by the British, many Chinese merchants moved their businesses to Hong Kong, under the correct assumption that trade would be given a freer hand there than on the mainland. Eventually, Hong Kong became the home of mammoth trading conglomerates. The laborers in these profitable enterprises came to Hong Kong, for the most part, as political refugees from mainland China in the early 1900s. Another wave of immigrants arrived in the 1930s upon the invasion of Manchuria by the Japanese, and yet another influx came after the Communists took over China in 1949. Thus, like Taiwan, Hong Kong became a place of refuge for those in economic or political trouble on the mainland.

Overcrowding plus a favorable climate for doing business have produced extreme social and economic inequalities. Some of the richest people on Earth live in Hong Kong, alongside some of the most wretchedly poor, notable among whom are recent refugees from China and Southeast Asia

(Photo by Lisa Clyde Nielsen)

Land is so expensive in Hong Kong that most residences and businesses today are located in skyscrapers. While the buildings are thoroughly modern, construction crews typically erect bamboo scaffolding as the floors mount up, protecting the netting. The skyscrapers that appear darker in color in the forefront of this photo are being built using this technique.

(more than 300,000 Vietnamese sought refuge in Hong Kong after the Communists took over South Vietnam, although many have now been repatriated—some forcibly). Some of these refugees have joined the traditionally poor boat peoples living in Aberdeen Harbor. Although surrounded by poverty, many of Hong Kong's economic elites have not found it inappropriate to indulge in ostentatious displays of wealth, such as riding in chauffeured, pink Rolls-Royces or wearing full-length mink coats. Some Hong Kong residents are listed among the wealthiest people in the world: Li Ka-Shing worth US$12.4 billion, the Kwok brothers, worth US$11.4 billion; Lee Shau Kee, worth US$6.3 billion; and Michael Kadoorie, worth US$3.9 billion, to name a few.

Workers are on the job six days a week, morning and night, yet the average pay for a worker in industry is only about $5,000 per year. With husband, wife, and older children all working, families can survive; some even make it into the ranks of the fabulously wealthy. Indeed, the desire to make money was the primary reason why Hong Kong was settled in the first place. That fact is not lost on anyone who lives there today. Noise and air pollution (which have worsened dramatically in the past several years), traffic congestion, and dirty and smelly streets do not deter people from abandoning the countryside in favor of the consumptive and sparkling lifestyle of one

FREEDOM

Hong Kong was an appendage to one of the world's foremost democracies for 160 years. Thus, its residents enjoyed the civil liberties guaranteed by British law. Under the new Basic Law of 1997, the Chinese government has agreed to maintain the capitalist way of life and other freedoms for 50 years, although many residents believe that, slowly, democracy is being compromised by China's repressive policies.

Yet materialism has not wholly effaced the cultural arts and social rituals that are essential to a cohesive society. Indeed, with the vast majority of Hong Kong's residents hailing originally from mainland China, the spiritual beliefs and cultural heritage of China's long history abound. Some residents hang small, eight-sided mirrors outside windows to frighten away malicious spirits, while others burn paper money in the streets each August to pacify the wandering spirits of deceased ancestors. Business owners carefully choose certain Chinese characters for the names of their companies or products, which they hope will bring them luck. Even skyscrapers are designed following ancient Chinese customs so that their entrances are in balance with the elements of nature.

Buddhist and Taoist beliefs remain central to the lives of many residents. In the back rooms of many shops, for example,

small religious shrines are erected; joss sticks burning in front of these shrines are thought to bring good fortune to the proprietors. Elaborate festivals, such as those at New Year's, bring the costumes, art, and dance of thousands of years of Chinese history to the crowded streets of Hong Kong. And the British legacy may be found in the cricket matches, ballet troupes, philharmonic orchestras, English-language radio and television broadcasts, and the legal system under which capitalism flourished.

THE END OF AN ERA

Britain was in control of this tiny speck of Asia for nearly 160 years. Except during World War II, when the Japanese occupied Hong Kong for about four years, the territory was governed as a Crown colony of Great Britain, with a governor appointed by the British sovereign. In 1997, China recovered control of Hong Kong from the British. The events happened in this way: In 1984, British prime minister Margaret Thatcher and Chinese leader Deng Xiaoping concluded two years of acrimonious negotiations over the fate of Hong Kong upon the expiration of the New Territories' lease in 1997. Great Britain claimed the right to control Hong Kong Island and Kowloon forever—a claim disputed by China, which argued that the treaties granting these lands to Britain had been imposed by military force. Hong Kong Island and Kowloon,

(Photo by Lisa Clyde Nielsen)

A fishing family lives on this houseboat in Aberdeen Harbor. On the roof, strips of fish are hung up to dry.

however, constituted only about 10 percent of the colony; the other 90 percent was to return automatically to China at the expiration of the lease. The various parts of the colony having become fully integrated, it seemed desirable to all parties to keep the colony together as one administrative unit. Moreover, it was felt that Hong Kong Island and Kowloon could not survive alone.

HEALTH/WELFARE

Schooling is free and compulsory in Hong Kong through junior high school. The government has devoted large sums for low-cost housing, aid for refugees, and social services such as adoption. Housing, however, is cramped and inadequate.

The British government had hoped that the People's Republic of China would agree to the status quo, or that it would at least permit the British to maintain administrative control over the colony should it be returned to China. Many Hong Kong Chinese felt the same way, since they had, after all, fled to Hong Kong to escape the Communist regime in China. For its part, the P.R.C. insisted that the entire colony be returned to its control by 1997. After difficult negotiations, Britain agreed to return the entire colony to China as long as China would grant important concessions. Foremost among these were that the capitalist economy and lifestyle, including private-property ownership and basic human rights, would not be changed for 50 years. The P.R.C. agreed to govern Hong Kong as a

"Special Administrative Region" (SAR) within China and to permit British and local Chinese to serve in the administrative apparatus of the territory. The first direct elections for the 60-member Legislative Council were held in September 1991, while the last British governor, Christopher Patten, attempted to expand democratic rule in the colony as much as possible before the 1997 Chinese takeover—reforms that the Chinese dismantled to some extent after 1997.

The Joint Declaration of 1984 was drafted by top governmental leaders, with very little input from the people of Hong Kong. This fact plus fears about what P.R.C. control would mean to the free-wheeling lifestyle of Hong Kong's ardent capitalists caused thousands of residents, with billions of dollars in assets in tow, to abandon Hong Kong for Canada, Bermuda, Australia, the United States, and Great Britain. Surveys found that as many as one third of the population of Hong Kong wanted to leave the colony before the Chinese takeover. In the year before the change to Chinese rule, so many residents—16,000 at one point—lined up outside the immigration office to apply for British passports that authorities had to open up a nearby sports stadium to accommodate them. About half of Hong Kong residents already held British citizenship; but many of the rest, particularly recent refugees from China, wanted to secure their futures in case life under Chinese rule became repressive. Immigration officials received more than 100,000 applications for British passports in a single month in 1996!

Emigration and unease over the future have unsettled, but by no means ruined, Hong Kong's economy. According to the World Bank, Hong Kong is home to the world's eighth-largest stock market, the fifth-largest banking center and foreign-exchange market (and the second largest in Asia after Japan), and it per capita GDP at US$28,800 is second highest in Asia, just after Australia. As of 2005, it will also be home to Hong Kong Disneyland, Disney's third theme park outside the United States. The 766-acre project on Lantau Island is expected to attract up to 10 million visitors each year by 2020. Close to 9,000 multinational corporations have offices in Hong Kong. Despite the objections of the Chinese government, the outgoing British authorities embarked on several ambitious infrastructural projects that would allow Hong Kong to continue to grow economically in the future. Chief among these was the airport on Chek Lap Kok Island. At a cost of $21 billion, the badly needed airport was one of the largest construction projects in the Pacific Rim.

Opinion surveys showed that despite fears of angering the incoming Chinese government, most Hong Kong residents supported efforts to improve the economy and to democratize the government by lowering the voting age and allowing direct election rather than appointment by Beijing of more officials. In the 1998 elections, more than 50 percent of registered voters cast ballots—more than voted in Hong Kong's last election under British rule. In 1999, the opposition Democratic

Party gained a substantial number of seats in district-level elections. These indicate a desire by the people of Hong Kong for more democracy.

ACHIEVEMENTS

Hong Kong has the capacity to hold together a society where the gap between rich and poor is enormous. The so-called boat people have long been subjected to discrimination, but most other groups have found social acceptance and opportunities for economic advancement.

Unfortunately, that desire may be under attack. Although in 1997 the Chinese did not impose massive changes to the structure of democracy in Hong Kong, little by little the instances of restrictions are piling up. For example, the Beijing bureau chief for the *South China Morning Post*, a Hong Kong newspaper, was fired after complaining about restrictions on freedom of the press. The Hong Kong government continues to feel tremendous pressure from Beijing to clamp down on the Falun Gong religious movement in Hong Kong, as Beijing did on the mainland. One of the biggest flashpoints since reversion to the mainland was the proposed anti-subversion law, which would have given the mainland-influenced government more power to define ordinary journalists, labor activists, and even academics as subversives. Street demonstrations in 2003 in which 500,000 protestors marched, were successful in preventing the proposal from becoming law. Another flashpoint is the election in 2007 of the Hong Kong Chief Executive. Hong Kong residents want the right to freely and directly elect the successor to Tung Chee-hwa, rather than have the position controlled by China. They also want to democratically elect all lawmakers in 2008. Once again, residents took to the streets when it was announced by Beijing that direct elections would not be allowed. In 2004, voters also gave pro-democracy parties 3 more seats in Legco, Hong Kong's legislature, although 30 seats are still appointed by a China-controlled group of business leaders. Annual protests—sometimes involving more than 70,000 people—also occur in Hong Kong on the anniversary of the Tiananmen Square massacre in Beijing in which hundreds or perhaps thousands of unarmed students were killed by the Chinese military.

These protests may suggest that the people of Hong Kong would have preferred to remain a British colony. However, while there were large British and American communities in Hong Kong, and although English has been the medium of business and government for many years (China is now proposing the elimination of English as a language of instruction in most schools), many residents over the years had little or no direct emotional involvement with British culture and no loyalty to the British Crown. They asserted that they were, first and foremost, Chinese. This, of course, does not amount to a popular endorsement of Beijing's rule, but it does imply that some residents of Hong Kong feel that if they have to be governed by others, they would rather it be by the Chinese. Moreover, some believe that the Chinese government may actually help rid Hong Kong of financial corruption and allocate more resources to the poor—although with tourism down since the handover to China and the Hong Kong stock market still suffering the effects of the general Asian financial problems of 1998 and 1999, there may be fewer resources to distribute in the future.

Hong Kong's natural links with China had been expanding steadily for years before the handover. In addition to a shared language and culture, there are in Hong Kong thousands of recent immigrants with strong family ties to China. In 2001, 4.4 million Chinese tourists visited Hong Kong, spending more per person than tourists from any other country. And there are increasingly important commercial ties. Hong Kong has always served as south China's entrepôt to the rest of the world for both commodity and financial exchanges. For instance, for years Taiwan has circumvented its regulations against direct trade with China by transshipping its exports through Hong Kong. Commercial trucks plying the highways between Hong Kong and the P.R.C. form a bumper-to-bumper wall of commerce between the two regions. Sixty percent of Hong Kong's re-exports now originate in China. Despite a tightening of laws regarding human rights, the P.R.C. realizes that Hong Kong needs to remain more or less as it is—therefore, the transition to Chinese rule may be less jarring to residents than was expected. Most people think that Hong Kong will remain a major financial and trading center for Asia, but adjustments are under way as Hong Kong copes with its new status as just one of China's holdings. For instance, the sluggish world economy has pushed up unemployment, and in 2001, some 13,000 people—including some former millionaires—declared bankruptcy. Competition from China's less expensive docking facilities and from the booming, Hong Kong–like metropolis of Shanghai has contributed to some of the downturn. Perhaps the biggest drain on the economy is the devaluation of property values. Still expensive by world standards, Hong Kong property now suffers from "negative equity"; that is, some property is worth less than the mortgage that was used to buy it. There may be more such adjustments in the future, but the people of Hong Kong are hard-working and say that they are determined to make the new system work.

Timeline: PAST

A.D. 1839–1842
The British begin to occupy and use Hong Kong Island; the first Opium War

1842
The Treaty of Nanking cedes Hong Kong to Britain

1856
The Chinese cede Kowloon and Stonecutter Island to Britain

1898
England gains a 99-year lease on the New Territories

1898–1900
The Boxer Rebellion

1911
Sun Yat-sen overthrows the emperor of China to establish the Republic of China

1941
The Japanese attack Pearl Harbor and take Hong Kong

1949
The Communist victory in China produces massive immigration into Hong Kong

1980s
Great Britain and China agree to the return of Hong Kong to China

1990s
China resumes control of Hong Kong on July 1, 1997; prodemocracy politicians sweep the 1998 elections

PRESENT

2000s
Hong Kong's economy continues its recovery from the Asian financial crisis

Signs emerge of increasing P.R.C. control

Half a million protestors successfully defeat a proposed anti-subversion law

China refuses to allow democratic election of the Chief Executive in 2007 elections

Indonesia (Republic of Indonesia)

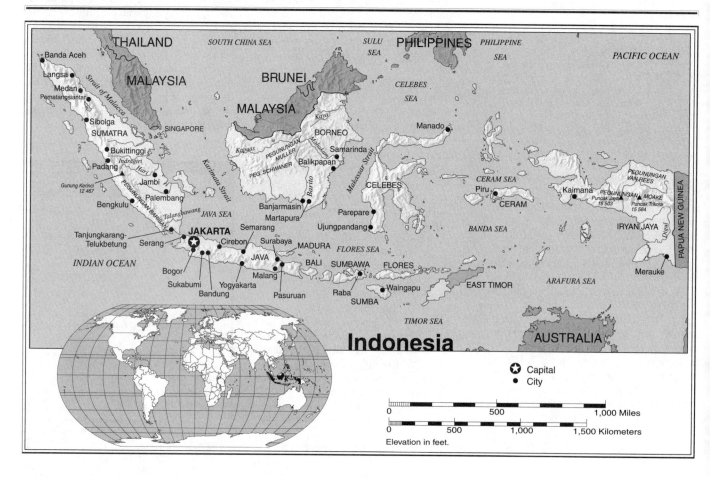

Indonesia Statistics

GEOGRAPHY

Area in Square Miles (Kilometers):
740,903 (1,919,440) (nearly 3 times the size of Texas)

Capital (Population): Jakarta (11,429,000)

Environmental Concerns: air and water pollution; sewage; deforestation; smoke and haze from forest fires

Geographical Features: the world's largest archipelago; coastal lowlands; larger islands have interior mountains

Climate: tropical; cooler in highlands

PEOPLE

Population

Total: 238,452,952
Annual Growth Rate: 1.49%
Rural/Urban Population Ratio: 60/40

Major Languages: Bahasa Indonesian; English; Dutch; Javanese; many others

Ethnic Makeup: 45% Javanese; 14% Sundanese; 7.5% Madurese; 7.5% coastal Malay; 26% others

Religions: 88% Muslim; 8% Christian; 4% Hindu, Buddhist, and others

Health

Life Expectancy at Birth: 66 years (male); 71 years (female)
Infant Mortality: 39.4/1,000 live births
Physicians Available: 1/6,570 people
HIV/AIDS in Adults: 0.05%

Education

Adult Literacy Rate: 87%
Compulsory (Ages): 7–16

COMMUNICATION

Telephones: 7.75 million main lines

Daily Newspaper Circulation: 20/1,000 people
Televisions: 145/1,000 people
Internet Users: 4,400,000 (2002)

TRANSPORTATION

Highways in Miles (Kilometers): (1999) 212,474 (342,700)
Railroads in Miles (Kilometers): 3,875 (6,450)
Usable Airfields: 490
Motor Vehicles in Use: 4,800,000

GOVERNMENT

Type: republic
Independence Date: December 27, 1949 (legally; from the Netherlands)
Head of State/Government: President Susilo Bambang Yudhoyono is both head of state and head of government

Political Parties: Golkar; Indonesia Democracy Party-Struggle; Development Unity Party; Crescent Moon and Star Party; National Awakening Party; others

Suffrage: universal at 17; married persons regardless of age

MILITARY

Military Expenditures (% of GDP): 1.3%

Current Disputes: territorial disputes with Malaysia, others; internal strife

ECONOMY

Currency ($ U.S. Equivalent): 8,577 rupiahs = $1
Per Capita Income/GDP: $3,200/$758 billion
GDP Growth Rate: 4.1%
Inflation Rate: 11.5%
Unemployment Rate: 8.7%%
Labor Force by Occupation: 45% agriculture; 39% services; 16% industry
Population Below Poverty Line: 27%
Natural Resources: petroleum; tin; natural gas; nickel; timber; bauxite; copper; fertile soils; coal; gold; silver
Agriculture: rice; cassava; peanuts; rubber; cocoa; coffee; copra; other

tropical; livestock products; poultry; beef; pork; eggs

Industry: petroleum; natural gas; textiles; mining; cement; chemical fertilizers; food; rubber; wood

Exports: $63.8 billion (primary partners Japan, United States, Singapore)

Imports: $40.2 billion (primary partners Japan, United States, Singapore)

SUGGESTED WEBSITES

http://www.cia.gov/cia/
publications/factbook/geos/
id.html
http://www.indonesia.elga.net.id

Indonesia Country Report

Present-day Indonesia is a kaleidoscope of some 300 languages and more than 100 ethnic groups. Beginning about 5000 B.C., people of Mongoloid stock settled the islands that today constitute Indonesia, in successive waves of migration from China, Thailand, and Vietnam. Animism—the nature-worship religion of these peoples—was altered substantially (but never completely lost) about A.D. 200, when Hindus from India began to settle in the area and wield the dominant cultural influence. Several hundred years later, Buddhist missionaries and settlers began converting Indonesians in a proselytizing effort that produced strong political and religious antagonisms. In the thirteenth century, Muslim traders began the Islamization of the Indonesian people; today, 87 percent of the population claim the Muslim faith—meaning that there are more Muslims in Indonesia than in any other country of the world, including the states of the Middle East. Commingling with all these influences are cultural inputs from the islands of Polynesia and colonial remnants from the Portuguese and the Dutch.

DEVELOPMENT

Indonesia continues to be hamstrung by its heavy reliance on foreign loans, a burden inherited from the Sukarno years. Current Indonesian leaders speak of "stabilization" and "economic dynamism," but there are always obstacles—government corruption and such natural disasters as the devastating tsunami of 2004—that hamper smooth economic improvement.

The real roots of the Indonesian people undoubtedly go back much further than any of these historic cultures. In 1891, the fossilized bones of a hominid who used stone tools, camped around a fire, and probably had a well-developed language were found on the island of Java. Named *Pithecanthropus erectus* ("erect ape-man"), these important early human fossils, popularly called Java Man, have been dated at about 750,000 years of age. Fossils similar to Java Man have been found in Europe, Africa, and Asia.

Modern Indonesia was sculpted by the influence of many outside cultures. Portuguese Catholics, eager for Indonesian spices, made contact with Indonesia in the 1500s and left 20,000 converts to Catholicism, as well as many mixed Portuguese–Indonesian communities and dozens of Portuguese "loan words" in the Indonesian-style Malay language. In the following century, Dutch Protestants established the Dutch East India Company to exploit Indonesia's riches. Eventually the Netherlands was able to gain complete political control; it reluctantly gave it up, in the face of insistent Indonesian nationalism, only as recently as 1950. Before that, however, the British briefly controlled one of the islands, and the Japanese ruled the country for three years during the 1940s.

Indonesians, including then-president Sukarno, initially welcomed the Japanese as helpers in their fight for independence from the Dutch. Everyone believed that the Japanese would leave soon. Instead, the Japanese military forced farmers to give food to the Japanese soldiers, made everyone worship the Japanese emperor, neglected local industrial development in favor of military projects, and took 270,000 young men away from Indonesia

to work elsewhere as forced laborers (fewer than 70,000 survived to return home). Military leaders who attempted to revolt against Japanese rule were executed. Finally, in August 1945, the Japanese abandoned their control of Indonesia, according to the terms of surrender with the Allied powers.

FREEDOM

Demands for Western-style human rights are frequently heard, but until recently, only the army has had the power to impose order on the numerous and often antagonistic political groups. However, some progress is evident: in 2004, Indonesians directly elected their president for the first time in decades.

Consider what all these influences mean for the culture of modern Indonesia. Some of the most powerful ideologies ever espoused by humankind—supernaturalism, Islam, Hinduism, Buddhism, Christianity, mercantilism, colonialism, and nationalism—have had an impact on Indonesia. Take music, for example. Unlike Western music, which most people just listen to, Indonesian music, played on drums and gongs, is intended as a somewhat sacred ritual in which all members of a community are expected to participate. The instruments themselves are considered sacred. Dances are often the main element in a religious service whose goal might be a good rice harvest, spirit possession, or exorcism. Familiar musical styles can be heard here and there around the country. In the eastern part of Indonesia, the Nga'dha peoples, who were converted to Christianity in the early

1900s, sing Christian hymns to the accompaniment of bronze gongs and drums. On the island of Sumatra, Minang Kabau peoples, who were converted to Islam in the 1500s, use local instruments to accompany Islamic poetry singing. Communal feasts in Hindu Bali, circumcision ceremonies in Muslim Java, and Christian baptisms among the Bataks of Sumatra all represent borrowed cultural traditions. Thus, out of many has come one rich culture.

But the faithful of different religions are not always able to work together in harmony. For example, in the 1960s, when average Indonesians were trying to distance themselves from radical Communists, many decided to join Christian faiths. Threatened by this tilt toward the West and by the secular approach of the government, many fundamentalist Muslims resorted to violence. They burned Christian churches, threatened Catholic and Baptist missionaries, and opposed such projects as the construction of a hospital by Baptists. As we have seen, Indonesia has more Muslims than any other country in the world, and the hundreds of Islamic socioreligious, political, and paramilitary organizations intend to keep it that way. Some 9,000 people in the eastern provincial capital of Ambon in the Maluku islands, many of them Christians desiring independence from Indonesia, were killed in Muslim-Christian sectarian violence in 2001, and gangs from both religions fought street battles there again in 2004, leaving more than two dozen dead and scores wounded. In addition to exploding bombs and hacking people to death with swords, the gangs set fire to churches and destroyed a United Nations office. Clearly, Indonesians have a long way to go in developing mutual respect and tolerance for the diversity of cultures in their midst.

A LARGE LAND, LARGE DEBTS

Unfortunately, Indonesia's economy is not as rich as its culture. Three quarters of the population live in rural areas; more than half of the people engage in fishing and small-plot rice and vegetable farming. The average income per person is only US$3,200 a year, based on gross domestic product. A 1993 law increased the minimum wage in Jakarta to $2.00 *per day.*

HEALTH/WELFARE

Indonesia has one of the highest birth rates in the Pacific Rim. Many children will grow up in poverty, never learning even to read or write their national language, Bahasa Indonesian.

Also worrisome is the level of government debt. Indonesia is blessed with large oil reserves (Pertamina is the state-owned oil company) and minerals and timber of every sort (also state-owned), but to extract these natural resources has required massive infusions of capital, most of it borrowed. In fact, Indonesia has borrowed more money than any other country in Asia. The country must allocate 40 percent of its national budget just to pay the interest on loans. Low oil prices in the 1980s made it difficult for the country to keep up with its debt burden. Extreme political unrest and an economy that contracted nearly 14 percent (!) in 1998 have seriously exacerbated Indonesia's economic headaches in recent years.

An example of the difficulty that Indonesia has in moving forward its economy is the experience of ExxonMobil Corporation. The U.S. company operates a large oil and gas refinery in Aceh Province in north Sumatra Island. But many of the Acehnese are locked in a bitter battle with the Indonesian government for independence. Wanting more revenue from the company, Acehnese rebels started a series of pipe bombings, kidnappings for ransom, bus hijackings, and even mortar and grenade attacks. In order to keep business going, the government had to send in 3,000 troops to guard the refinery day and night. Troops line the company airport for every take-off and landing; and, to protect the safety of its staff en route, the company has purchased 16 armor-plated Land Rovers. Thus, Indonesia's problems always involve both business and political entanglements.

To cope with these problems, Indonesia has relaxed government control over foreign investment and banking, and it seems to be on a path toward privatization of other parts of the economy. Still, the gap between the modernized cities and the traditional countryside continues to plague the government.

Indonesia's financial troubles seem puzzling, because in land, natural resources, and population, the country appears quite well-off. Indonesia is the second-largest country in Asia (after China). Were it superimposed on a map of the United States, its 13,677 tropical islands would stretch from California, past New York, and out to Bermuda in the Atlantic Ocean. Oil and hardwoods are plentiful, and the population is large enough to constitute a viable internal consumer market. But transportation and communication are problematic and costly in archipelagic states. Before the Asian financial crisis of the late 1990s hit, Indonesia's national airline, Garuda Indonesia, had hoped to launch a $3.6 billion development program that would have

brought into operation 50 new aircraft stopping at 13 new airports. New seaports were also planned. But the contraction of the economy and the enormous cost of linking together some 6,000 inhabited islands made most of the projects unworkable. Moreover, exploitation of Indonesia's amazing panoply of resources is drawing the ire of more and more people around the world who fear the destruction of one more part of the world's fragile ecosystem.

Illiteracy and demographic circumstances also constrain the economy. Indonesia's population of 238.5 million is one of the largest in the world, but 12 percent of adults (17 percent of females) cannot read or write. Only about 600 people per 100,000 attend college, as compared to 3,580 in nearby Philippines. Moreover, since almost 70 percent of the population reside on or near the island of Java, on which the capital city, Jakarta, is located, educational and development efforts have concentrated there, at the expense of the communities on outlying islands. Many children in the out-islands never complete the required six years of elementary school. Some ethnic groups, in the remote provinces of Papua (formerly called Irian Jaya) on the island of New Guinea and Kalimantan (Borneo), for example, continue to live isolated in small tribes, much as they did thousands of years ago. By contrast, the modern city of Jakarta, with its classical European-style buildings, is home to millions of people. Over the past 20 years, poverty has been reduced from 60 percent (the current poverty rate is about 27 percent), but Indonesia was seriously damaged by the Asian financial crisis and by a series of natural disasters in 2004, including the mammoth tsunami that devastated Aceh province, and experts expect that the economy will not return to normal in the near future.

ACHIEVEMENTS

Balinese dancers' glittering gold costumes and unique choreography epitomize the "Asian-ness" of Indonesia as well as the Hindu roots of some of its communities.

A big blow to Indonesia's tourism revenues came in October 2002, when terrorists linked to the al-Qaeda network bombed a nightclub in the popular resort community of Bali. Killing dozens of locals and tourists—many of them Australians—the bombings inflicted major damage to the tourism industry and convinced the world that Indonesia was not doing enough to support the international war on terrorism. Muslim Jemaah Islamiyah militants from Malaysia detonated a bomb outside the Australian embassy in Jakarta in 2004, wounding 173 people and killing 9. Other foreign interests, such as the Jakarta Marriott Hotel, have also been targets of deadly bomb attacks, and although the government has arrested several suspects in these and other bombings, militants can often escape to one of the thousands of remote islands that make up the country, where they find refuge among like-minded residents.

With 2.3 million new Indonesians entering the labor force every year, and with half the population under age 20, serious efforts must be made to increase employment opportunities. For the 1990s, the government earmarked millions of dollars to promote tourism. Nevertheless, the most pressing problem was to finish the many projects for which World Bank and Asian Development Bank loans had already been received. With considerable misgivings, the World Bank, the Asian Development Bank, and the government of Japan agreed in 2000 to provide more than $4 billion in additional aid to Indonesia to alleviate poverty and help decentralize government authority.

MODERN POLITICS

Establishing the current political and geographic boundaries of the Republic of Indonesia has been a bloody and protracted task. So fractured is the culture that many people doubt whether there really is a single country that one can call Indonesia. During the first 15 years of independence (1950–1965), there were revolts by Muslims and pro-Dutch groups, indecisive elections, several military coups, battles against U.S.–supported rebels, and serious territorial disputes with Malaysia and the Netherlands. In 1966, nationalistic President Sukarno, who had been a founder of Indonesian independence, lost power to Army General Suharto. (Many Southeast Asians had no family names until influenced by Westerners; Sukarno and Suharto have each used only one name.) Anti-Communist feeling grew during the 1960s, and thousands of suspected members of the Indonesian Communist Party (PKI) and other Communists were killed before the PKI was banned in 1966.

In 1975, ignoring the disapproval of the United Nations, President Suharto invaded and annexed East Timor, a Portuguese colony. Although the military presence in East Timor was subsequently reduced, separatists were beaten and killed by the Indonesian Army as recently as 1991; and in 1993, a separatist leader was sentenced to 20 years in prison. In late 1995, Amnesty International accused the Indonesian military of raping and executing human-rights activists in East Timor, while the 20th anniversary of the Indonesian takeover was marked by Timorese storming foreign embassies and demanding asylum and redress for the kidnapping and killing of protesters. In 1996, antigovernment rioting in Jakarta resulted in the arrest of more than 200 opposition leaders and the disappearance of many others. The rioters were supporters of the Indonesian Democracy Party and its leader, Megawati Sukarnoputri, daughter of Sukarno, and the woman who would shortly become president of the country.

Suharto's so-called New Order government ruled with an iron hand, suppressing student and Muslim dissent and controlling the press and the economy. With the economy in serious trouble in 1998, and with the Indonesian people tired of government corruption and angry at the control of Suharto and his six children over much of the economy, rioting broke out all over the country. Some 15,000 people took to the streets, occupied government offices, burned cars, and fought with police. The International Monetary Fund suspended vital aid because it appeared that Suharto would not conform to the belt-tightening required of IMF aid recipients. With unemployed migrant workers streaming back to Indonesia from Malaysia and surrounding countries, with the government unable to control forest fires burning thousands of acres and producing a haze all over Southeast Asia, with even his own lifetime political colleagues calling for him to step down, Suharto at last resigned, ending a 32-year dictatorship. The new leader, President Bacharuddin Jusuf Habibie, pledged to honor IMF commitments and restore dialogue on the East Timor dispute. But protests dogged Habibie, because he was seen as too closely allied with the Suharto leadership. In the first democratic elections in years, a respected Muslim cleric, Abdurrahman Wahid, was elected president, with Megawati Sukarnoputri (daughter of Indonesia's founding father, Sukarno) as vice-president.

In August 1999, the 800,000 residents of East Timor, the majority of whom were Catholic Christians, voted overwhelmingly (78.5 percent) for independence from Indonesia, in a peaceful referendum. Unfortunately, anti-independence Muslim militias, together with Indonesian troops, launched a hellish drive to prevent East Timor from separating from Indonesia. They drove thousands of residents from their homes, beat and killed them, and then burned their homes and businesses. The militias virtually obliterated the capital city of Dili. The violence became so severe that nearby Australia felt compelled to send in some 8,000 troops to prevent a wholesale bloodbath. Eventually East Timor obtained the independence it desired, but not before

Timeline: PAST

750,000 B.C.
Java Man lived here

A.D. 600
Buddhism gains the upper hand

1200
Muslim traders bring Islam to Indonesia

1509
The Portuguese begin to trade and settle in Indonesia

1596
Dutch traders begin to influence Indonesian life

1942
The Japanese defeat the Dutch

1949–1950
Indonesian independence from the Netherlands; President Sukarno retreats from democracy and the West

1966
General Suharto takes control of the government from Sukarno and establishes his New Order, pro-Western government

1975
Indonesia annexes East Timor

1990s
Suharto steps down after 32 years in power; East Timor votes for independence; violence erupts in Borneo

PRESENT

2000s
The economy remains stalled

East Timor obtains independence

Aceh Province signs a peace agreement ending a 130-year armed rebellion

A peace agreement in separatist Aceh Province mutes, but does not eliminate, some of the violence there

Moderate Susilo Bambang Yudhoyono, age 55, wins a landslide election to become the first Indonesian president directly elected by the voters since the end of the Suharto dictatorship

A series of natural disasters, including a volcanic eruption in northeastern Indonesia, rain-triggered landslides in Sumatra, powerful earthquakes in Papua, and a mammoth tsunami in Aceh kill tens of thousands and add further woes to the struggling economy

Indonesia received worldwide criticism for its ineffectual response to the rampage. Other separatist movements are also challenging the government, and as of this writing, over half a million Indonesians are displaced from their homes due to government attempts to suppress separatists.

The East Timor violence was not the first-time that Indonesians had run "amok" (a Malay word meaning to erupt in violent rage). In 1965, for example, an abortive communist coup precipitated a violent purge of communists by the Army. Within a few months over 250,000 citizens had been killed. In 1975, General Suharto invaded East Timor; his troops tortured, raped, and mutilated their victims. After four years, some 200,000 people had perished.

CRISIS IN BORNEO

In early 1999, violence broke out on the Indonesian island of Borneo. By the time it ended months later, 50,000 people had fled the island, and hundreds of men, women, and children had been killed, their decapitated heads displayed in towns and villages. The attackers even swaggered through towns victoriously holding up the dismembered body parts of their victims.

What caused this violence? It started with overpopulation and poverty. Some 50 years ago, the Indonesian government decided that it had to do something about overpopulation on the soil-poor island of Madura. Located near the island of Java, Madura, with its white-sand beaches and its scores of related islets, was the home of the Madurese, Muslims who had long since found it difficult to survive by farming the rocky ground. The government thus decided to move some of the Madurese to the island of Borneo, where land was more plentiful and fertile. It seemed like a logical solution, but the government failed to consider the ethnic context—that is, the island of Borneo was inhabited by the Dayaks, a people who considered the island their

tribal homeland. They had little or no interest in the government in Jakarta, regarding themselves as Dayaks first and Indonesians a distant second, if at all.

Almost immediately the Dayaks began harassing the newcomers, whom they judged to be "hot-headed" and crude. They resented both the loss of their land and, later, the loss of jobs in the villages. Over the years, hundreds of people were killed in sporadic attacks, but the violence in 1999 was worse than ever, in part because the Dayaks were, for the first time, joined by the Malays in "cleansing" the island of the hated Madurese. At first, the Indonesian government seemed to do little to stop the carnage. Eventually, when the Madurese had been chased from their homes, President Wahid promised aid money and assistance with relocation. But he wanted the Madurese to return to Borneo. The Dayaks, with little regard for the Indonesian government, responded that they would kill any who returned. As of this writing, thousands of Madurese remain as refugees on the island of Java.

In mid-2000, in the face of mounting criticism of his seeming inability to restore order and jump-start the economy, Wahid, who was nearly blind due to a serious eye disease, announced that he was turning over day-to-day administration to his vice-president. Wahid was subsequently censured, impeached, and then replaced. The new female president, Megawati Sukarnoputri, made some attempts to clean up the leftover corruption of the Suharto

years, and she encouraged foreign investment, particularly from Japan, which buys 23 percent of Indonesia's produce. In the past, companies like Toyota had invested millions in Indonesia's ASTRA automobile company, while Japanese banks had supported the expansion of Indonesia's tourist industry. Closer links with the West, particularly the United States, are difficult due to the Iraq War, against which thousands of Indonesians have protested in mass rallies in Jakarta.

In the 2004 elections, in which 24 parties vied for 14,000 seats at all levels of government and fielded 450,000 (!) candidates, Megawati, whom the U.S. accused of having made illegal oil deals with Iraq's Saddam Hussein, was swept from office in a landslide victory by 55-year-old guitar-playing, poetry-reading Susilo Bambang Yudhoyono, the former security minister in Megawati's cabinet. Voters had great hope that Yudhoyono would be more approachable than Megawati, but barely had he taken office when the country was hit by a devastating natural disaster, the December 2004 tsunami. Caused by a mammoth earthquake in the Indian Ocean, a massive wave over five stories tall and traveling at speeds of 500 mph smashed into northern Sumatra. Over 100,000 Indonesians were confirmed dead, and many thousands more were unaccounted for. Yudhoyono's approach to this tragedy will likely determine his ability to solve some of Indonesia's many other pressing problems.

Laos (Lao People's Democratic Republic)

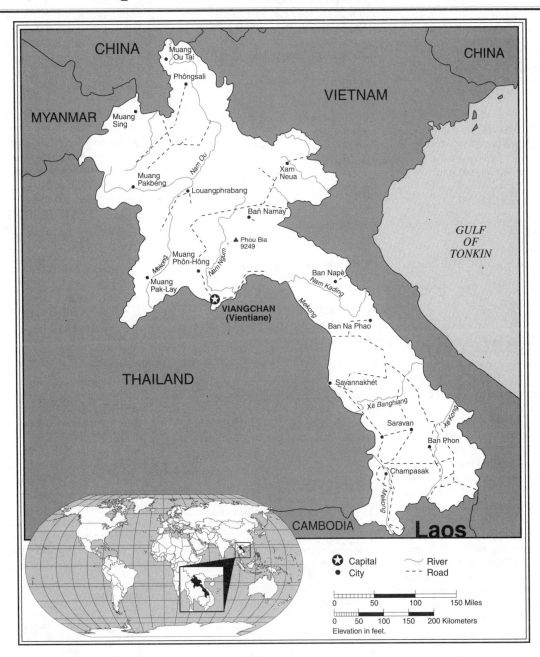

Laos Statistics

GEOGRAPHY

Area in Square Miles (Kilometers):
91,400 (236,800) (about the size of Utah)
Capital (Population): Viangchan
(Vientiane) (663,000)
Environmental Concerns: unexploded
ordnance; deforestation; soil erosion;
lack of access to potable water
Geographical Features: mostly rugged
mountains; some plains and plateaus
Climate: tropical monsoon

PEOPLE

Population

Total: 6,068,117
Annual Growth Rate: 2.4%
Rural/Urban Population (Ratio): 77/23
Major Languages: Lao; French; English;
ethnic languages
Ethnic Makeup: 50% Lao; 20% tribal
Thai; 15% Phoutheung (Kha); 15% Meo,
Hmong, Yao, and others

Religions: 60% Buddhist; 40% indigenous
beliefs and others

Health

Life Expectancy at Birth: 52 years (male);
56 years (female)

Infant Mortality Rate: 91/1,000 live births

Physicians Available: 1/3,555 people

HIV/AIDS Rate in Adults: 0.05%

Education
Adult Literacy Rate: 57%
Compulsory (Ages): for 5 years between ages 6–15

COMMUNICATION
Telephones: 28,500 main lines
Televisions: 17/1,000 people
Internet Users: 10,000 (2002)

TRANSPORTATION
Highways in Miles (Kilometers): 8,400 (14,000)
Railroads in Miles (Kilometers): none
Usable Airfields: 51
Motor Vehicles in Use: 18,000

GOVERNMENT
Type: Communist state
Independence Date: July 19, 1949 (from France)

Head of State/Government: President Khamtai Siphandon; Prime Minister Boungnang Volachit
Political Parties: Lao People's Revolutionary Party; other parties proscribed
Suffrage: universal at 18

MILITARY
Military Expenditures (% of GDP): 4.2%
Current Disputes: internal strife; indefinite border with Thailand; disputes with Thailand and Vietnam over squatters

ECONOMY
Currency ($ U.S. Equivalent): 7,600 kip = $1
Per Capita Income/GDP: $1,700/$10.3 billion
GDP Growth Rate: 5.5%
Inflation Rate: 10%

Unemployment Rate: 5.7%
Labor Force by Occupation:: 80% agriculture
Population Below Poverty Line: 40%
Natural Resources: timber; hydropower; gypsum; tin; gold; gemstones
Agriculture: rice; sweet potatoes; vegetables; coffee; tobacco; sugarcane; cotton; livestock
Industry: mining; timber; garments; electric power; agricultural processing; construction
Exports: $332 million (primary partners Thailand, France, Germany, Vietnam)
Imports: $492 million (primary partners Thailand, China, Vietnam)

SUGGESTED WEBSITES
http://www.global.lao.net/
laoVL.html
http://www.cia.gov/cia/
publications/factbook/geos/
la.html

Laos Country Report

Laos seems a sleepy place. Almost everyone lives in small villages where the only distraction might be the Buddhist temple gong announcing the day. Water buffalo plow quietly through centuries-old rice paddies, while young Buddhist monks in saffron-colored robes make their silent rounds for rice donations. Villagers build their houses on stilts for safety from annual river flooding, and top them with thatch or tin. Barefoot children play under the palm trees or wander to the village Buddhist temple for school in the outdoor courtyard. Mothers work at home, weaving brightly colored cloth for the family and preparing meals—on charcoal or wood stoves—of rice, bamboo shoots, pork, duck, and snakes seasoned with hot peppers and ginger. Even the 133,000-person capital city of Viangchan (the name means Sandalwood City) seems laid-back, with chickens wandering the downtown streets.

Below this serene surface, however, Laos is a nation divided. The name Laos is taken from the dominant ethnic group, but there are actually about 70 ethnic groups in the country. Over the centuries, they have battled one another for supremacy, for land, and for tribute money. The constant feuding has weakened the nation and served as an invitation for neighboring countries to annex portions of Laos forcibly, or to align themselves with one or another of the Laotian royal families or generals for material gain. China, Burma

(today called Myanmar), Vietnam, and especially Thailand—with which Laotian people share many cultural and ethnic similarities—have all been involved militarily in Laos.

Historically, jealousy among members of the royal family caused most of Laos's bloodshed. More recently, Laos has been seen as a pawn in the battle of the Western powers for access to the rich natural resources of Southeast Asia, or as a "domino" that some did and others did not want to fall to communism. Former members of the royal family continue to find themselves on the opposite sides of many issues.

The results of these struggles have been devastating. Laos is now one of the poorest countries in the world, with a per capita income of only US$1,700. With electricity available in only a few cities, there are few industries in the country, so most people survive by subsistence farming and fishing,

raising or catching just what they need to eat rather than growing food to sell. In fact, some "hill peoples" (about two thirds of the Laotian people live in the mountains) in the long mountain range that separates Laos from Vietnam continue to use the most ancient farming technique known, slash-and-burn farming, an unstable method of land use that allows only a few years of good crops before the soil is depleted and the farmers must move to new ground. Today, soil erosion and deforestation pose significant threats to economic growth.

Even if all Laotian farmers used the most modern techniques and geared their production to cash crops, it would still be difficult to export food (or, for that matter, anything else) because of Laos's woefully inadequate transportation network. There are no railroads, and muddy, unpaved roads make many mountain villages completely inaccessible by car or truck. Only one bridge in Laos, the Thai-Lao Friendship Bridge near the capital city of Viangchan (Vientiane), spans the famous Mekong River. Moreover, Laos is land-locked. In a region of the world where wealth flows toward those countries with the best ports, having no direct access to the sea is a serious impediment to economic growth.

In addition, for years the economy has been strictly controlled by the government. Foreign investment and trade have not been welcomed; tourists were not allowed into the country until 1989. But the economy began to open up in the late 1980s. The government's "New Economic Mechanism" (NEP) in 1986 called for foreign investment in all sectors and anticipated gross domestic product growth of 8 percent per year (although growth in 2004 was actually about 5 percent). The year 1999 was declared "Visit Laos Year," and the government set a goal of drawing 1 million tourists. With its technological infrastructure abysmally underdeveloped, tourism seemed like the easiest way for the country to gain some foreign currency and create jobs. In Luan Prabang province, where the U.S. is helping locals preserve historical artifacts, tourists can visit a large statue of Buddha and observe a lifestyle that has changed little in centuries. In Viangchan, tourists can visit the mammoth gold Buddhist stupa of Pha That Luang around which saffron-robed monks pray and chant to early-morning drums and then depart for the streets with their begging bowls until the heat becomes too oppressive. Breakfast (and lunch and dinner for that matter) for Loatians often consists of beef soup mixed with rice noodles, fish sauce, bean sprouts, and chili powder. The lingering French influence can be seen in the Fountain Circle area, where foreign restaurants continue to serve French dishes.

Some economic progress has been made in the past decade. Laos is now self-sufficient in its staple crop, rice; and surplus electricity generated from dams along the Mekong River is sold to Thailand to earn foreign exchange. Laos imports various commodity items from Thailand, Vietnam, Singapore, Japan, and other countries, and it has received foreign aid from the Asian Development Bank and other organiza-

tions, including a $40.2 million loan in 2001 from the International Monetary Fund. Exports to Thailand, China, and the United States include teakwood, tin, and various minerals. In 1999, a state-owned bank in Vietnam agreed to set up a joint venture with the Laotian Bank for Foreign Trade. The new bank will mainly deal with imports and exports, particularly between Vietnam and Laos. In 2000, leaders of the world's wealthiest nations, the G-7 group, agreed to help Laos by offering various types of debt relief, but it will be many years before the country can claim that its economy is solid.

Despite the 1995 "certification" by the United States that Laos is cooperating in the world antidrug effort, Laos continues to supply opium, cannabis, and heroin to users in Europe and North America. The Laotian government is now trying to prevent hill peoples from cutting down valuable forests for opium-poppy cultivation.

HISTORY AND POLITICS

The Laotian people, originally migrating from south China, settled Laos in the thirteenth century A.D., when the area was controlled by the Khmer (Cambodian) Empire. Early Laotian leaders expanded the borders of Laos through warfare with Cambodia, Thailand, Burma, and Vietnam. Internal warfare, however, led to a loss of autonomy in 1833, when Thailand forcibly annexed the country (against the wishes of Vietnam, which also had designs on Laos). In the 1890s, France, determined to have a part of the lucrative Asian trade and to hold its own against growing British strength in Southeast Asia, forced Thailand to give up its hold on Laos. Laos, Vietnam, and Cambodia were combined into a new political entity, which the French termed *Indochina.* Between these French possessions and the British possessions of Burma and Malaysia (then called Malaya) lay Thailand; thus, France, Britain, and Thailand effectively controlled mainland Southeast Asia for several decades.

There were several small uprisings against French power, but these were easily suppressed until the Japanese conquest of Indochina in the 1940s. The Japanese, with their "Asia for Asians" philosophy,

convinced the Laotians that European domination was not a given. In the Geneva Agreement of 1949, Laos was granted independence, although full French withdrawal did not take place until 1954.

Prior to independence, Prince Souphanouvong (who died in 1995 at age 82) had organized a Communist guerrilla army, with help from the revolutionary Ho Chi Minh of the Vietnamese Communist group Viet Minh. This army called itself *Pathet Lao* (meaning "Lao Country"). In 1954, it challenged the authority of the government in the Laotian capital. Civil war ensued, and by 1961, when a cease-fire was arranged, the Pathet Lao had captured about half of Laos. The Soviet Union supported the Pathet Lao, whose strength was in the northern half of Laos, while the United States supported a succession of pro-Western but fragile governments in the south. A

Timeline: PAST

A.D. 1300s
The first Laotian nation is established

1890s
Laos is under French control

1940s
The Japanese conquer Southeast Asia

1949
France grants independence to Laos

1971
South Vietnamese troops, with U.S. support, invade Laos

1975
Pathet Lao Communists gain control of the government

1977
Laos signs military and economic agreements with Vietnam

1980s
The government begins to liberalize some aspects of the economy

1990s
The Pathet Lao government maintains firm control; efforts to maintain high GDP growth are threatened by deforestation and soil erosion

Funded by Australia, a "Friendship Bridge" connecting Laos and Thailand across the Mekong River opens for commerce.

PRESENT

2000s
Laos works for closer economic ties to Vietnam

The G-7 nations promise debt relief for Laos

coalition government consisting of Pathet Lao, pro-Western, and neutralist leaders was installed in 1962, but it collapsed in 1965, when warfare once again broke out.

During the Vietnam War, U.S. and South Vietnamese forces bombed and invaded Laos in an attempt to disrupt the North Vietnamese supply line known as the Ho Chi Minh Trail. Americans flew nearly 600,000 bombing missions over Laos (many of the small cluster bombs released during those missions remain unexploded in fields and villages and present a continuing danger). Communist battlefield victories in Vietnam encouraged and aided the Pathet Lao Army, which became the dominant voice in a new coalition government established in 1974. The Pathet Lao controlled the government exclusively by 1975. In the same year, the government proclaimed a new "Lao People's Democratic Republic." It abolished the 622-year-old monarchy and sent the king and the royal family to a detention center to learn Marxist ideology.

Vietnamese Army support and flight by many of those opposed to the Communist regime have permitted the Pathet Lao to maintain control of the government. The ruling dictatorship is determined to prevent the democratization of Laos: In 1993, several cabinet ministers were jailed for 14 years for trying to establish a multiparty democracy. In 2001, the government deported five activists, including a Belgian member of the European Parliament, for handing out prodemocracy leaflets on the streets of Viangchan. A year before, students doing the same thing disappeared; they have never been seen since.

The Pathet Lao government was sustained militarily and economically by the Soviet Union and other East bloc nations for more than 15 years. However, with the end of the Cold War and the collapse of the Soviet Union, Laos has had to look elsewhere, including non-Communist countries, for support. In 1992, Laos signed a friendship treaty with Thailand to facilitate trade between the two historic enemy countries. In 1994, the Australian government, continuing its plan to integrate itself more fully into the strong Asian economy, promised to provide Laos with more than $33 million in aid.

In 1995, Laos joined with ASEAN nations to declare the region a nuclear-free zone.

Trying to teach communism to a devoutly Buddhist country has not been easy. Popular resistance has caused the government to retract many of the regulations it has tried to impose on the Buddhist Church (technically, the Sangara, or order of the monks—the Buddhist equivalent of a clerical hierarchy). As long as the Buddhist hierarchy limits its activities to helping the poor, it seems to be able to avoid running afoul of the Communist leadership.

Intellectuals, especially those known to have been functionaries of the French administration, have fled Laos, leaving a leadership and skills vacuum. As many as 300,000 people are thought to have left Laos for refugee camps in Thailand and elsewhere. Many have taken up permanent residence in foreign countries. Among these are some 15,000 Hmongs, an ethnic group that helped the United States during the Vietnam War. Having been persecuted ever since, the Hmongs are now being resettled in the U.S.

Macau (Macau Special Administrative Region)

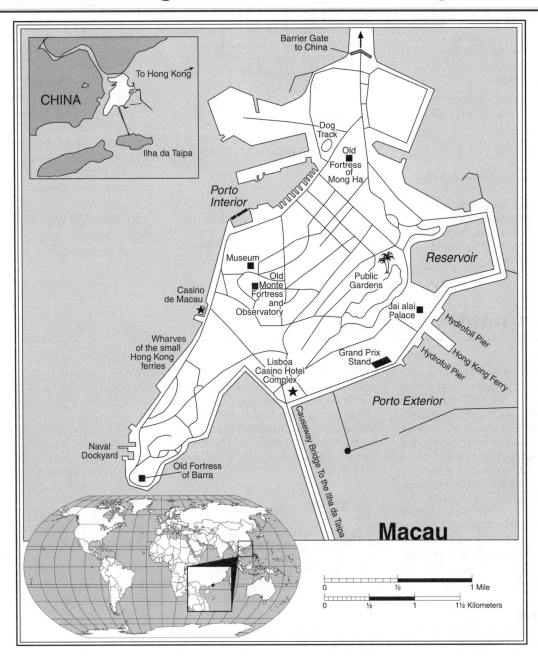

Macau Statistics

GEOGRAPHY

Area in Square Miles (Kilometers): 8 (21) (about 1/10 the size of Washington, D.C.)

Capital (Population): Macau (445,600)

Environmental Concerns: air and water pollution

Geographical Features: generally flat

Climate: subtropical; marine with cool winters, warm summers

PEOPLE

Population

Total: 445,286
Annual Growth Rate: 0.87%
Rural/Urban Population Ratio: 0/100
Major Languages: Portuguese; Cantonese

Ethnic Makeup: 95% Chinese; 5% Macanese, Portuguese, and others
Religions: 50% Buddhist; 15% Roman Catholic; 35% unaffiliated or other

Health

Life Expectancy at Birth: 79 years (male); 85 years (female)
Infant Mortality: 4.4/1,000 live births
Physicians Available: 1/2,470 people

Education

Adult Literacy Rate: 94%

COMMUNICATION

Telephones: 180,000 main lines
Internet Users: 102,000 (2002)

TRANSPORTATION

Highways in Miles (Kilometers): 31 (50)
Railroads in Miles (Kilometers): none
Usable Airfields: 1

GOVERNMENT

Type: Special Administrative Region of China
Independence Date: none; Special Administrative Region of China
Head of State/Government: President (of China) Hu Jintao; Chief Executive Edmund Ho Hau-Wah

Political Parties: there are no formal political parties; civic associations are used instead
Suffrage: direct election at 18; universal for permanent residents living in Macau for 7 years; indirect election limited to organizations registered as "corporate voters" and a 300-member Election Committee

MILITARY

Military Expenditures (% of GDP): defense is the responsibility of China
Current Disputes: none

ECONOMY

Currency ($ U.S. Equivalent): 8.03 pataca = $1 (tied to Hong Kong dollar)
Per Capita Income/GDP: $19,400/$9.1 billion

GDP Growth Rate: 4%
Inflation Rate: -2%
Unemployment Rate: 6.5%
Labor Force by Occupation: 26% restaurants and hotels; 20% manufacturing; 54% other services and agriculture
Natural Resources: fish
Agriculture: rice; vegetables
Industry: clothing; textiles; toys; tourism; electronics; footwear; gambling
Exports: $2.4 billion (primary partners United States, China, Germany)
Imports: $2.5 billion (primary partners China, Hong Kong, Japan)

SUGGESTED WEBSITE

http://www.cia.gov/cia/
publications/factbook/geos/
mc.html

Macau Country Report

Just 17 miles across the Pearl River estuary from Hong Kong lies the world's most densely populated territory: the former Portuguese colony of Macau (sometimes spelled Macao). Consisting of only six square miles of land, the peninsula and two tiny islands are home to nearly half a million people, 95 percent of whom are Chinese, with the remainder being Portuguese or Portuguese/Asian mixtures. Until December 20, 1999, when Macau reverted to China as a "Special Administrative Region" (SAR), it had been the oldest outpost of European culture in the Far East, with a 442-year history of Portuguese administration.

DEVELOPMENT

The development of industries related to gambling and tourism (tourists are primarily from Hong Kong) has been very successful. Most of Macau's foods, energy, and fresh water are imported from China; Japan and Hong Kong are the main suppliers of raw materials.

Macau was frequented by Portuguese traders as early as 1516, but it was not until 1557 that the Chinese agreed to Portuguese settlement of the land; unlike Hong Kong, however, China did not acknowledge Portuguese sovereignty. Indeed, the Chinese government did not recognize the Portuguese demand for "perpetual occupation" until 1887.

Macau's population has varied over the years, depending on conditions in China. During the Japanese occupation of parts of China during the 1940s, for instance, Macau's Chinese population is believed to have doubled, and more refugees streamed in when the Communists took over China in 1949.

FREEDOM

Under the "Basic Law," China agreed to maintain Macau's separate legal, political, and economic system. The Legislature is partly elected and partly appointed.

In 1987, Chinese and Portuguese officials signed an agreement, effective December 20, 1999, to end European control of the first—and last—colonial outpost in China. Actually, Portugal had offered to return Macau on two earlier occasions (in 1967, during the Chinese Cultural Revolution; and in 1974, after the coup in Portugal that ended the dictatorship there), but China refused. The 1999 transition went smoothly, although Portugal's president announced that he would not attend the hand-over ceremony if China sent in troops before the official transition date. China relented, but when the troops arrived on December 20, they were generally cheered by crowds who hoped that they would bring some order to the gang-infested society.

The transition agreement was similar to that signed by Great Britain and China over the fate of Hong Kong. China agreed to allow Macau to maintain its capitalist way of life for 50 years, to permit local elections, and to allow its residents to travel freely without Chinese intervention. Unlike Hong Kong residents, who staged massive demonstrations against future Chinese rule or emigrated from Hong Kong before its return to China, Macau residents—some of whom have been openly pro-Communist—have not seemed bothered by the new arrangements. Indeed, businesses in Macau (as well as Hong Kong) have contributed to a de facto merging with the mainland by investing more than $20 billion in China since the mid-1990s.

HEALTH/WELFARE

Macau has very impressive quality-of-life statistics. It has a low infant mortality rate and very high life expectancy for both males and females. Literacy is close to 95 percent.

Since it was established in the sixteenth century as a trading colony with interests in oranges, tea, tobacco, and lacquer, Macau has been heavily influenced by Roman Catholic priests of the Dominican and Jesuit orders. Christian churches, interspersed with Buddhist temples, abound. With Buddhist immigrants from China re-

Timeline: PAST

A.D. 1557
A Portuguese trading colony is established at Macau

1849
Portugal declares sovereignty over Macau

1887
China signs a treaty recognizing Portuguese sovereignty over Macau

1940s
Immigrants from China flood into the colony

1967
Pro-Communist riots in Macau

1970s
Portugal begins to loosen direct administrative control over Macau

1976
Macau becomes a Chinese territory but is still administered by Portugal

1987
China and Portugal sign an agreement scheduling the return of Macau to Chinese control

1990s
Portugal sends troops to help tamp down gambling-related crime; Macau reverts to Chinese control on December 20, 1999

PRESENT

2000s
The reversion to Chinese control goes smoothly

Macau fights to recover from the Asian financial crisis

ducing the proportion of Christians, about 15 percent of the population now claims to be Christian, and about 50 percent claims to be Buddhist. The name of Macau itself reflects its deep and enduring religious roots; the city's official name is "City of the Name of God in China, Macau, There Is None More Loyal." Macau has perhaps the highest density of churches and temples per square mile in the world.

A HEALTHY ECONOMY

Macau's modern economy is a vigorous blend of light industry, fishing, tourism, and gambling. Revenues from the latter source—casino gambling—are impressive, accounting for almost 40 percent of gross domestic product as well as one third of all jobs. The government imposes a 32 percent tax on casinos. Major infrastructural improvements, such as the new airport, are funded by gambling revenues. There are five major casinos and many other gambling opportunities in Macau, which, along with the considerable charms of the city itself, attract more than 5 million foreign visitors a year, more than 80 percent of them Hong Kong Chinese with plenty of money to spend (gambling is illegal on the mainland). For 40 years, Macau's gaming industry was run by Stanley Ho and his Macau Travel and Amusement Company, which held monopoly rights on all gambling. Not everyone was pleased with that arrangement, and rival street gangs, striving to control part of the lucrative business, launched a crime spree in 1998. Military troops had to be sent from Portugal to restore order. In 2002, the government diversified control of the industry by giving one license to Ho and two to U.S. gaming interests. The three licensees had to agree to invest US$500 million in new casino facilities—a sum not particularly onerous for Ho who, in 2004, was listed as not only the wealthiest person in Macau, but one of the richest persons in the world, with assets over US$1 billion and operations in several Asian countries. He employs over 10,000 people in Macau alone.

Export earnings derived from light-industry products such as textiles, fireworks, plastics, and electronics are also critical to the colony. Macau's leading export markets are the United States, China, Germany, France, and Hong Kong; ironically, Portugal consumes only about 3 percent of Macau's exports.

As might be expected, the general success of the economy has a downside. In Macau's case, hallmarks of modernization—crowded apartment blocks, packed casino hotels, and bustling traffic—are threatening to eclipse the remnants of the old, serene, Portuguese-style seaside town.

Malaysia

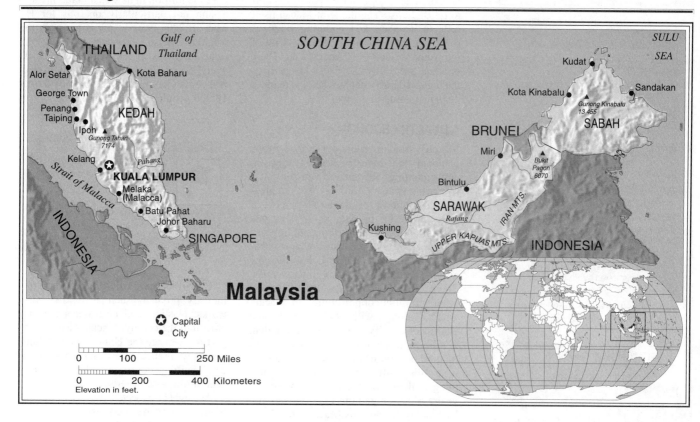

Malaysia Statistics

GEOGRAPHY

Area in Square Miles (Kilometers):
121,348 (329,750) (slightly larger than
New Mexico)
Capital (Population): Kuala Lumpur
(1,410,000)
Environmental Concerns: air and water
pollution; deforestation; smoke/haze
from Indonesian forest fires
Geographical Features: coastal plains
rising to hills and mountains
Climate: tropical; annual monsoons

PEOPLE

Population

Total: 23,522,482
Annual Growth Rate: 1.83%
Rural/Urban Population Ratio: 43/57
Major Languages: Peninsular Malaysia:
Bahasa Malaysia, English, Chinese
dialects, Tamil; Sabah: English, Malay,
numerous tribal dialects, Mandarin and
Hakka dialects; Sarawak: English,
Malay, Mandarin, numerous tribal
dialects, Arabic, others
Ethnic Makeup: 58% Malay and other
indigenous; 26% Chinese; 7% Indian;
9% others
Religions: Peninsular Malaysia: Malays
nearly all Muslim, Chinese mainly
Buddhist, Indians mainly Hindu; Sabah:
33% Muslim, 17% Christian, 45%
others; Sarawak: 35% traditional
indigenous, 24% Buddhist and
Confucian, 20% Muslim, 16% Christian,
5% others

Health

Life Expectancy at Birth: 69 years (male);
74 years (female)
Infant Mortality: 19.6/1,000 live births
Physicians Available: 1/2,153 people
HIV/AIDS in Adults: 0.42%

Education

Adult Literacy Rate: 88%
Compulsory (Ages): 6–16; free

COMMUNICATION

Telephones: 4,600,000 main lines
Daily Newspaper Circulation: 139/1,000
people
Televisions: 454/1,000 people
Internet Users: 5,700,000

TRANSPORTATION

Highways in Miles (Kilometers): 38,803
(64,672)
Railroads in Miles (Kilometers): 1,116
(1,800)
Usable Airfields: 116
Motor Vehicles in Use: 3,948,000

GOVERNMENT

Type: constitutional monarchy
Independence Date: August 31, 1957
(from the United Kingdom)
Head of State/Government: Paramount
Ruler Tuanku Syed Sirajuddin ibni
Almarhum Tuanku Syed Putra

Jamalullail; Prime Minister Abdullah bin Ahmad Badawi

Political Parties: Peninsular Malaysia: National Front and others; Sabah: National Front and others; Sarawak: National Front and others

Suffrage: universal at 21

MILITARY

Military Expenditures (% of GDP): 2%

Current Disputes: complex dispute over the Spratly Islands; Sabah is claimed by the Philippines; other territorial disputes

ECONOMY

Currency ($ U.S. Equivalent): 3.80 ringgits = $1

Per Capita Income/GDP: $9,000/$207 billion

GDP Growth Rate: 5.2%

Inflation Rate: 1.5%

Unemployment Rate: 3.6%

Population Below Poverty Line: 8%

Natural Resources: tin; petroleum; timber; natural gas; bauxite; iron ore; copper; fish

Agriculture: rubber; palm oil; rice; coconut oil; pepper; timber

Industry: rubber and palm oil manufacturing and processing; light manufacturing; electronics; tin mining and smelting; logging and timber processing; petroleum; food processing

Exports: $98.4 billion (primary partners United States, Singapore, Japan)

Imports: $74.4 billion (primary partners Japan, United States, Singapore)

SUGGESTED WEBSITE

http://ianchai.50megs.com/ malaysia.html

Malaysia Country Report

About the size of Japan and famous for its production of natural rubber and tin, Malaysia sounds like a true political, economic, and social entity. Although it has all the trappings of a modern nation-state, Malaysia is one of the most fragmented nations on Earth.

DEVELOPMENT

Efforts to move the economy away from farming and toward industrial production have been very successful. Manufacturing now accounts for 30% of GDP, and Malaysia is the third-largest producer of semiconductors in the world. With Thailand, Malaysia will build a $1.3 billion, 530-mile natural-gas pipeline.

Consider its land. West Malaysia, wherein reside 86 percent of the population, is located on the Malay Peninsula between Singapore and Thailand; but East Malaysia, with 60 percent of the land, is located on the island of North Borneo, some 400 miles of ocean away.

Similarly, Malaysia's people are divided along racial, religious, and linguistic lines. Fifty-eight percent are Malays and other indigenous peoples, many of whom adhere to the Islamic faith or animist beliefs; 26 percent are Chinese, most of whom are Buddhist, Confucian, or Taoist; 7 percent are Indians and 9 percent are Pakistanis and others, some of whom follow the Hindu faith. Bahasa Malaysia is the official language, but English, Arabic, two forms of Chinese, Tamil, and other languages are also spoken. Thus, although the country is called Malaysia (a name adopted only 35 years ago), many people living in Kuala Lumpur, the capital, or in the many villages in the countryside have a stronger identity with their ethnic group or village than with the country of Malaysia per se.

Malaysian culture is further fragmented because each ethnic group tends to replicate the architecture, social rituals, and norms of etiquette peculiar to itself. The Chinese, whose ancestors were imported in the 1800s from south China by the British to work the rubber plantations and tin mines, have become so economically powerful that their cultural influence extends far beyond their actual numbers. Like the Malays and the Hindus, they surround themselves with the trappings of their original, "mother" culture.

Malaysian history is equally fragmented. Originally controlled by numerous sultans who gave allegiance to no one, or only reluctantly to various more powerful states in surrounding regions, Malaysia first came to Western attention in A.D. 1511, when the prosperous city of Malacca, which had been founded on the west coast of the Malay Peninsula about a century earlier, was conquered by the Portuguese. The Dutch took Malacca away from the Portuguese in 1641. The British seized it from the Dutch in 1824 (the British had already acquired an island off the coast and had established the port of Singapore). By 1888, the British were in control of most of the area that is now Malaysia.

However, British hegemony did not mean total control, for each of the many sultanates—the origin of the 13 states that constitute Malaysia today—continued to act more or less independently of the British, engaging in wars with one another and maintaining an administrative apparatus apart from the British. Some groups, such as the Dayaks, an indigenous people living in the jungles of Borneo, remained more or less aloof from the various intrigues of modern state-making and developed little or no identity of themselves as citizens of any modern nation.

FREEDOM

Malaysia is attempting to govern according to democratic principles. Ethnic rivalries, however, severely hamper the smooth conduct of government and limit such individual liberties as the right to form labor Unions. Evidence of un-democratic tactics, such as the government's treatment and imprisonment of ex-deputy Prime Minister Anwar Ibrahim, bring out large numbers of protestors.

It is hardly surprising, then, that Malaysia has had a difficult time emerging as a nation. Indeed, it is not likely that there would have been an independent Malaysia had it not been for the Japanese, who defeated the British in Southeast Asia during World War II and promulgated their alluring doctrine of "Asia for Asians."

After the war, Malaysian demands for independence from European domination grew more persuasive; Great Britain attempted in 1946 to meet these demands by proposing a partly autonomous Malay Union. However, ethnic rivalries and power-sensitive sultans created such enormous tension that the plan was scrapped. In an uncharacteristic display of cooperation, some 41 different Malay groups organized the United Malay National Organization (UMNO) to oppose the British plan. In 1948, a new Federation of Malaya was attempted. It granted considerable freedom within a framework of British supervision, allowed sultans to retain power over their own regions, and placed certain restrictions on the power of the Chinese living in the country.

Opposing any agreement short of full independence, a group of Chinese Communists, with Indonesian support, began a guerrilla war against the government and against capitalist ideology. Known as "The Emergency," the war lasted more than a decade and involved some 250,000 government troops. Eventually, the insurgents withdrew.

HEALTH/WELFARE

City dwellers have ready access to educational, medical, and social opportunities, but the quality of life declines dramatically in the countryside. Malaysia has one of the highest illiteracy rates in the Pacific Rim. It spends only a small percentage of its GDP on education.

The three main ethnic groups—Malayans, represented by UMNO; Chinese, represented by the Malayan Chinese Association, or MCA; and Indians, represented by the Malayan Indian Congress, or MIC—were able to cooperate long enough in 1953 to form a single political party under the leadership of Abdul Rahman. This party demanded and received complete independence for the Federation in 1957, although some areas, such as Brunei, refused to join. Upon independence, the Federation of Malaya (not yet called Malaysia), excluding Singapore and the territories on the island of Borneo, became a member of the British Commonwealth of Nations and was admitted to the United Nations. In 1963, a new Federation was proposed that included Singapore and the lands on Borneo. Again, Brunei refused to join. Singapore joined but withdrew in 1965. Thus, what we call Malaysia today acquired its current form in 1966.

Political troubles stemming from the deep ethnic divisions in the country, however, remain a constant feature of Malaysian life. With nine of the 13 states controlled by independent sultans, every election is a test of the ability of the National Front, a multiethnic coalition of 11 different parties. As of the 2004 elections, the National Front held almost 80 percent of the 219-seat federal Parliament and scored victories in regional races as well. The strong showing was apparently in reaction to the efforts of Islamic fundamentalists to win control and transform Malaysia into an Islamic state. In addition, voters seemed pleased with the efforts of their secular prime minister, Abdullah Ahmad Badawi, whose low-key, rational approach to politics was a refreshing break from the combative nationalism of his predecessor, Mahathir Mohamad. Despite the current success of the National Front coali-

tion, it will continue to be difficult for any government to maintain political stability. Particularly troublesome has been the state of Sabah (an area claimed by the Philippines), many of whose residents have wanted independence or, at least, greater autonomy from the federal government. In recent years, the National Front was able to gain a slight majority in Sabah elections, indicating the growing confidence that people have in the federal government's economic development policies.

ECONOMIC DEVELOPMENT

For years, Malaysia's "miracle" economy kept social and political instability in check. Although it had to endure normal fluctuations in market demand for its products, the economy grew at 5 to 8 percent per year from the 1970s to the late 1990s, making it one of the world's top 20 exporters/importers. The manufacturing sector developed to such an extent that it accounted for 70 percent of exports. Then, in 1998, a financial crisis hit. Malaysia was forced to devalue its currency, the ringgit, making it more difficult for consumers to buy foreign products, and dramatically slowing the economy. The government found it necessary to deport thousands of illegal Indonesian and other workers (dozens of whom fled to foreign embassies to avoid deportation) in order to find jobs for Malaysians. In the 1980s and early 1990s, up to 20 percent of the Malaysian workforce had been foreign workers, but the downturn of the late 1990s produced "Operation Get Out," in which at least 850,000 "guest workers" were deported to their home countries. The deportations were made more urgent in 2002, when imported workers, many of them Indonesians and Filipinos working as menial construction and rubber-plantation workers, rioted in protest of a new drug-testing law. Penalties for illegal or law-breaking immigrant workers included a fine, three months in prison, or six lashes with a cane. The deportations caused consternation in Indonesia and the Philippines, especially when it was learned that 17,000 laborers were being detained in holding camps and that more than a dozen had died in detention.

ACHIEVEMENTS

Malaysia has made impressive economic advancements, and its New Economic Policy has resulted in some redistribution of wealth to the poorer classes. Malaysia has been able to recover from the Asian financial crisis of the late 1990s and now expects solid GDP growth. The country has also made impressive social and political gains.

Timeline: PAST

A.D. 1403
The city of Malacca is established; it becomes a center of trade and Islamic conversion

1511
The Portuguese capture Malacca

1641
The Dutch capture Malacca

1824
The British obtain Malacca from the Dutch

1941
Japan captures the Malay Peninsula

1948
The British establish the Federation of Malaya; a Communist guerrilla war begins, lasting for a decade

1957
The Federation of Malaya achieves independence under Prime Minister Tengku Abdul Rahman

1963
The Federation of Malaysia, including Singapore but not Brunei, is formed

1965
Singapore leaves the Federation of Malaysia

1980s
Malaysia attempts to build an industrial base

1990s
The NEP is replaced with Vision 2020; economic crisis

PRESENT

2000s
The economy rebounds; the environment suffers

Former deputy prime minister Anwar Ibrahim is arrested and convicted under questionable circumstances

The World Court awards two tiny Celebes Sea islands to Malaysia in a dispute with Indonesia

Abdullah Ahmad Badawi replaces Mahathir Mohamad as prime minister

Ex-deputy prime minister Anwar Ibrahim's conviction is overturned; he vows to push for democratic reforms

Malaysia continues to be rich in raw materials; therefore, it is not likely that the crisis of the late 1990s will permanently cripple its economy. Moreover, the Malaysian government has a good record of active planning and support of business ventures—directly modeled after Japan's export-oriented strategy. Malaysia launched a "New Economic Policy" (NEP) in the 1970s that welcomed foreign direct investment and sought to diversify the economic base. Japan, Taiwan, and the United States invested heavily in Malaysia. So successful was this strategy that economic-growth targets set for the mid-1990s were actually achieved several years early. In 1991, the government replaced NEP with a new plan, "Vision 2020." Its goal was to bring Malaysia into full "developed nation" status by the year 2020. Sectors targeted for growth included the aerospace industry, biotechnology, microelectronics, and information and energy technology. The government expanded universities and encouraged the creation of some 170 industrial and research parks, including "Free Zones," where export-oriented businesses were allowed duty-free imports of raw materials. Some of Malaysia's most ambitious projects, including a $6 billion hydroelectric dam (strongly opposed by environmentalists), have been shelved, at least until the full effects of the Asian financial crisis are overcome. That may not be long, for while the economy nosedived, the growth rate has picked up and occasionally even exceeded the average world growth rate of 4 percent.

Despite Malaysia's substantial economic successes, serious social problems remain. They stem not from insufficient revenues but from inequitable distribution of wealth. The Malay portion of the population in particular continues to feel economically deprived as compared to the more affluent Chinese and Indian segments. Furthermore, most Malays are farmers, and rural areas have not benefited from Malaysia's economic boom as much as urban areas have.

In the 1960s and 1970s, riots involving thousands of college students were headlined in the Western press as having their basis in ethnicity. This was true to some degree, but the core issue was economic inequality. Included in the economic master plan of the 1970s were plans (similar to affirmative action in the United States) to change the structural barriers that prevented many Malays from fully enjoying the benefits of the economic boom. Under the leadership of Prime Minister Datuk Mahathir bin Mohamad, plans were developed that would assist Malays until they held a 30 percent interest in Malaysian businesses. In 1990, the government announced that the figure had already reached an impressive 20 percent. Unfortunately, many Malays have insufficient capital to maintain ownership in businesses, so the government has been called upon to acquire many Malay businesses in order to prevent their being purchased by non-Malays. In addition, the system of preferential treatment for Malays has created a Malay elite, detached from the Malay poor, who now compete with the Chinese and Indian elites; interethnic and interracial goodwill is still difficult to achieve. Nonetheless, social goals have been attained to a greater extent than most observers have thought possible. Educational opportunities for the poor have been increased, farmland development has proceeded on schedule, and the poverty rate has dropped below 10 percent.

THE LEADERSHIP

In a polity so fractured as Malaysia's, one would expect rapid turnover among political elites, but for over a decade, Malaysia was run, sometimes ruthlessly, by Malay prime minister Mahathir Mohamad of the United Malay Naitonal Organization. Most Malaysians were relieved that his aggressive nationalistic rhetoric was usually followed by more moderate behavior vis-à-vis other countries. The Chinese Democratic Action Party (DAP) was sometimes able to reduce his political strength in Parliament, but his successful economic strategies muted most critics.

Malaysia's economic success, symbolized perhaps in one of the world's tallest buildings, the mammoth Petronas Twin Towers in Luala Lumpur, has not been achieved without some questionable practices. The government under Mahathir seemed unwilling to regulate economic growth, even though strong voices were raised against industrialization's deleterious effects on the old-growth teak forests and other parts of the environment. The environmentalists' case was substantially strengthened in 1998 when forest and peat bog fires in Malaysia and Indonesia engulfed Kuala Lumpur in a thick haze for weeks. The government, unable to snuff out the fires, resorted to installing sprinklers atop the city's skyscrapers to settle the dust and lower temperatures and tempers. In 2004, a chemical plant in Prai, a town across from Penang island, exploded, sending noxious smoke into the community and requiring the evacuation of hundreds of people. And, like other Asian nations, Malaysia has had to cope with several outbreaks of avian flu. Other problems involve workers.

Blue-collar workers who are the muscle behind Malaysia's economic success are prohibited from forming labor unions, and outspoken critics have been silenced. The most outspoken critic was Anwar Ibrahim. He had been the deputy prime minister and heir-apparent to Mahathir; but when he challenged Mahathir's policies, he was fired, arrested, beaten, and eventually sent to prison for 14 years on various charges. Protesters frequently took to the streets in his defense, but they were beaten by police and sprayed with tear gas and water cannons. The largest opposition newspaper came to Anwar's defense (for which the editor was charged with sedition), and Anwar's wife started a new political party to challenge the government. The scandal severely tarnished Mahathir's reputation, and he resigned in late 2003. Another Malay, Abdullah Ahmad Badawi, whose resistance to the creation of an Islamic state won him respect from the majority of the people, including many moderate Muslims, replaced him. In 2004, the conviction of ex-deputy prime minister Ibrahim was overturned, and immediately upon leaving prison, he vowed to return to politics and push for greater respect for human rights.

Myanmar (Union of Myanmar; formerly Burma)

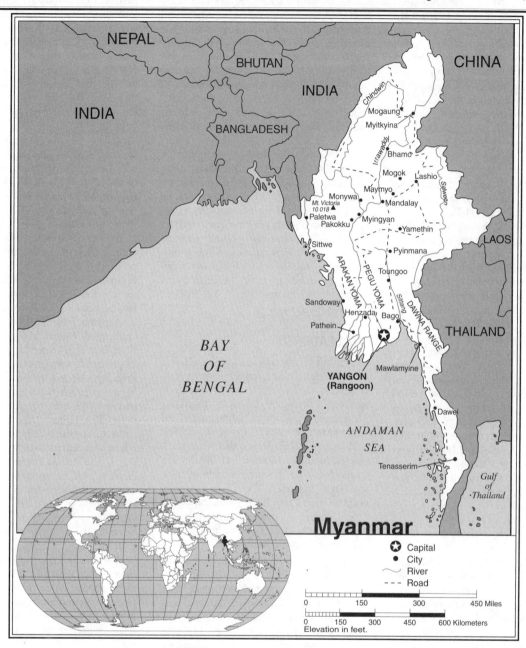

Myanmar Statistics

GEOGRAPHY

Area in Square Miles (Kilometers):
261,901 (678,500) (slightly smaller than Texas)

Capital (Population): Yangon (Rangoon) (4,504,000)

Environmental Concerns: deforestation; air, soil, and water pollution; inadequate sanitation and water treatment

Geographical Features: central lowlands ringed by steep, rugged highlands

Climate: tropical monsoon

PEOPLE

Population

Total: 42,720,196

Annual Growth Rate: 0.47%

Rural/Urban Population Ratio: 73/27

Major Languages: Burmese; various minority languages

Ethnic Makeup: 68% Burman; 9% Shan; 7% Karen; 4% Rakhine; 3% Chinese; 9% Mon, Indian, and others

Religions: 89% Buddhist; 4% Muslim; 4% Christian; 3% others

Health

Life Expectancy at Birth: 54 years (male); 57 years (female)

Infant Mortality: 72.2/1,000 live births

Physicians Available: 1/3,554 people

HIV/AIDS Rate in Adults: 2%

Education
Adult Literacy Rate: 85% (official)
Compulsory (Ages): 5–10; free

COMMUNICATION
Telephones: 250,000 main lines
Daily Newspaper Circulation: 23/1,000 people
Televisions: 22/1,000 people
Internet Users: 10,000 (2002)

TRANSPORTATION
Highways in Miles (Kilometers): 17,484 (28,200)
Railroads in Miles (Kilometers): 2,474 (3,991)
Usable Airfields: 80
Motor Vehicles in Use: 69,000

GOVERNMENT
Type: military regime

Independence Date: January 4, 1948 (from the United Kingdom)
Head of State/Government: Prime Minister and Chairman of the State Peace and Development Council (General) Than Shwe is both head of state and head of government
Political Parties: Union Solidarity and Development Association; National League for Democracy; National Unity Party; others
Suffrage: universal at 18

MILITARY
Military Expenditures (% of GDP): 2.1%
Current Disputes: internal strife; border conflicts with Thailand

ECONOMY
Currency ($ U.S. Equivalent): 6.20 kyat = $1
Per Capita Income/GDP: $1,800/$74 billion

GDP Growth Rate: -0.5%
Inflation Rate: 20%
Unemployment Rate: 4.2%
Labor Force by Occupation: 65% agriculture; 25% services; 10% industry
Population Below Poverty Line: 25%
Natural Resources: petroleum; tin; timber; antimony; zinc; copper; tungsten; lead; coal; marble; limestone; precious stones; natural gas
Agriculture: rice; corn; oilseed; sugarcane; pulses; hardwood
Industry: agricultural processing; textiles; footwear; wood and wood products; petroleum refining; mining; construction materials; pharmaceuticals; fertilizer
Exports: $1.8 billion (primary partners United States, India, Thailand, China)
Imports: $2.07 billion (primary partners China, Singapore, Thailand)

SUGGESTED WEBSITE
http://www.myanmar.com

Myanmar (Burma) Country Report

For four decades, Myanmar (as Burma was officially renamed in 1989) has been a tightly controlled society. Telephones, radio stations, railroads, and many large companies have been under the direct control of a military junta that has brutalized its opposition and forced many to flee the country. For many years, tourists were allowed to stay only two week (for awhile the limit was 24 hours), were permitted accommodations only at military-approved hotels, and were allowed to visit only certain parts of the country. Citizens too were highly restricted: They could not leave their country by car or train to visit nearby countries because all the roads were sealed off by government decree, and rail lines terminated at the border. Even Western-style dancing was declared illegal. Until a minor liberalization of the economy was achieved in 1989, all foreign exports—every grain of rice, every peanut, every piece of lumber—though generally owned privately, had to be sold to the government rather than directly to consumers.

Observers attribute this state of affairs to military commanders who overthrew the legitimate government in 1962, but the roots of Myanmar's political and economic dilemma actually go back to 1885, when the British overthrew the Burmese government and declared Burma a colony of Britain. In the 1930s, European-educated Burmese college students organized strikes and demonstrations against the British. Seeing that the Japanese Army had successfully toppled other European colonial governments in Asia, the students decided to assist the Japanese during their invasion of the country in 1941. Once the British had been expelled, however, the students organized the Anti-Fascist People's Freedom League (AFPFL) to oppose Japanese rule.

DEVELOPMENT

Primarily an agricultural nation, Myanmar has a poorly developed industrial sector. Until recently, the government forbade foreign investment and severely restricted tourism. In 1989, recognizing that the economy was on the brink of collapse, the government permitted foreign investment and signed contracts with Japan and others for oil exploration.

When the British tried to resume control of Burma after World War II, they found that the Burmese people had given their allegiance to U Aung San, one of the original student leaders. He and the AFPFL insisted that the British grant full independence to Burma, which they reluctantly did in 1948. So determined were the Burmese to remain free of foreign domination that, unlike most former British colonies, they refused to join the British Commonwealth of Nations. This was the first of many decisions that would have the effect of isolating Burma from the global economy.

FREEDOM

Myanmar is a military dictatorship. Until 1989, only the Burma Socialist Program Party was permitted. Other parties, while now legal, are intimidated by the military junta. The democratically elected National League for Democracy has not been permitted to assume office. The government has also restricted the activities of Buddhist monks and has carried out "ethnic cleansing" against minorities.

Unlike Japan, with its nearly homogeneous population and single national language, Myanmar is a multiethnic state; in fact, only about 60 percent of the people speak Burmese. The Burman people are genetically related to the Tibetans and the Chinese; the Chin are related to peoples of nearby India; the Shan are related to Thais; and the Mon migrated to Burma from Cambodia. In general, these ethnic groups live in separate political states within Myanmar—the Kachin State, the Shan State, the Karen State, and so on; and for hundreds of years, they have warred against one another for dominance. Upon the withdrawal of the British in 1948, some ethnic groups, particularly the Kachins, the Karens, and the Shans, embraced the Communist ideology of change through violent revolution. Their rebellion against the government in the capital city of Yangon (the new name of Rangoon) had the effect of removing from government control large portions of the country. Headed by U Nu (U Aung San and several of the original government leaders having been assassinated shortly before independence), the government considered its position precarious and determined that to align itself with the Communist forces then ascendant in China and other parts of Asia would strengthen the hand of the ethnic separatists, whereas to form alliances with the capitalist world would invite a repetition of decades of Western domination. U Nu thus attempted to steer a decidedly neutral course during the cold war era and to be as tolerant as possible of separatist groups within Burma. Burma refused U.S. economic aid, had very little to do with the warfare afflicting Vietnam and the other Southeast Asian countries, and was not eager to join the Southeast Asian Treaty Organization or the Asian Development Bank.

HEALTH/WELFARE

The Myanmar government provides free health care and pensions to citizens, but the quality and availability of these services are erratic, to say the least. Malnourishment and preventable diseases are common, and infant mortality is high. Overpopulation is not a problem; Myanmar is one of the most sparsely populated nations in Asia.

Some factions of Burmese society were not pleased with U Nu's relatively benign treatment of separatist groups. In 1958, a political impasse allowed Ne Win, a military general, to assume temporary control of the country. National elections were held in 1962, and a democratically elected government was installed in power. Shortly thereafter, however, Ne Win staged a military coup. The military has controlled Myanmar ever since. Under Ne Win, competing political parties were banned, the economy was nationalized, and the country's international isolation became even more pronounced. In what is surely a proof of the saying that "violence begets violence," the current military junta recently arrested Ne Win and his daughter and have charged his son-in-law and three grandsons with treason—that is, with plotting the overthrow of the government. The head of the air force as well as the chief of police were also purged for their involvement in the alleged plot.

Years of ethnic conflict, inflexible socialism, and self-imposed isolation have severely damaged economic growth in Myanmar. In 1987, despite Burma's abundance of valuable teak and rubber trees in its forests, sizable supplies of minerals in the mountains to the north, onshore oil, rich farmland in the Irrawaddy Delta, and a reasonably well-educated population, the United Nations declared Burma one of the least-developed countries in the world (it had once been the richest country in Southeast Asia). Debt incurred in the 1970s exacerbated the country's problems, as did the government's fear of foreign investment. The per capita annual income of US$1,800 makes Myanmar the third poorest country in Asia, after Laos and East Timor.

ACHIEVEMENTS

Myanmar is known for the beauty of its Buddhist architecture. Pagodas and other Buddhist monuments and temples dot many of the cities, especially Pagan, one of Burma's earliest cities. Politically, it is notable that the country was able to remain free of the warfare that engulfed much of Indochina during the 1960s and 1970s.

Myanmar's industrial base is still very small; about two thirds of the population of 42 million make their living by farming (rice is a major export) and by fishing. The tropical climate yields abundant forest cover, where some 250 species of valuable trees abound. Good natural harbors and substantial mineral deposits of coal, natural gas, and others also bless the land. Only about 10 percent of gross domestic product comes from the manufacturing sector (as compared to, for example, approximately 45 percent in wealthy Taiwan). In the absence of a strong economy, black-marketeering has increased, as have other forms of illegal economic transactions. It is estimated that 80 percent of the heroin smuggled into New York City comes from the jungles of Myanmar and northern Thailand. Methamphetamines are also a major export from the region, with profits going to the army. Indeed, so heavy is the drug traffic in the area that, in response to Chinese complaints of drug activity along its border, Myanmar was forced to resettle some 120,000 Wah ethnics away from the border. Figures from 2003 and 2004 show a drop in opium and heroin production, perhaps as a result of a United Nations program to control drugs.

Over the years, the Burmese have been advised by economists to open up their country to foreign investment and to develop the private sector of the economy. They have resisted the former idea because

of their deep-seated fear of foreign domination; they have similar suspicions of the private sector because it was previously controlled almost completely by ethnic minorities (the Chinese and Indians). The government has relied on the public sector to counterbalance the power of the ethnic minorities.

Beginning in 1987, however, the government began to admit publicly that the economy was in serious trouble. To counter massive unrest in the country, the military authorities agreed to permit foreign investment from countries such as Malaysia, South Korea, Singapore, and Thailand and to allow trade with China and Thailand. In 1989, the government signed oil-exploration agreements with South Korea, the United States, the Netherlands, Australia, and Japan. Both the United States and West Germany withdrew foreign aid in 1988, but Japan did not; in 1991, Japan supplied $61 million—more than any other country—in aid to Myanmar. Ironically, such aid is not welcomed by all. Many Burmese outside the country believe that economic assistance of this kind just helps the military stay in power. One U.S.–based group, the Free Burma Coalition, works to dissuade companies from trading with Myanmar. Their pleas are being heard overseas where, in 2004, the European Union approved tighter sanctions against the military dictatorship: no investment in state-run companies; no loans to Myanmar; and no visas granted to any high-level general. For years, the United States government has kept Myanmar (along with China and North Korea) on its list of countries where freedom of religion is not recognized.

POLITICAL STALEMATE

For many years, the people of Myanmar have been in a state of turmoil caused by governmental repression. In 1988, thousands of students participated in six months of demonstrations to protest the lack of democracy in the country and to demand multiparty elections. General Saw Maung brutally suppressed the demonstrators, imprisoning many students—and killing more than 3,000 of them. He then took control of the government and reluctantly agreed to multiparty elections. About 170 political parties registered for the elections, which were held in 1990—the first elections in 30 years. Among these were the National Unity Party (a new name for the Burma Socialist Program Party, the only legal party since 1974) and the National League for Democracy, a new party headed by Aung San Suu Kyi, daughter of slain national hero U Aung San.

This village leader looks over a field where farmers have replaced opium poppies with tea and other crops under the direction of the UN Drug Control Program (UNDCP).

The campaign was characterized by the same level of military control that had existed in all other aspects of life since the 1960s. Martial law, imposed in 1988, remained in effect; all schools and universities were closed; opposition-party workers were intimidated; and, most significantly, the three most popular opposition leaders were placed under house arrest and barred from campaigning. The United Nations began an investigation of civil-rights abuses during the election and, once again, stu-

dents demonstrated against the military government. Several students even hijacked a Burmese airliner to demand the release of Aung San Suu Kyi, who had been placed under house arrest.

As the votes were tallied, it became apparent that the Burmese people were eager to end military rule; the National League for Democracy won 80 percent of the seats in the National Assembly. But military leaders did not want the National League for Democracy in control of the government, so they refused to allow the elected leaders to take office. Under General Than Shwe, who replaced General Saw Maung in 1992, the military has organized various operations against Karen rebels and has so oppressed Muslims that some 40,000 to 60,000 of them have fled to Bangladesh. Hundreds of students who fled the cities during the 1988 crackdown on student demonstrations have now joined rural guerrilla organizations, such as the Burma Communist Party and the Karen National Union, to continue the fight against the military dictatorship. Among those most vigorously opposed to military rule are Buddhist monks. Five months after the elections, monks in the capital city of Yangon boycotted the government by refusing to conduct religious rituals for soldiers. Tens of thousands of people joined in the boycott. The government responded by threatening to shut down monasteries in Yangon and Mandalay.

The military government calls itself the State Law and Order Restoration Council (SLORC) and appears determined to stay in power. SLORC has kept Aung San Suu Kyi, the legally elected leader of the country, under house arrest off and on for years, watching her every move. For several years, even her husband and children were forbidden to visit her. In 1991, she was awarded the Nobel Peace Prize; in 1993, several other Nobelists gathered in nearby Thailand to call for her release from house arrest—a plea ignored by SLORC. The United Nations showed its displeasure with the military junta by substantially cutting development funds, as did the United States (which, on the basis of Myanmar's heavy illegal-drug activities, has disqualified the country from receiving most forms of economic aid). In 2002, the military released Aung San Suu Kyi from house arrest and also released some 600 other political prisoners (albeit keeping another 1,600 behind bars). These high-profile moves were intended to satisfy the United States and others that Myanmar was willing to do what was necessary to qualify for economic assistance. But shortly thereafter, Aung San Suu Kyi was put under house arrest again.

Timeline: PAST

800 B.C.
Burman people enter the Irrawaddy Valley from China and Tibet

A.D. 1500s
The Portuguese are impressed with Burmese wealth

1824–1826
The First Anglo-Burmese War

1852
The Second Anglo-Burmese War

1885
The Third Anglo-Burmese War results in the loss of Burmese sovereignty

1948
Burma gains independence of Britain

1962
General Ne Win takes control of the government in a coup

1980s
Economic crisis; the pro-democracy movement is crushed; General Saw Maung takes control of the government

1989
Burma is renamed Myanmar (though most people prefer the name Burma)

1990s
The military refuses to give up power; Aung San Suu Kyi's activities remain restricted

PRESENT

2000s
The world community increasingly registers its disapproval of the Myanmar junta

Aung San Suu Kyi is released from house arrest, but in May 2003 she was detained once again and universities and colleges were closed

But perhaps the greatest pressure on the dictatorship is from within the country itself. Despite brutal suppression, the military seems to be losing control of the people. Both the Kachin and Karen ethnic groups have organized guerrilla movements against the regime; in some cases, they have coerced foreign lumber companies to pay them protection money, which they have used to buy arms to fight against the junta. Opponents of SLORC control one third of Myanmar, especially along its eastern borders with Thailand and China and in the north alongside India. With the economy in shambles, the military appears to be involved with the heroin trade as a way of acquiring needed funds; it reportedly engages in bitter battles with drug lords periodically for control of the trade. To ease economic pressure, the military rulers have ended their monopoly of some businesses and have legalized the black market, making products from China, India, and Thailand available on the street.

Pressure from the European Union and the United States seems to be paying off: in 2004, SLORC released nearly 10,000 thousand prisoners, including some who were high-profile members of the National League for Democracy (although party leader Aung San Suu Kyi remained under house arrest). The National League for Democracy was allowed to re-open its party headquarters, and the junta even proposed a multi-party national constitutional convention. The NLD, however, announced it would boycott the meetings until their leader was released from house arrest. The political softening suffered a setback in late 2004 when the relatively moderate premier, Khin Nyunt (along with some of his supporters) was suddenly removed from his position, held in house arrest, and replaced by the more hard-line Lt. Gen. Soe Win. At this point, it is difficult to say how sincere the move toward political liberalization really is, although some progress appears to be developing.

Still, for ordinary people, especially those in the countryside, life is anything but pleasant. A 1994 human-rights study found that as many as 20,000 women and girls living in Myanmar near the Thai border had been abducted to work as prostitutes in Thailand. For several years, SLORC has carried out an "ethnic-cleansing" policy against villagers who have opposed their rule; thousands of people have been carried off to relocation camps, forced to work as slaves or prostitutes for the soldiers, or simply killed. Some 400,000 members of ethnic groups have fled the country, including 300,000 Arakans who escaped to Bangladesh and 5,000 Karenni, 12,000 Mon, and 50,000 Karens who fled to Thailand. Food shortages plague certain regions of the country, and many young children are forced to serve in the various competing armies rather than acquire an education or otherwise enjoy a normal childhood. Indeed,

warfare and violence are the only reality many youth know. In 1999, a group of "Burmese Student Warriors" seized the Myanmar Embassy in Thailand, taking hostages and demanding talks between the military junta and Aung San Suu Kyi. The next year, a youth group calling itself God's Army and led by 12-year-old twin brothers, seized a hospital in Thailand and held 800 patients and staff hostage. The boys were from the Karen people, a Christian subculture long persecuted in Burma. The boys attacked Thai residents to protest their villages having been shelled by the Thai military, which is increasingly uneasy with the large number of Burmese refugees inside and along its border.

THE CULTURE OF BUDDHA

Although Myanmar's most famous religious buildings are of Hindu origin, Buddhism, representing the beliefs of 89 percent of the population, has been the dominant religion for decades. In fact, for a brief period in the 1960s, Buddhism was the official state religion of Burma. The government quickly repealed this appellation in order to weaken the power of the Buddhist leadership, or *Sangha*, vis-à-vis the polity. Still, Buddhism remains the single most important cultural force in the country. Even the Burmese alphabet is based, in part, on Pali, the sacred language of Buddhism. Buddhist monks joined with college students after World War II to pressure the British government to withdraw from Burma, and they have brought continual pressure to bear on the current military junta.

Historically, so powerful has been the Buddhist Sangha in Burma that four major dynasties have fallen because of it. This has not been the result of ideological antagonism between church and state (indeed, Burmese rulers have usually been quite supportive of Buddhism) but, rather, because Buddhism soaks up resources that might otherwise go to the government or to economic development. Believers are willing to give money, land, and other resources to the religion, because they believe that such donations will bring them

spiritual merit; the more merit one acquires, the better one's next life will be. Thus, all over Myanmar, but especially in older cities such as Pagan, one can find large, elaborate Buddhist temples, monuments, or monasteries, some of them built by kings and other royals on huge, untaxed parcels of land. These monuments drained resources from the government but brought to the donor unusual amounts of spiritual merit. As Burmese scholar Michael Aung-Thwin explained it: "One built the largest temple because one was spiritually superior, and one was spiritually superior because one built the largest temple."

Today, the Buddhist Sangha is at the forefront of the opposition to military rule. This is a rather unusual position for Buddhists, who generally prefer a more passive attitude toward "worldly" issues. Monks have joined college students in peaceful-turned-violent demonstrations against the junta. Other monks have staged spiritual boycotts against the soldiers by refusing to accept merit-bringing alms from them or to perform weddings and funerals. The junta has retaliated by banning some Buddhist groups altogether and purging many others of rebellious leaders. The military regime now seems to be relaxing its intimidation of the Buddhists, has reopened universities, and has invited some foreign investment. Efforts by the junta to enhance tourism have been opposed by the National League for Democracy, on the grounds that any improvement in the economy would strengthen the hand of the military rulers. All 15 European Union members as well as the national U.S. Chamber of Commerce supported a 2000 Massachusetts state law that would have penalized companies for doing business with Myanmar; the law was thrown out by the U.S. Supreme Court, but the sentiment of broad opposition to the Myanmar dictatorship remained. Although the Japanese have invested in Myanmar throughout the military dictatorship, some potential investors from other countries refuse to invest in the brutal regime.

New Zealand

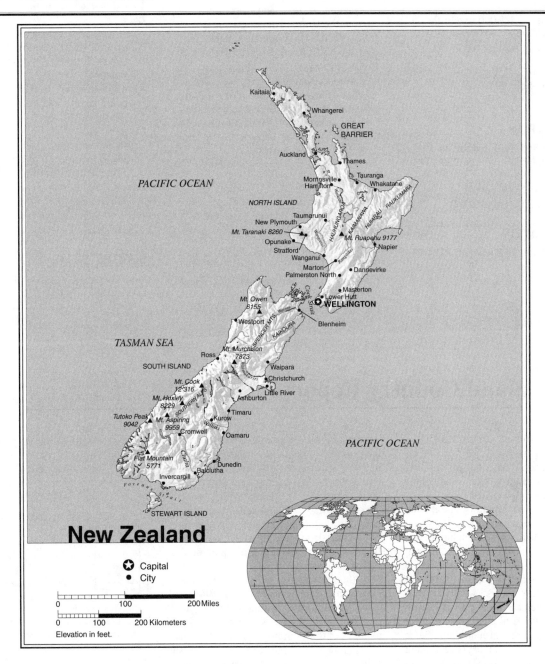

New Zealand

- ⊛ Capital
- ● City

0   100   200 Miles

0   100   200 Kilometers

Elevation in feet.

New Zealand Statistics

GEOGRAPHY

Area in Square Miles (Kilometers):
98,874 (268,680) (about the size of Colorado)

Capital (Population): Wellington (345,000)

Environmental Concerns: deforestation; soil erosion; damage to native flora and fauna from outside species

Geographical Features: mainly mountainous with some large coastal plains

Climate: temperate; sharp regional contrasts

PEOPLE

Population

Total: 3,993,817

Annual Growth Rate: 1.05%

Rural/Urban Population (Ratio): 14/86

Major Languages: English; Maori

Ethnic Makeup: 80% New Zealand and other European; 10% Maori; 4% Pacific islander; 6% Asian and others

Religions: 67% Christian; 33% unspecified or none

Health

Life Expectancy at Birth: 75 years (male);
  81 years (female)
Infant Mortality: 6.1/1,000 live births
Physicians Available: 1/318 people
HIV/AIDS Rate in Adults: 0.06%

Education

Adult Literacy Rate: 99%
Compulsory (Ages): 6–16; free

COMMUNICATION

Telephones: 1.765 million main lines
Daily Newspaper Circulation: 239/1,000
  people
Televisions: 514/1,000 people
Internet Users: 2,060,000 (2002)

TRANSPORTATION

Highways in Miles (Kilometers): 57,164
  (92,200)
Railroads in Miles (Kilometers): 2,383
  (3,813)
Usable Airfields: 106
Motor Vehicles in Use: 2,053,000

GOVERNMENT

Type: parliamentary democracy
Independence Date: September 26, 1907
  (from the United Kingdom)
Head of State/Government: Queen
  Elizabeth II; Prime Minister Helen Clark
Political Parties: ACT; Alliance (a
  coalition); National Party; New Zealand
  Labour Party; New Zealand First Party;
  Democratic Party; New Zealand Liberal
  Party; Green Party; Mana Motuhake;
  others
Suffrage: universal at 18

MILITARY

Military Expenditures (% of GDP): 1%
Current Disputes: disputed territorial
  claim in Antarctica

ECONOMY

Currency ($ U.S. Equivalent): 1.72 New
  Zealand dollars = $1
Per Capita Income/GDP: $21,600/$85.3
  billion
GDP Growth Rate: 3.5%
Inflation Rate: 2.6%
Unemployment Rate: 4.7%
Labor Force by Occupation: 65%
  services; 25% industry; 10% agriculture
Natural Resources: natural gas; iron ore;
  sand; coal; timber; hydropower; gold;
  limestone
Agriculture: wool; beef; dairy products;
  wheat; barley; potatoes; pulses; fruits;
  vegetables; fishing
Industry: food processing; wood and paper
  products; textiles; machinery;
  transportation equipment; banking;
  insurance; tourism; mining
Exports: $15.9 billion (primary partners
  Australia, Japan, United States)
Imports: $16.1 billion (primary partners
  Australia, United States, Japan)

SUGGESTED WEBSITES

http://www.cia.gov/cia/
  publications/factbook/geos/
  nz.html
http://dir.yahoo.com/regional/
  countries/New_Zealand/

New Zealand Country Report

Thanks to the filming in New Zealand of the blockbuster movie series *Lord of the Rings*, many people have had an opportunity to see some of the natural beauty of the country; but few have visited there, owing, no doubt, to its geographic isolation from the rest of the world. No doubt, the isolation, along with a sparse population, has helped preserve the incredible natural beauty of the country. Glacier-carved snow-capped mountains, lush green valleys, crystal-clear rivers, waterfalls (including Sutherland Falls, the world's fifth highest), and white plumes of hot steam from numerous underground lava fields make New Zealand a country-size version of North America's Yellowstone National Park. Rare animals, such as the yellow-eyed penguin, the Kea or alpine parrot, and the royal albatross, add to the dazzling beauty of the country, as do the miles of spectacular coastline; one can never get more than 80 miles from the ocean. Urban areas, both large and small, are also unusually clean and tidy.

New Zealand's pristine beauty is not the only quality that distinguishes it from other Pacific Rim countries. As with Australia, the majority of the people (80 percent) are of European descent (primarily British), English is the official language, and most people, even many of the original Maori

inhabitants, are Christians. Britain claimed the beautiful, mountainous islands officially in 1840, after agreeing to respect the property rights of Maoris, most of whom lived on the North Island. Blessed with a temperate climate and excellent soils for crop and dairy farming, New Zealand—divided into two main islands, North Island and South Island—became an important part of the British Empire.

Although largely self-governing since 1907 and fully independent as of 1947, the country has always maintained very close ties with the United Kingdom and is a member of the Commonwealth of Nations. It has, in fact, attempted to re-create British culture—customs, architecture, even vegetation—in the Pacific. English sheep, which literally transformed the economy, were first imported to the South Island in 1773. Some towns seem very, very English, such as Christchurch, the largest city on the South Island, through which flows the Avon River and which was founded in part by the Church of England. Others seem more Scottish, such as Dunedin. Everywhere one can find town names derived from British life: Wellington, Nelson, Queenstown, Stirling. So close were the links with Great Britain in the 1940s that England purchased fully 88 percent of New Zealand's exports (mostly agricultural and

dairy products), while 60 percent of New Zealand's imports came from Britain. As a part of the British Empire, New Zealand always sided with the Western nations on matters of military defense.

Efforts to maintain a close cultural link with Great Britain do not stem entirely from the common ethnicity of the two nations; they also arise from New Zealand's extreme geographical isolation from the centers of European and North American activity. Even Australia is more than 1,200 miles away. Therefore, New Zealand's policy—until the 1940s—was to encourage the British presence in Asia and the Pacific, by acquiring more lands or building up naval bases, to make it more likely that Britain would be willing and able to defend New Zealand in a time of crisis. New

(The Peabody Museum of Salem photo)

Maoris occupied New Zealand long before the European settlers moved there. The Maoris quickly realized that the newcomers were intent on depriving them of their land, but it was not until the 1920s that the government finally regulated unscrupulous land-grabbing practices. Today, the Maoris pursue a lifestyle that preserves key parts of their traditional culture while incorporating skills necessary for survival in the modern world.

Zealand had involved itself somewhat in the affairs of some nearby islands in the late 1800s and early 1900s, but its purpose was not to provide development assistance or defense. Rather, its aim was to extend the power of the British Empire and put New Zealand in the middle of a mini-empire of its own. To that end, New Zealand annexed the Cook Islands in 1901 and took over German Samoa in 1914. In 1925, it assumed formal control over the atoll group known as the Tokelau Islands.

REGIONAL RELIANCE

During World War II (or, as the Japanese call it, the Pacific War), Japan's rapid conquest of the Malay Archipelago, its seizure of many Pacific islands, and its plans to attack Australia demonstrated to New Zealanders the futility of relying on the British to guarantee their security. After the war, and for the first time in its history, New Zealand began to pay serious attention to the real needs and ambitions of the peoples nearby rather than to focus on Great Britain. In 1944 and again in 1947, New Zealand joined with Australia and other colonial nations to create regional associations on behalf of the Pacific islands. One of the organizations, the South Pacific Commission, has itself spawned many regional subassociations dealing with trade, education, migration, and cultural and economic development. Although it had neglected the islands that it controlled during its imperial phase, in the early 1900s, New Zealand cooperated fully with the United Nations in the islands' decolonization during the 1960s (although Tokelau, by choice, and the Ross dependency remain under New Zealand's control), while at the same time increasing development assistance. New Zealand's first alliance with Asian nations came in 1954, when it joined the Southeast Asian Treaty Organization. New Zealand's continuing involvement in regional affairs is demonstrated by its sending of troops to help maintain the peace in East Timor and its agreement to take in residents ("environmental refugees") from the island of Tuvalu, an island that is being literally engulfed by the Pacific as global warming produces higher sea levels.

New Zealand's new international focus certainly did not mean the end of cooperation with its traditional allies, however. In fact, the common threat of the Japanese during World War II strengthened cooperation with the United States to the extent that, in 1951, New Zealand joined a three-way, regional security agreement known as ANZUS (for Australia, New Zealand, and the United States). Moreover, because the United States was, at war's end, a Pacific/Asian power, any agreement with the United States was likely to bring New Zealand into more, rather than less, contact with Asia and the Pacific. Indeed, New Zealand sent troops to assist in all of the United States' military involvements in Asia: the occupation of Japan in 1945, the Korean War in 1950, and the Vietnam War in the 1960s. And, as a member of the British Commonwealth, New Zealand sent troops in the 1950s and 1960s to fight Malaysian Communists and Indonesian insurgents.

FREEDOM

New Zealand partakes of the democratic heritage of English common law and subscribes to all the human-rights protections that other Western nations have endorsed. Maoris, originally deprived of much of their land, are now guaranteed the same legal rights as whites. Social discrimination against Maoris is much milder than with many other colonized peoples, but tensions remain.

A NEW INTERNATIONALISM

Beginning in the 1970s, especially when the Labour Party of Prime Minister Norman Kirk was in power, New Zealand's orientation shifted even more markedly toward its own region. Under the Labour Party, New Zealand defined its sphere of interest and responsibility as the Pacific, where it hoped to be seen as a protector and benefactor of smaller states. Of immediate concern to many island nations was the issue of nuclear testing in the Pacific. Both the United States and France had undertaken tests by exploding nuclear devices on tiny Pacific atolls. In the 1960s, the United States ceased these tests, but France continued. On behalf of the smaller islands, New Zealand argued before the United Nations against testing, but France still did not stop. Eventually, the desire to end testing congealed into the more comprehensive position that the entire Pacific should be declared a nuclear-free zone. Not only testing but also the transport of nuclear weap-

ons through the area would be prohibited under the plan.

New Zealand's Labour government issued a ban on the docking of ships with nuclear weapons in New Zealand, despite the fact that such ships were a part of the ANZUS agreement. When the National Party regained control of the government in the late 1970s, the nuclear ban was revoked, and the foreign policy of New Zealand tipped again toward its traditional allies. The National Party government argued that, as a signatory to ANZUS, New Zealand was obligated to open its docks to U.S. nuclear ships. However, under the subsequent Labour government of Prime Minister David Lange, New Zealand once again began to flex its muscles over the nuclear issue. Lange, like his Labour Party predecessors, was determined to create a foreign policy based on moral rather than legal rationales. In 1985, a U.S. destroyer was denied permission to call at a New Zealand port, even though its presence there was due to joint ANZUS military exercises. Because the United States refused to say whether or not its ship carried nuclear weapons, New Zealand insisted that the ship could not dock. Diplomatic efforts to resolve the standoff were unsuccessful; and in 1986, New Zealand, claiming that it was not fearful of foreign attack, formally withdrew from ANZUS.

The issue of use of the Pacific for nuclear-weapons testing by superpowers is still of major concern to the New Zealand government. The nuclear test ban treaty signed by the United States in 1963 has limited U.S. involvement in that regard, but France has continued to test atmospheric weapons, and at times both the United States and Japan have proposed using uninhabited Pacific atolls to dispose of nuclear waste. In 1995, when France ignored the condemnation of world leaders and detonated a nuclear device in French Polynesia, New Zealand recalled its ambassador to France out of protest.

In the early 1990s, a new issue came to the fore: nerve-gas disposal. With the end of the Cold War, the U.S. military proposed disposing of most of its European stockpile of nerve gas on an atoll in the Pacific. The atoll is located within the trust territory granted to the United States at the conclusion of World War II. The plan is to burn the gas away from areas of human habitation, but those islanders living closest (albeit hundreds of miles away) worry that residues from the process could contaminate the air and damage humans, plants, and animals. The religious leaders of Melanesia, Micronesia, and Polynesia have condemned the plan, not only on environmental grounds

but also on grounds that outside powers should not be permitted to use the Pacific region without the consent of the inhabitants there—a position with which the Labour government of New Zealand strongly concurs.

In addition to a strong stand on the nuclear issue, the Labour government has successfully implemented laws that have made New Zealand more socially "liberal" than even some European nations, and certainly more so than the increasingly conservative United States. Abortion is not only legal, it is paid for by the National Health Service, and young girls, even preteens, are not required to inform their parents if they have an abortion. The use of birth control is so common that the birth rate is falling to just about natural replacement level. Prostitution has been legalized, and bills that would allow euthanasia for the elderly and civil unions for homosexuals have been seriously considered by Parliament. Some religious leaders are speaking out against such proposals, but many New Zealanders seem comfortable with the trend and continue to vote in liberal candidates for office. In fact, New Zealand was the first country in the world, in the 1970s, to have an environmentalist Green Party, and the Labour/Green coalition constantly reveals itself in policies such as New Zealand's signing of a fishing-regulations treaty to slow depletion of world fish stocks.

HEALTH/WELFARE

New Zealand established pensions for the elderly as early as 1898. Child-welfare programs were started in 1907, followed by the Social Security Act of 1938, which augmented the earlier benefits and added a minimum-wage requirement and a 40-hour work week. A national health program was begun in 1941. The government began dispensing free birth-control pills to all women in 1996 in an attempt to reduce the number of abortions, although, recently, the abortion rate has been rising.

ECONOMIC CHALLENGES

The New Zealand government's new foreign-policy orientation has caught the attention of observers around the world, but more urgent to New Zealanders themselves is the state of their own economy. Until the 1970s, New Zealand had been able to count on a nearly guaranteed export market in Britain for its dairy and agricultural products. Moreover, cheap local energy supplies as well as inexpensive oil from the Middle East had produced several decades of steady improvement in the standard of living. Whenever the economy showed signs of being sluggish, the government would artificially protect certain industries to ensure full employment.

All of this came to a halt in 1973, when Britain joined the European Union (then called the European Economic Community) and when the Organization of Petroleum Exporting Countries sent the world into its first oil shock. New Zealand actually has the potential of near self-sufficiency in oil, but the easy availability of Middle East oil over the years has prevented the full development of local oil and natural-gas reserves. As for exports, New Zealand had to find new outlets for its agricultural products, which it did by contracting with various countries throughout the Pacific Rim. Currently, about one third of New Zealand's trade occurs within the Pacific Rim. In the transition to these new markets, farmers complained that the manufacturing sector—intentionally protected by the government as a way of diversifying New Zealand's reliance on agriculture—was getting unfair, favorable treatment. Subsequent changes in government policy toward industry resulted in a new phenomenon for New Zealand: high unemployment. Moreover, New Zealand had constructed a rather elaborate social-welfare system since World War II, so, regardless of whether economic growth was high or low, social-welfare checks still had to be sent. This untenable position has made for a difficult political situation, for, when the National Party cut some welfare benefits and social services, it lost the support of many voters. The welfare issue, along with a change to a mixed member proportional voting system that enhanced the influence of smaller parties, threatened the National Party's political power. Thus, in order to remain politically dominant, in 1996 the National Party was forced to form a coalition with the United Party—the first such coalition government in more than 60 years.

ACHIEVEMENTS

New Zealand is notable for its efforts on behalf of the smaller islands of the Pacific. In addition to advocating a nuclear-free Pacific, New Zealand has promoted interisland trade and has established free-trade agreements with Western Samoa, the Cook Islands, and Niue. It provides educational and employment opportunities to Pacific islanders who reside within its borders.

In the 1970s, for the first time in its history, New Zealand's standard of living began to drop when compared to other Pacific Rim nations. Still high by world standards, New Zealand's per capita gross

domestic product of US$21,600 puts it in fifth place in the Pacific Rim, behind Australia, Hong Kong, Japan, and Singapore, with Taiwan close to overtaking New Zealand. The government realized that the country's dependence agriculture to support the economy was no longer feasible and began a program of economic diversification. Food processing, wood and paper, and many other industries were encouraged. The effort seems to be paying dividends; the country's annual economic growth rate is now 3.5 percent, and unemployment has dropped to 4.7 percent. In fact, government revenue surpluses were prompting calls for tax cuts. Still, in 2000, the Labour Party prime minister, Helen Clark, had to cancel a large purchase of F-16 fighter jets from the United States because she said the budget could not handle it. The bankruptcy of Air New Zealand's Australian subsidiary also caused concern.

New Zealanders are well aware of the economic strength of countries in North Asia such as Japan and China, and they see the potential for benefiting their own economy through joint ventures, loans, and trade. Yet they also worry that Japanese and Chinese wealth may come to constitute a symbol of New Zealand's declining strength as a culture. For instance, in the 1980s, as Japanese tourists began traveling en masse to New Zealand, complaints were raised about the quality of New Zealand's hotels. Unable to find the funds for a massive upgrading of the hotel industry, New Zealand agreed to allow Japan to build its own hotels; it reasoned that the local construction industry could use an economic boost and that the better hotels would encourage well-heeled Japanese to spend even more tourist dollars in the country. However, they also worried that, with the Japanese owning the hotels, New Zealanders might be relegated to low-level jobs.

Concern about their status vis-à-vis nonwhites had never been much of an issue to many Anglo-Saxon New Zealanders; they always simply assumed that nonwhites were inferior. Many settlers of the 1800s believed in the Social Darwinistic philosophy that the Maori and other brown- and black-skinned peoples would gradually succumb to their European "betters." It did not take long for the Maoris to realize that, land guarantees notwithstanding, the whites intended to deprive them of their land and culture. Violent resistance to these intentions occurred in the 1800s, but Maori landholdings continued to be gobbled up, usually deceptively, by white farmers and sheep herders. Government control of these unscrupulous practices was lax until the 1920s. Since that time, many Maoris (whose population has in-

Timeline: PAST

A.D. 1300s
Maoris, probably from Tahiti, settle the islands

1642
New Zealand is "discovered" by Dutch navigator A. J. Tasman

1769
Captain James Cook explores the islands

1840
Britain declares sovereignty over New Zealand

1865
A gold rush attracts new immigrants

1907
New Zealand becomes an almost independent dominion of Great Britain

1941
Socialized medicine is implemented

1947
New Zealand becomes fully independent within the British Commonwealth of Nations

New Zealand backs creation of the South Pacific Commission

1950s
Restructuring of export markets

1970s
The National Party takes power; New Zealand forges foreign policy more independent of traditional allies

1980s
The Labour Party regains power; New Zealand withdraws from ANZUS

1990s
New Zealanders consider withdrawing from the Commonwealth; Maoris and white New Zealanders face economic challenges from other Pacific Rimmers

PRESENT

2000s
New Zealand is led by its second woman prime minister, Helen Clark

New Zealand champions cultural and environmental goals

Parliament debates the merits of liberal social legislation such as a parent-leave law, a euthanasia law, and a civil union law for homosexuals.

creased to about 260,000) have intentionally sought to create a lifestyle that preserves key parts of traditional culture while incorporating the skills necessary for survival in a white world. In 1999, Maoris on the South Island, using land-loss funds provided by the government, made such a large land purchase (nearly 300,000 acres) that they became the largest land owner on the island. In 2004, Maori's started their own television station, broadcasting 50 percent in English and 50 percent in Maori. Maoris continue to insist that the Treaty of Waitangi, which they signed with the British, and which guaranteed them rights to their own lands, be honored in full. Some members of Parliament, such as those affiliated with the New Zealand First Party and the National Party, want to end what they see as "special treatment" of Maoris, i.e., entitlements such as health care and government paid education. But the Maoris are, literally, fighting back, throwing mud at the National Party leader and marching 10,000-strong, in what sometimes turn into violent demonstrations. The Maoris have generally accommodated themselves to those who rule over them, but they are increasingly educated and increasingly vocal. They recently had a stand-off with the Japanese over net fishing in the oceans and the damage the Japanese were inflicting on the Maori fishing industry, and in 2004, they created their own political party—the first ever Maori party.

More so than the Australians, New Zealanders are attempting to rectify the historic discrimination against the country's indigenous peoples. The current cabinet, for example, includes four Maoris and a member of Pacific Island descent (as well as 11 women and 1 Muslim). The new Labour/ Green Party coalition government has also attempted to eliminate some of the colonial cobwebs from their society by abolishing knighthoods bestowed by the British Crown. Local honors are awarded instead. Outspoken Prime Minister Helen Clark, defining the queen of England's role in New Zealand's affairs as "absurd," says it is only a matter of time until New Zealand will make a clean break from its old British Empire orientation and become a republic. Naturally, opinions on such matters differ; Clark's Labour Party has never won enough seats in Parliament to govern without allying itself with others; and in the most recent elections, the far-right, anti-immigration "New Zealand First" party was supported by 11 percent of the voters, giving them more seats in Parliament than before.

New Zealand will have to assert itself with vigor if it wants to prevent a slide to mediocrity when compared to the economic strength of some of its Asian neighbors. Still, given the favorable land-to-population ratio, the beauty of the land, the high quality of the infrastructure, and the general ease of life, New Zealanders can, in the main, be said to have a lifestyle that is the envy of most other peoples in the Pacific Rim—indeed, in the world.

North Korea (Democratic People's Republic of Korea)

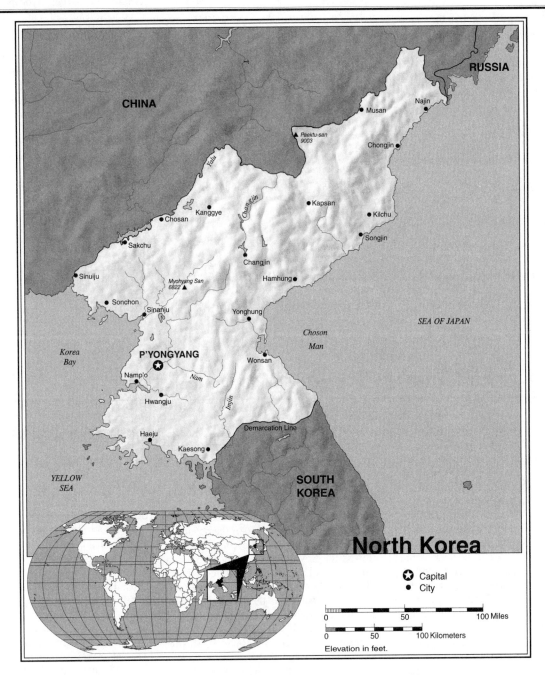

North Korea Statistics

GEOGRAPHY

Area in Square Miles (Kilometers):
44,358 (120,540) (about the size of Mississippi)

Capital (Population): P'yongyang (3,197,000)

Environmental Concerns: air and water pollution; insufficient potable water

Geographical Features: mostly hills and mountains separated by deep, narrow valleys; coastal plains

Climate: temperate

PEOPLE*

Population

Total: 22,697,553

Annual Growth Rate: 0.98%

Rural/Urban Population (Ratio): 40/60

Major Language: Korean

Ethnic Makeup: homogeneous Korean

Religions: mainly Buddhist and Confucianist (autonomous religious activities now almost nonexistent)

Health

Life Expectancy at Birth: 68 years (male); 74 years (female)
Infant Mortality: 22.8/1,000 live births
Physicians Available: 1/370 people

Education

Adult Literacy Rate: 99%
Compulsory (Ages): 6–17; free

COMMUNICATION

Telephones: 1,100,000 main lines
Daily Newspaper Circulation: 213/1,000 people
Televisions: 85/1,000 people

TRANSPORTATION

Highways in Miles (Kilometers): 19,345 (31,200)
Railroads in Miles (Kilometers): 3,000 (4,800)
Usable Airfields: 87

GOVERNMENT

Type: authoritarian socialist; one-man dictatorship
Independence Date: August 15, 1945 (from Japan)
Head of State/Government: President Kim Jong-Il; Premier Hong Song-nam
Political Parties: Korean Workers' Party; Korean Social Democratic Party; Chondoist Chongu Party
Suffrage: universal at 17

MILITARY

Military Expenditures (% of GDP): 31.3%
Current Disputes: Demarcation Line with South Korea; unclear border with China

ECONOMY

Currency ($ U.S. Equivalent): 2.20 won = $1
Per Capita Income/GDP: $1,300/$29 billion
GDP Growth Rate: 1%

Labor Force by Occupation: 64% nonagricultural; 36% agricultural
Natural Resources: hydropower; iron ore; copper; lead; salt; zinc; coal; magnesite; gold; tungsten; graphite; pyrites; fluorspar
Agriculture: rice; corn; potatoes; soybeans; pulses; livestock
Industry: machinery; military products; electric power; chemicals; mining; metallurgy; textiles; food processing; tourism
Exports: $1.04 billion (primary partners South Korea, China, Japan)
Imports: 2.04 billion (primary partners China, Thailand, Japan)

---

*Note: Statistics for North Korea are generally estimated due to unreliable official information.

SUGGESTED WEBSITES

```
http://www.skas.org
http://memory.loc.gov/frd/cs/
    kptoc.html
```

North Korea Country Report

The area that we now call North Korea has, at different times in Korea's long history, been separated from the South. In the fifth century A.D., the Koguryo Kingdom in the North was distinct from the Shilla, Paekche, and Kaya Kingdoms of the South. Later, the Parhae Kingdom in the North remained separate from the expanded Shilla Kingdom in the South. Thus, the division of Korea in 1945 into two unequal parts was not without precedent. Yet this time, the very different paths of development that the North and South chose rendered the division more poignant and, to those separated with little hope of reunion, more emotionally painful.

DEVELOPMENT

Already more industrialized than South Korea at the time of the Korean War, North Korea built on this foundation with massive assistance from China and the Soviet Union. Heavy industry was emphasized to the detriment of consumer goods. Economic isolation presages more negative growth ahead.

Beginning in 1945, Kim Il-Song, with the strong backing of the Soviet Union, pursued a hard-line Communist policy for both the political and economic develop-

ment of North Korea. The Soviet Union's involvement on the Korean Peninsula arose from its opportunistic entry into the war against Japan, just eight days before Japan's surrender to the Allies. Thus, when Japan withdrew from its long colonial rule over Korea, the Soviets, an allied power, were in a position to be one of the occupying armies. Reluctantly, the United States allowed the Soviet Union to move troops into position above the 38th Parallel, a temporary dividing line for the respective occupying forces. It was the Soviet Union's intention to establish a Communist buffer state between itself and the capitalist West. Therefore, it moved quickly to establish the area north of the 38th Parallel as a separate political entity. The northern city of P'yongyang was established as the capital.

When United Nations representatives arrived in 1948 to oversee elections and ease the transition from military occupation and years of Japanese rule to an independent Korea, the Soviets would not cooperate. Kim Il-Song took over the reins of power in the North. Separate elections were held in the South, and the beginning of separate political systems got underway. The 38th Parallel came to represent not only the division of the Korean Peninsula but also the boundary between the worlds of capitalism and communism.

THE KOREAN WAR (1950–1953)

Although not pleased with the idea of division, the South, without a strong army, resigned itself to the reality of the moment. In the North, a well-trained military, with Soviet and Chinese help, began preparations for a full-scale invasion of the South. The North attacked in June 1950, a year after U.S. troops had vacated the South, and quickly overran most of the Korean Peninsula. The South Korean government requested help from the United Nations, which dispatched personnel from 19 nations, under the command of U.S. general Douglas MacArthur. U.S. president Harry Truman ordered American troops to Korea just a few days after the North attacked.

MacArthur's troops advanced against the North's armies and by October were in control of most of the peninsula. However, with massive Chinese help, the North once again moved south. In response, UN troops bombed the North, inflicting heavy destruction. Whereas South Korea was primarily agricultural, North Korea was the industrialized sector of the peninsula. Bombing of the North's industrial targets severely damaged the economy, forcing several million North Koreans to flee south to escape both the war and the Communist dictatorship under which they found themselves.

Eventually, U.S. troops recaptured South Korea's capital, Seoul. Realizing that further fighting would lead to an expanded Asian war, the two sides agreed to cease-fire talks. They signed a truce that established a 2.5-mile-wide "demilitarized zone" (DMZ) for 155 miles across the peninsula and more or less along the former 38th Parallel division. The Korean War took the lives of more than 54,000 American soldiers, 58,000 South Koreans, and 500,000 North Koreans—but when it was over, both sides occupied about the same territory as they had at the beginning. Yet because neither side has ever declared peace, the two countries remain officially in a state of war.

FREEDOM

The mainline Communist approach has meant that the human rights commonplace in the West have never been enjoyed by North Koreans. Through suppression of dissidents, a controlled press, and restrictions on travel, the regime has kept North Koreans isolated from the world.

The border between North and South has been one of the most volatile places in Asia. The North staged military exercises along the border in 1996, firing into the DMZ, and thus breaking the cease-fire of 1953 (over the years there have many shooting and other incidents across the border). The South responded by raising its intelligence-monitoring activities and by requesting U.S. AWACS surveillance planes to monitor military movements in the North.

Scholars are still debating whether the Korean War should be called the United States' first losing war and whether or not the bloodshed was really necessary. To understand the Korean War, one must remember that in the eyes of the world, it was more than a civil war among different kinds of Koreans. The United Nations, and particularly the United States, saw North Korea's aggression against the South as the first step in the eventual communization of the whole of Asia. Just a few months before North Korea attacked, China had fallen to the Communist forces of Mao Zedong, and Communist guerrilla activity was being reported throughout Southeast Asia. The "Red Scare" frightened many Americans, and witchhunting for suspected Communist sympathizers—a college professor who might have taught about Karl Marx in class or a news reporter who might have praised the educational reforms of a Communist country—became the everyday preoccupation of such groups

as the John Birch Society and the supporters of U.S. senator Joseph McCarthy.

In this highly charged atmosphere, it was relatively easy for the U.S. military to promote a war whose aim it was to contain communism. Containment rather than defeat of the enemy was the policy of choice, because the West was weary after battling Germany and Japan in World War II. The containment policy also underlay the United States' approach to Vietnam. Practical though it may have been, this policy denied Americans the opportunity of feeling satisfied in victory, since there was to be no victory, just a stalemate. Thus, the roots of the United States' dissatisfaction with the conduct of the Vietnam War actually began in the policies shaping the response to North Korea's offensive in 1950. North Korea was indeed contained, but the communizing impulse of the North remained. In a way, the war continues, not only because a peace accord has never been signed, but also because remnants of the war keep coming back, as in the case of the remains of an American soldier that were found as recently as 2004 in North Korea and repatriated to the United States.

COLLECTIVE CULTURE

With Soviet backing, North Korean leaders moved quickly to repair war damage and establish a Communist culture. The school curriculum was rewritten to emphasize nationalism and equality of the social classes. Traditional Korean culture, based on Confucianism, had stressed strict class divisions, but the Communist authorities refused to allow any one class to claim privileges over another (although eventually the families of party leaders came to constitute a new elite). Higher education at the more than 600 colleges and training schools was redirected to technical rather than analytical subjects. Industries were nationalized; farms were collectivized into some 3,000 communes; and the communes were invested with much of the judicial and executive powers that other countries grant to cities, counties, and states. To overcome labor shortages, nearly all women were brought into the workforce, and the economy slowly returned to prewar levels. Of course, while the North was economically normalizing, the South was creating itself into an economic dynamo that soon eclipsed the North.

Today, many young people bypass formal higher education in favor of service in the military. Although North Korea has not published economic statistics for nearly 30 years, it is estimated that military expenses consume one quarter to one third of the entire national budget—this despite near-

starvation conditions in many parts of the country.

With China and the former Communist-bloc nations constituting natural markets for North Korean products, and with substantial financial aid from both China and the former Soviet Union in the early years, North Korea was able to regain much of its former economic, and especially industrial, strength. Today, North Korea successfully mines iron and other minerals and exports such products as cement and cereals. China has remained North Korea's only reliable ally; trade between the two countries is substantial. In one Chinese province, more than two thirds of the people are ethnic Koreans, most of whom take the side of the North in any dispute with the South.

Tensions with the South have remained high since the war. Sporadic violence along the border has left patrolling soldiers dead, and the assassination of former South Korean president Park Chung Hee and attempts on the lives of other members of the South Korean government have been attributed to North Korea, as was the bombing of a Korean Airlines flight in 1987. Both sides have periodically accused each other of attempted sabotage. In 1996, North Korea tried to send spies to the South via a small submarine; the attempt failed, and most of the spies were killed. More worrisome have been clashes with the North Korean Navy. Dozens of North Korean sailors were killed in 1999 while attempting to penetrate the South; and in 2002, two North Korean patrol boats fired on and sank a South Korean ship in the Yellow Sea. Thirty North Korean sailors were killed in the attack. To most observers, such attacks seem pointless and irrational.

HEALTH/WELFARE

Under the Kim Il-Song government, illiteracy was greatly reduced. Government housing is available at low cost, but shoppers are often confronted with empty shelves and low-quality goods. Malnutrition is widespread, and mass starvation has been reported in some regions.

The North has long criticized the South for its suppression of dissidents. Although the North's argument is bitterly ironic, given its own brutal suppression of human rights, it is nonetheless accurate in its view that the government in the South has been blatantly dictatorial. To suppress opponents, the South Korean government has, among other things, abducted its own students from Europe, abducted opposition leader Kim Dae Jung from Japan, tortured dissidents, and violently silenced demonstrators. All of this was said to be necessary because of the need for unity in the face of the threat from the North; as pointed out by scholar Gavan McCormack, the South used the North's threat as an excuse for maintaining a rigid dictatorial system. (Recent South Korean leaders have endeavored to bring democracy to the country and to open serious dialogue with the North.)

Under these circumstances, it is not surprising that the formal reunification talks, begun in 1971 with much fanfare, have only recently shown any promise. Visits of residents separated by the war were approved in 1985—the first time in 40 years that an opening of the border had even been considered. In 1990, in what many saw as an overture to the United States, North Korea returned the remains of five American soldiers killed during the Korean War. But real progress came in late 1991, when North Korean premier Yon Hyong Muk and South Korean premier Chung Won Shik signed a nonaggression and reconciliation pact, whose goal was the eventual declaration of a formal peace treaty between the two governments. In 1992, the governments established air, sea, and land links and set up mechanisms for scientific and environmental cooperation. North Korea also signed the nuclear nonproliferation agreement with the International Atomic Energy Agency. This move placated growing concerns about North Korea's rumored development of nuclear weapons and opened the way for investment by such countries as Japan, which had refused to invest until they received assurances on the nuclear question.

ACHIEVEMENTS

North Korea has developed its resources of aluminum, cement, and iron into solid industries for the production of tools and machinery while developing military superiority over South Korea, despite a population numbering less than half that of South Korea.

THE NUCLEAR ISSUE FLARES UP

The goodwill deteriorated quickly in 1993 and 1994, when North Korea refused to allow inspectors from the International Atomic Energy Agency (IAEA) to inspect its nuclear facilities, raising fears in the United States, Japan, and South Korea that the North was developing a nuclear bomb, despite their promise not to. When pressured to allow inspections, the North responded by threatening to withdraw from the IAEA and expel the inspectors. Tensions mounted, with all parties engaging in military threats and posturing and the United States, South Korea, and Japan (whose

shores could be reached in minutes by the North's new ballistic missiles) threatening economic sanctions. Troops in both Koreas were put on high alert. Former U.S. president Jimmy Carter helped to defuse the issue by making a private goodwill visit to Kim Il-Song in P'yongyang, the unexpected result of which was a promise by the North to hold a first-ever summit meeting with the South. Then, in a near-theatrical turn of events, Kim Il-Song, at five decades the longest national office-holder in the world, died, apparently of natural causes. The summit was canceled and international diplomacy was frozen while the North Korean government mourned the loss of its "Great Leader" and informally selected a new one, "Dear Leader" Kim Jong-Il, Kim Il-Song's son. Eventually, the North agreed to resume talks, a move interpreted as evidence that, for all its bravado, the North wanted to establish closer ties with the West. In the 1994 Agreed Framework with the United States, North Korea agreed to dismantle its graphite-moderated nuclear reactors in exchange for oil and help in building two light-water reactors. The world community breathed easier for a while. Unfortunately, tensions increased when North Korea launched a missile over Japan in 1998 and promised to keep doing so.

In the midst of these developments, a most amazing breakthrough occurred: the North agreed to a summit. In June 2000, President Kim Dae Jung of the South flew to P'yongyang in the North for talks with President Kim Jung-Il. Unlike anything the South had expected, the South Korean president was greeted with cheers from the crowds lining the streets, and was feted at a state banquet. An agreement was reached that seemed to pave the way for peace and eventual reunification. To reward the North for this dramatic improvement in relations, the United States immediately lifted trade sanctions that had been in place for 50 years. Both the North and the South suspended propaganda broadcasts and began plans for reuniting families long separated by the fortified DMZ. The North even went so far as to establish diplomatic ties with Australia and Italy.

Yet the West still has grave misgivings about the sincerity of North Korea. For example, in 2004, a large explosion sent a mushroom cloud high into the air in the North. Was it a nuclear test? North Korea said it was not, but many people were not convinced, especially as it occurred very near a military base where North Korea's Taepo-dong ballistic missiles are stored. Likewise, the North's motives have been questioned when, after requesting food relief for its starving peoples, it refused to allow observers from the United Nations

World Food Program to monitor the distribution of aid. Moreover, despite saying it will cooperate with the six-nation talks that would bring resolution to many of its problems, the North always seems to find a reason to halt talks with the West once they have started.

THE CHANGING INTERNATIONAL LANDSCAPE

North Korea has good reason to promote better relations with the West, because the world of the 1990s is not the world of the 1950s. In 1989, for instance, several former Soviet-bloc countries cut into the

Timeline: PAST

A.D. 1945
Kim Il-Song comes to power

1948
The People's Democratic Republic of Korea is created

1950
The Korean War begins

1953
A truce is arranged between North Korea and UN troops

1968
A U.S. spy boat, the *Pueblo*, is seized by North Korea

1969
A U.S. spy plane is shot down over North Korea

1971
Reunification talks begin

1990s
A nonaggression pact is signed with the South; North and South are granted seats in the UN; the world fears that North Korea is developing nuclear weapons; Kim Il-Song dies and is succeeded by his son, Kim Jong-Il

PRESENT

2000s
Famine causes thousands to flee the country

North and South Korea meet in a dramatic summit in 2000; North and South Koreans march together in the 2000 Summer Olympics

Tensions increase with the Bush administration

The government admits that it is still developing a nuclear-weapons program

A massive train explosion kills nearly 200 and destroyed thousands of homes

Six-nation talks to resolve the nuclear issue continue intermittently

North's economic monopoly by welcoming trade initiatives from South Korea; some even established diplomatic relations. At the same time, the disintegration of the Soviet Union meant that North Korea lost its primary political and military ally. Perhaps most alarming to the North is its declining economy; it has suffered negative growth for several years. Severe flooding in 1995 destroyed much of the rice harvest and forced the North to do the unthinkable: accept rice donations from the South. Hundreds of North Koreans have defected to the South (over 500 in the year 2004 alone), all of them complaining of near-famine conditions. Even more, perhaps as many as 2,000 per month have fled to China. With the South's economy consistently booming and the example of the failed economies of Eastern Europe as a danger signal, the North appears to understand that it must break out of its decades of isolation or lose its ability to govern. Nevertheless, it is not likely that North Koreans will quickly retreat from the Communist model of development that they have espoused for so long.

Kim Il-Song, who controlled North Korea for nearly 50 years, promoted the development of heavy industries, the collectivization of agriculture, and strong linkages with the then–Communist bloc. Governing with an iron hand, Kim denied basic civil rights to his people and forbade any tendency of the people to dress or behave like the "decadent" West. He kept tensions high by asserting his intention of communizing the South. His son, Kim Jong-Il, who had headed the North Korean military but was barely known outside his country, was eventually named successor to his father—the first dynastic power transfer in the Communist world. An enigmatic leader, the younger Kim has apparently approved actions that have increased

tensions with the South while doing little to improve the wrecked economy. When communism was introduced in North Korea in 1945, the government nationalized major companies and steered economic development toward heavy industry. In contrast, the South concentrated on heavy industry to balance its agricultural sector until the late 1970s but then geared the economy toward meeting consumer demand. Thus, the standard of living in the North for the average resident remains far behind that of the South. Indeed, Red Cross, United Nations, and other observers have documented widespread malnutrition and starvation in North Korea, conditions that are likely to continue well into the twenty-first century unless the North dramatically alters its current economic policies. Conditions are so bad in some areas that there is no electricity nor chlorine to run water-treatment plants, resulting in contaminated water supplies for about 60 percent of the population.

Even before the June 2000 summit, North Korea exhibited signs of liberalization. The government agreed to allow foreign companies to establish joint ventures inside the country, tourism was being promoted as a way of earning foreign currency, and two small Christian churches were allowed to be established. In an unusual move, North Korea openly requested aid from several foreign countries after nearly 200 people were killed and some 8,000 homes destroyed or damaged when a train exploded in the city of Ryongchon near the Chinese border in 2004. About the same time, the North even stopped broadcasting propaganda across the DMZ from large loudspeakers it had once erected to spread the philosophy of communism to Southerners across the border. Nevertheless, years of a totally controlled economy

in the North and shifting international alliances indicate many difficult years ahead for North Korea.

RECENT TRENDS

Although political reunification still seems to be years away, social changes are becoming evident everywhere as a new generation, unfamiliar with war, comes to adulthood, and as North Koreans are being exposed to outside sources of news and ideas. Many North Koreans now own radios that receive signals from other countries. South Korean stations are now heard in the North, as are news programs from the Voice of America. Modern North Korean history, however, is one of repression and control, first by the Japanese and then by the Kim government, who used the same police surveillance apparatus as did the Japanese during their occupation of the Korean Peninsula. It is not likely, therefore, that a massive push for democracy will be forthcoming soon from a people long accustomed to dictatorship.

Most alarming of all is the North's admission in 2002 that it did not dismantle its nuclear-bomb-making program. U.S. president George W. Bush was incensed at the rebuff and cut shipments of oil and other supplies. The North responded by evicting the United Nations inspectors who had been keeping an eye on its nuclear program. Defined by the West as a rogue nation (and one of Bush's "axis of evil"), North Korea is not likely to regain its standing (such as it was) in the world community until it makes bold initiatives away from aggression and toward economic and political reform—both of which are unlikely in the current climate.

Papua New Guinea (Independent State of Papua New Guinea)

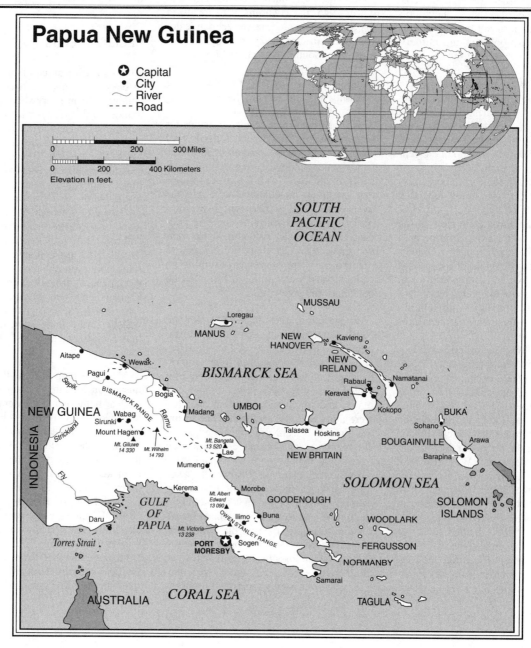

Papua New Guinea

- ⊛ Capital
- • City
- River
- - - - Road

Elevation in feet.

0 — 200 — 300 Miles
0 — 200 — 400 Kilometers

SOUTH PACIFIC OCEAN

MUSSAU

Loregau
MANUS

NEW HANOVER
Kavieng
NEW IRELAND
Namatanai

BISMARCK SEA

Aitape
Wewak
Pagui
Sepik
Bogia
BISMARCK RANGE
NEW GUINEA
Wabag
Ramu
Madang
UMBOI
Rabaul
Keravat
Kokopo
Sirunki
Mount Hagen
Mt. Giluwe 14 330
Mt. Wilhelm 14 793
Strickland
INDONESIA
Talasea
Hoskins
BUKA
Sohano
Mt. Bangeta 13 520
Lae
NEW BRITAIN
BOUGAINVILLE
Arawa
Barapina
Mumeng

Fly
Kerema
Mt. Albert Edward 13 090
Morobe
GOODENOUGH
SOLOMON SEA
SOLOMON ISLANDS

GULF OF PAPUA
Daru
Ilimo
Buna
WOODLARK
OWEN STANLEY RANGE
Mt. Victoria 13 238
PORT MORESBY
Sogen
FERGUSSON
NORMANBY

Torres Strait

AUSTRALIA
CORAL SEA
Samarai
TAGULA

Papua New Guinea Statistics

GEOGRAPHY

Area in Square Miles (Kilometers):
178,612 (461,690) (about the size of California)

Capital (Population): Port Moresby (259,000)

Environmental Concerns: deforestation; pollution from mining projects; drought

Geographical Features: mostly mountains; coastal lowlands and rolling foothills

Climate: tropical monsoon

PEOPLE

Population

Total: 5,420,280

Annual Growth Rate: 2.3%

Rural/Urban Population Ratio: 83/17

Major Languages: English; New Guinea Pidgin; Motu; 850 indigenous languages

Ethnic Makeup: predominantly Melanesian and Papuan; some Negrito, Micronesian, and Polynesian

Religions: 66% Christian; 34% indigenous beliefs

Health

Life Expectancy at Birth: 62 years (male); 68 years (female)
Infant Mortality: 53.15/1,000 live births
Physicians Available: 1/5,584 people
HIV/AIDS Rate in Adults: 0.6%

Education

Adult Literacy Rate: 65%

COMMUNICATION

Telephones: 61,200 main lines
Daily Newspaper Circulation: 15/1,000 people
Televisions: 23/1,000 people
Internet Users: 13,500 (2001)

TRANSPORTATION

Highways in Miles (Kilometers): 11,904 (19,200)
Railroads in Miles (Kilometers): none
Usable Airfields: 559
Motor Vehicles in Use: 99,000

GOVERNMENT

Type: parliamentary democratic/monarchy
Independence Date: September 16, 1975 (from the Australian-administered UN trusteeship)
Head of State/Government: Queen Elizabeth II; Prime Minister Michael Somare
Political Parties: Black Party; People's Democratic Movement; People's Action Party; People's Progress Party; United Party; Papua Party; National Party; Melanesian Alliance; others
Suffrage: universal at 18

MILITARY

Military Expenditures (% of GDP): 1.4%
Current Disputes: none

ECONOMY

Currency ($ U.S. Equivalent): 3.57 kina = $1

Per Capita Income/GDP: $2,200/$11.48 billion
GDP Growth Rate: 1.4%
Inflation Rate: 10.3%
Labor Force by Occupation: 85% agriculture
Population Below Poverty Line: 37%
Natural Resources: gold; copper; silver; natural gas; timber; petroleum; fisheries
Agriculture: coffee; cocoa; coconuts; palm kernels; tea; rubber; sweet potatoes; fruit; vegetables; poultry; pork
Industry: copra crushing; palm oil processing; wood processing and production; mining; construction; tourism
Exports: $1.938 billion (primary partners Australia, Japan, China)
Imports: $967 million (primary partners Australia, Singapore, New Zealand)

SUGGESTED WEBSITE

http://www.cia.gov/cia/
publications/factbook/geos/
pp.html

Papua New Guinea Country Report

Papua New Guinea is an independent nation and a member of the British Commonwealth. The capital is Port Moresby, where, in addition to English, the Motu language and a hybrid language known as New Guinea Pidgin are spoken. Occupying the eastern half of New Guinea (the second-largest island in the world) and many outlying islands, today Papua New Guinea is probably the most overlooked of all the nations in the Pacific Rim.

DEVELOPMENT

Agriculture (especially coffee and copra) is the mainstay of Papua New Guinea's economy. Copper, gold, and silver mining are also important, but large-scale development of other industries is inhibited by rough terrain, illiteracy, and a huge array of spoken languages—more than 700. There are substantial reserves of untapped oil.

It was not always overlooked, however. Spain claimed the vast land in the mid-sixteenth century, followed by Britain's East India Company in 1793. The Netherlands laid claim to part of the island in the 1800s and eventually came to control the western half (known for many years as Irian Jaya but now called Papua, a province of the Republic of Indonesia). In the 1880s, German settlers occupied the northeastern part of the island; and in 1884, Britain signed a treaty with Germany, which gave the British about half of what is now Papua New Guinea. In 1906, Britain gave its part of the island to Australia. Australia invaded and quickly captured the German area in 1914. Eventually, the League of Nations and, later, the United Nations gave the captured area to Australia to administer as a trust territory.

During World War II, the northern part of New Guinea was the scene of bitter fighting between a large Japanese force and Australian and U.S. troops. The Japanese had apparently intended to use New Guinea as a base for the military conquest of Australia. Australia resumed control of the eastern half of the island after Japan's defeat, and it continued to administer Papua New Guinea's affairs until 1975, when it granted independence.

STONE-AGE PEOPLES

Early Western explorers found the island's resources difficult to reach. The coastline and some of the interior are swampy and mosquito- and tick-infested, and the high, snow-capped mountainous regions are densely forested and hard to traverse. But perhaps most daunting to early would-be settlers and traders were the local inhabitants. Consisting of hundreds of sometimes warring tribes with totally different languages and customs, the New Guinea populace was determined to prevent outsiders from settling the island. Many adventurers were killed, their heads displayed in villages as victory trophies. The origins of the Papuan people are unknown, but some tribes share common practices with Melanesian islanders. Others appear to be Negritos, and some may be related to the Australian Aborigines. More than 700 languages are spoken in Papua New Guinea.

Australians and other Europeans found it beneficial to engage in trade with coastal tribes who supplied them with unique tropical lumbers, such as sandalwood and bamboo, and foodstuffs such as sugarcane, coconut, and nutmeg. Rubber and tobacco were also traded. Tea, which grows well in the highland regions, is an important cash crop.

But the resource that was most important for the economic development of Papua New Guinea was gold. It was discovered there in 1890 and produced two major gold rushes, in 1896 and 1926. Most prospectors were not Papuans; rather, they came from outside, most often Australia, and their presence antagonized the locals. Some prospectors were killed and cannibalized. A large number of airstrips in the otherwise unde-

(Photo UN/Witlin)

The interior of Papua New Guinea is very difficult to reach. Achieving easier access to the country's valuable minerals and exotic timber have caused a push for the development of transportation services. The island has nearly 500 airstrips, some in very isolated areas, along with increasing development of a road network. The negative impact of this development on the environment is of great concern.

veloped interior eventually were built by miners who needed a safe and efficient way to receive supplies. Today, copper is more important than gold—copper is, in fact, the largest single earner of export income for Papua New Guinea.

FREEDOM

 Papua New Guinea is a member of the British Commonwealth and officially follows the English heritage of law. However, in the country's numerous, isolated small villages, effective control is wielded by village elites with personal charisma; tribal customs take precedence over national law—of which many inhabitants are virtually unaware.

Meanwhile, pollution from mining is increasingly of concern to environmentalists, as is deforestation of Papua New Guinea's spectacular rain forests. A diplomatic flap occurred in 1992, when Australian environmentalists protested that a copper and gold mine in Papua New Guinea was causing enormous environmental damage. They called for the mine to be shut down. The Papuan government strongly resented the verbal intrusion into its sovereignty and reminded the protesters and the Australian government that it alone would establish environmental standards for companies operating inside its borders. The Papuan government holds a 20 percent interest in the mining company.

A similar flap occurred in the late 1990s and early 2000s over proposed logging of the forests near Collingwood Bay. The Prime Minister wanted to allow clear-cutting of the forests, sell the logs to China, replace the forests with coconut sap plantations, and create 50,000 jobs in the process. The Maisin people who live in the area issued a formal declaration opposing the logging, but the timber companies, including one from the Philippines, proceeded anyway. The Maisins believed that the development of painted tapa cloth industries would be better suited to their culture and protect the forests at the same time. Eventually, the matter ended up in the courts, where it remains today.

HEALTH/WELFARE

 Three quarters of Papua New Guinea's population have no formal education. Daily nutritional intake falls far short of recommended minimums, and tuberculosis and malaria are common diseases.

The tropical climate that predominates in all areas except the highest mountain peaks produces an impressive variety of plant and animal life. Botanists and other naturalists have been attracted to the island for scientific study for many years. Despite extensive contacts with these and other outsiders over the past century, and despite the establishment of schools and a university by the Australian government, some inland mountain tribes continue to live much as they probably did in the Stone

Age. Thus, the country lures not only miners and naturalists but also anthropologists and archaelogists looking for clues to humankind's early lifestyles. One of the most famous of these was Bronislaw Malinowski, the Polish-born founder of the field of social anthropology. In the early 1900s, he spent several years studying the cultural practices of the tribes of Papua New Guinea, particularly those of the Trobriand Islands.

Most of the 5 million Papuans live by subsistence farming. Agriculture for commercial trade is limited by the absence of a good transportation network: Most roads are unpaved, and there is no railway system. Travel on tiny aircraft and helicopters is common, however; New Guinea boasts nearly 500 airstrips, most of them unpaved and dangerously situated in mountain valleys. The harsh conditions of New Guinea life have produced some unique ironies. For instance, Papuans who have never ridden in a car or truck may have flown in a plane dozens of times.

ACHIEVEMENTS

Papua New Guinea, lying just below the equator, is world-famous for its astoundingly varied and beautiful flora and fauna, including orchids, birds of paradise, butterflies, and parrots. Dense forests cover 70 percent of the country. Some regions receive as much as 350 inches of rain a year.

In 1998, 23-foot-high tidal waves caused by offshore, undersea earthquakes inundated dozens of villages along the coast and drowned 6,000 people. A social earthquake, of sorts, occurred in the late 1980s when secessionist rebels on Bougainville Island, located about 500 miles from the capital of Port Moresby, launched a violent campaign to gain independence. Before the violence ended, some 20,000 had been killed, some with such armaments as stone-age spears and clubs. In 2001, the government signed a peace agreement with the rebels that granted them autonomy, including the right to operate their own police force. The government also agreed that residents could establish a provincial government and vote on independence within 10 to 15 years.

The government has had to handle other touchy issues in recent years. For instance, Prime Minister Bill Skate established diplomatic ties with Taiwan—which, as usual, raised a political storm from the mainland Chinese. The controversy eventually caused Skate's resignation as prime minister and the defection of several members of his party to the opposition. His successor, Sir Mekere Moruata, immediately reversed the decision over Taiwan, but barely had he been in office when he fired three of his cabinet ministers, claiming that they were plotting against him. Nevertheless, he moved quickly to privatize almost all state-owned enterprises, including the airlines, telecommunications, and others.

Moruata was replaced in the next election by Sir Michael Somare, the person most responsible for promoting independence from Australia and often called the nation's "founding father." It was Somare's third time as prime minister. But the election that brought him back to office was nothing short of chaotic. Forty-three political parties running 3,000 candidates contended for the right to govern the island nation. During the campaign, some 30 people were killed; ballot boxes were stolen and plundered, and gangs hired by tribal leaders intimidated voters in many outlying villages. There were so many voting irregularities that a final count was delayed for weeks, and several regions had to redo the vote. In the end, Somare's National Alliance had to ally with five other parties in order to gain a majority and constitute a government.

Given the differences in socialization of the Papuan peoples and the difficult conditions of life on their island, it will likely be many decades before Papua New Guinea, which joined the Asia Pacific Economic Cooperation group in 1993, is able to participate fully in the Pacific Rim community.

Timeline: PAST

A.D. 1511
The main island is sighted by Portuguese explorers

1828
The Dutch annex the west half of the island

1884
A British protectorate over part of the eastern half of the island; the Germans control the northeast

1890
Gold is discovered in Papua New Guinea

1906–1914
Australia assumes control of the island

1920
Australia is given the former German areas as a UN trust territory

1940s
Japan captures the northern part of the island; Australia resumes control in 1945

1988
Australia grants independence to Papua New Guinea

1988
A revolt against the government begins on the island of Bougainville

1990s
An economic blockade of Bougainville is lifted, but violence continues, claiming 3,000 lives

PRESENT

2000s
Bougainville independence referendum is set

The government moves to privatize industries

Philippines (Republic of the Philippines)

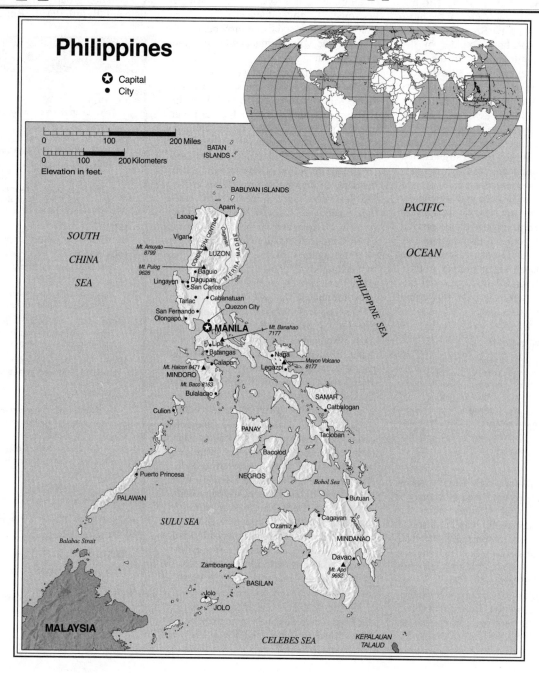

Philippines

★ Capital
● City

Philippines Statistics

GEOGRAPHY

Area in Square Miles (Kilometers):
110,400 (300,000) (about the size of Arizona)

Capital (Population): Manila (10,069,000)

Environmental Concerns: deforestation; air and water pollution; soil erosion; pollution of mangrove swamps

Geographical Features: mostly mountainous; coastal lowlands

Climate: tropical marine; monsoonal

PEOPLE

Population

Total: 86,241,697

Annual Growth Rate: 1.9%

Rural/Urban Population Ratio: 42/58

Major Languages: Pilipino (based on Tagalog); English

Ethnic Makeup: 95% Malay; 5% Chinese and others

Religions: 83% Roman Catholic; 9% Protestant; 5% Muslim; 3% Buddhist and others

Health

Life Expectancy at Birth: 66 years (male);
   72 years (female)
Infant Mortality: 24.2/1,000 live births
Physicians Available: 1/849 people
HIV/AIDS Rate in Adults: 0.07%

Education

Adult Literacy Rate: 92%
Compulsory (Ages): 7–12; free

COMMUNICATION

Telephones: 3,100,000 main lines
Daily Newspaper Circulation: 65/1,000
   people
Televisions: 125/1,000 people
Internet Users: 4,500,000 (2002)

TRANSPORTATION

Highways in Miles (Kilometers): 124,000
   (200,000)

Railroads in Miles (Kilometers): 499
   (800)
Usable Airfields: 275
Motor Vehicles in Use: 2,050,000

GOVERNMENT

Type: republic
Independence Date: July 4, 1946 (from the
   United States)
Head of State/Government: President
   Gloria Macapagal-Arroyo is both head of
   state and head of government
Political Parties: Democratic Action;
   Struggle of the Filipino Masses; Liberal
   Party; Lakas; People's Reform Party
Suffrage: universal at 18

MILITARY

Military Expenditures (% of GDP): 1.5%
Current Disputes: territorial disputes with
   China, Malaysia, Taiwan, Vietnam, and
   possibly Brunei

ECONOMY

Currency ($ U.S. Equivalent): 51.9
   Philippine pesos = $1
Per Capita Income/GDP: $4,600/$390
   billion
GDP Growth Rate: 4.5%
Inflation Rate: 6%
Unemployment Rate: 11.4%
Natural Resources: timber; petroleum;
   nickel; cobalt; silver; gold; salt; copper
Agriculture: rice; coconuts; corn;
   sugarcane; fruit; animal products; fish
Industry: food processing; chemicals;
   textiles; pharmaceuticals; wood
   products; electronics assembly;
   petroleum refining; fishing
Exports: $34.6 billion (primary partners
   United States, Japan, Hong Kong)
Imports: $35.9 billion (primary partners
   Japan, United States, Singapore)

Philippines Country Report

The Philippines has close historic ties to the West. Eighty-three percent of Filipinos are Roman Catholics, and most speak at least some English. Many use English daily in business and government; in fact, English is the language of instruction at school. Moreover, when they discuss their history as a nation, Filipinos will mention Spain, Mexico, the Spanish-American War, the United States, and cooperative Filipino–American attempts to defeat the Japanese in World War II. The country was even named after a European, King Philip II of Spain. (Currently, 4.4 percent of the United States' foreign-born population, second only to Mexicans, are from the Philippines.) If this does not sound like a typical Asian nation, it is because Philippine nationhood essentially began with the arrival of Westerners. That influence continues to dominate the political and cultural life of the country.

Yet the history of the islands certainly did not begin with European contact; indeed, there is evidence of human habitation in the area as early as 25,000 B.C. Beginning about 2,000 B.C., Austronesians, Negritos, Malays, and other tribal peoples settled many of the 7,107 islands that constitute the present-day Philippines. Although engaged to varying degrees in trade with China and Southeast Asia, each of these ethnic groups (nearly 60 distinct groups still exist) lived in relative isolation from one another, speaking different languages, adhering to different religions, and, for good or ill, knowing nothing of the concept of national identity.

DEVELOPMENT

The Philippines labors under more than $50 billion in foreign debt. Returns on investment in development projects have been so slow that about half of the earnings from all exports have to be used to service the debt. Still, the successful development of the former U.S. military bases into industrial zones is a bright spot in the economy.

Although 5 million ethnic peoples remain marginized from the mainstream today, for most islanders the world changed forever in the mid-1500s, when soldiers and Roman Catholic priests from Spain began conquering and converting the population. Eventually, the disparate ethnic groups came to see themselves as one entity, Filipinos, a people whose lives were controlled indirectly by Spain from Mexico—a fact that, unique among Asian countries, linked the Philippines with the Americas. Thus, the process of national-identity formation for Filipinos actually began in Europe and North America.

Members of some ethnic groups assimilated rather quickly, marrying Spanish soldiers and administrators and acquiring the language and cultural outlook of the West.

FREEDOM

Marcos's one-man rule meant that both the substance and structure of democracy were ignored. The Philippine Constitution is similar in many ways to that of the United States. President Corazon Aquino, who came to office in a "people power" revolution against Marcos, attempted to adhere to democratic principles; her successors pledged to do the same. The Communist Party was legalized in 1992.

The descendants of these mestizos ("mixed" peoples, including local/Chinese mixes) have become the cultural, economic, and political elite of the country. Others, particularly among the Islamic communities on the island of Mindanao, resisted assimilation right from the start and continue to challenge the authority of Manila. Indeed, the Communist insurgency, reported so often in the news, does not seem to go away. Several groups, among them the Moro Islamic Liberation Front, the Moro National Liberation Front, and the Islamic Abu Sayyaf Group, continue to take tourists and others as hostages, set off bombs, and inflict violence on villagers. For example, a bomb explosion at a basketball game in the southern Philippines in 2004 killed 11 people and wounded 40, including the town mayor, who was probably the target of the attack.

(United Nations photo by J.M. Micaud)

The Philippines has suffered from the misuse of funds entrusted to the government over the past several decades. The result has been a polarity of wealth, with many citizens living in severe poverty. Slums, such as Tondo in Manila, pictured above, are a common sight in many of the urban areas of the Philippines.

About the same time, an American businessperson was abducted and chained by his neck and feet for nearly a month. It is believe that some of these groups capture hostages (such as a Christian missionary from Kansas who was killed in 2002) and then send the ransom money to support the al-Qaeda terrorist network. On the day that nearly 3,000 U.S. troops arrived in the Philippines to help the local military in its efforts to eliminate the Abu Sayyaf rebels, bombs exploded at stores, a radio station, and a bus terminal, killing many innocent passersby. Such violence is, in part, connected to the worldwide rise of radical Islamism, but also it is an attempt by marginated ethnics and others to regain the cultural independence that their peoples lost some 400 years ago. The current government, led by President Gloria Macapagal-Arroyo, is hoping to revive stalled peace talks with these groups, and the United States is attempting to facilitate the process by promising to remove Philippine communist rebels from a list of terrorist organizations if they conclude a peace agreement and resume talks in Oslo, Norway.

As elsewhere in Asia, the Chinese community has played an important but controversial role in Philippine life. Dominating trade for centuries, the Philippine Chinese have acquired clout (and enemies) that far exceeds their numbers (fewer than 1 million). Former president Corazon Aquino was of part-Chinese ancestry, and some of the resistance to her presidency stemmed from her ethnic lineage. The Chinese-Philippine community, in particular, has been the target of ethnic violence—kidnappings

and abductions—because their wealth (relative to other Filipino groups) makes them easy prey.

FOREIGN INTERESTS

Filipinos occupy a resource-rich, beautiful land. Monsoon clouds dump as much as 200 inches of rain on the fertile, volcanic islands. Rice and corn grow well, as do hemp, coconut, sugarcane, and tobacco. Tuna, sponges, shrimp, and hundreds of other kinds of marine life flourish in the surrounding ocean waters. Part of the country is covered with dense tropical forests yielding bamboo and lumber and serving as habitat to thousands of species of plant and animal life. The northern part of Luzon Island is famous for its terraced rice paddies.

HEALTH/WELFARE

Quality of life varies considerably between the city and the countryside. Except for the numerous urban squatters, city residents generally have better access to health care and education. Most people still do not have access to safe drinking water. The gap between the upper-class elite and the poor is hugely pronounced, and growing.

Given this abundance, it is not surprising that several foreign powers have taken a serious interest in the archipelago. The Dutch held military bases in the country in the 1600s, the British briefly controlled Manila in the 1800s, and the Japanese overran the entire country in the 1940s. But

it was Spain, in control of most of the country for more than 300 years (1565–1898), that established the cultural base for the modern Philippines. Spain's interest in the islands—its only colony in Asia—was primarily material and secondarily spiritual. It wanted to take part in the lucrative spice trade and fill its galleon ships each year with products from Asia for the benefit of the Spanish Crown. It also wanted (or, at least, Rome wanted) to convert the so-called heathens (that is, nonbelievers) to Christianity. The friars were particularly successful in winning converts to Roman Catholicism because, despite some local resistance, there were no competing Christian denominations in the Philippines, and because the Church quickly gained control of the resources of the island, which it used to entice converts. Eventually, a Church-dominated society was established that mirrored in structure—social class divisions as well as religious and social values—the mother cultures of Spain and particularly Mexico.

Resisting conversion were the Muslims of the island of Mindanao, a group that continues to remain on the fringe of Philippine society but which signed a cease-fire with the government in 1994, after 20 years of guerrilla warfare. The cease-fire has been only partially observed, with violence flaring periodically, including immediately after the agreement, when 200 armed Muslims attacked and burned the town of Ipil on Mindanao Island.

Spanish rule in the Philippines came to an inglorious end in 1898, at the end of the Spanish-American War. Spain granted independence to Cuba and ceded the Philippines, Guam, and Puerto Rico to the United States. Filipinos hoping for independence were disappointed to learn that yet another foreign power had assumed control of their lives. Resistance to American rule cost several thousand lives in the early years, and the American public became outraged at the brutality exhibited by some of the U.S. soldiers against the Philippine people. Eventually, the majority of the people accepted their fate and even began to realize that, despite the deception and loss of independence, the U.S. presence was fundamentally different from that of Spain. The United States was interested in trade, and it certainly could see the advantage of having a military presence in Asia, but it viewed its primary role as one of tutelage. American officials believed that the Philippines should be granted independence, but only when the nation was sufficiently schooled in the process of democracy. Unlike Spain, the United States encouraged political parties and attempted to place Filipinos in positions of governmental authority.

Preparations were under way for independence when World War II broke out. The war and the occupation of the country by the Japanese undermined the economy, devastated the capital city of Manila, caused divisions among the political elite, and delayed independence. After Japan's defeat, the country was, at last, granted independence, on July 4, 1946. Manuel Roxas, a well-known politician, was elected president. Despite armed opposition from Communist groups, the country, after several elections, seemed to be maintaining a grasp on democracy.

MARCOS AND HIS AFTERMATH

Then, in 1965, Ferdinand E. Marcos, a Philippines senator and former guerrilla fighter with the U.S. armed forces, was elected president. He was reelected in 1969. Rather than addressing the serious problems of agrarian reform and trade, Marcos maintained people's loyalty through an elaborate system of patronage, whereby his friends and relatives profited from the misuse of government power and money. Opposition to his corrupt rule manifested itself in violent demonstrations and in a growing Communist insurgency. In 1972, Marcos declared martial law, arrested some 30,000 opponents, and shut down newspapers as well as the National Congress. Marcos continued to rule the country by personal proclamation until 1981. After the lifting of martial law, Marcos remained in power for another five years, and he and his wife, Imelda, and their extended family and friends increasingly were criticized for corruption. Finally, in 1986, after nearly a quarter-century of his rule, an uprising of thousands of dissatisfied Filipinos overthrew Marcos, who fled to Hawaii. He died there in 1990.

Taking on the formidable job of president was Corazon Aquino, the widow of murdered opposition leader Benigno Aquino. Aquino's People Power revolution had a heady beginning. Many observers believed that at last Filipinos had found a democratic leader around whom they could unite and who would end corruption and put the persistent Communist insurgency to rest. Aquino, however, was immediately beset by overwhelming economic, social, and political problems.

Opportunists and factions of the Filipino military and political elite still loyal to Marcos attempted numerous coups d'état in the years of Aquino's administration. Much of the unrest came from within the military, which had become accustomed to direct involvement in government during Marcos's martial-law era. A few Communist separatists, lured by Aquino's overtures of peace, turned in their arms, but most continued to plot violence against the government. Thus, the sense of security and stability that Filipinos needed in order to attract more substantial foreign investment and to reestablish the habits of democracy continued to elude them.

During the Aquino years, the anemic economy showed tentative signs of improvement. Some countries, particularly Japan and the United States and, more recently, Hong Kong, invested heavily in the Philippines, as did half a dozen international organizations. In fact, some groups complained that further investment was unwarranted, because already-allocated funds had not yet been fully utilized. Moreover, misuse of funds entrusted to the government—a serious problem during the Marcos era—continued, despite Aquino's promise to eradicate corruption.

ACHIEVEMENTS

Filipino women often run businesses or hold important positions in government. Two of the six presidents since independence have been women. Folk dancing is very popular, as is the *kundiman,* a unique blend of music and words found only in the Philippines.

A 1987 law, enacted after Corazon Aquino assumed the presidency, limited the president to one term in office. Half a dozen contenders vied for the presidency in 1992, including Imelda Marcos and other relatives of former presidents Marcos and Aquino; U.S. West Point graduate General Fidel Ramos, who had thwarted several coup attempts against Aquino and who thus had her endorsement, won the election. It was the first peaceful transfer of power in more than 25 years (although campaign violence claimed the lives of more than 80 people).

Timeline: PAST

25,000 B.C.
Negritos and others begin settling the islands

2,000 B.C.
Malays arrive in the islands

A.D. 400–1400
Chinese, Arabs, and Indians control parts of the economy and land

1542
The islands are named for the Spanish king Philip II

1890s
Local resistance to Spanish rule

1898
A treaty ends the Spanish-American War

1941
The Japanese attack the Philippines

1944
General Douglas MacArthur makes a triumphant return to Manila

1946
The United States grants complete independence to the Philippines

1947
Military-base agreements are signed with the United States

1965
Ferdinand Marcos is elected president

1972
Marcos declares martial law

1980s
Martial law is lifted; Corazon Aquino and her People Power movement drive Marcos into exile

1990s
The United States closes its military bases in the Philippines; economic crisis

PRESENT

2000s
The crippling "Love Bug" computer virus emanates from the Philippines

President Estrada is ousted for "economic plunder"

The Philippines withdraws more troops from Iraq in an effort to save the life of a Filipino hostage.

Incumbent Gloria Macapagal Arroyo is narrowly re-elected president of the Philippines in 2004 elections.

In the 1998 presidential campaign, some 83 candidates filed with the election commission, including Imelda Marcos. Despite the deaths of nearly 30 people and some bizarre moments, such as when a mayoral candidate launched a mortar attack on his opponents, the election was the most orderly in years. Former movie star Joseph Estrada won by a landslide. However, when he attempted to repeal the law limiting presidents to one six-year term, some 100,000 people took to the streets in protest.

Estrada had to handle such problems as the Mayon Volcano eruption, which forced 66,000 people from their homes, and the sometimes violent dispute over ownership of the Spratly Islands in the South China Sea. During the Estrada presidency, the Philippine Navy shot at and sank Chinese fishing vessels that were operating near the Islands. The Philippines demanded that China remove a pier that it had built on Mischief Reef. The dispute was not resolved until 2002, when (under Estrada's successor) the half-dozen claimants to the Spratlys reached agreement on a code of conduct for use of the resources in and under the sea. The Philippines is sure to watch the islands carefully, however, given that China appears to be building up its military capabilities there.

In 2001, growing evidence of "economic plunder" by the once-adored actor-turned-president led to popular demonstrations to remove Estrada from office. Investigators believe that Estrada siphoned away as much as $300 million in government money, hiding the funds in banks under various aliases. An attempt at impeachment failed, but public pressure drove Estrada from office. He was replaced by his vice-president, Gloria Macapagal-Arroyo, who filled out the remaining years of Estrada's term and then decided to stand for the presidency in her own right in 2004, which she won by narrow margins.

SOCIAL PROBLEMS

Much of the foreign capital coming into the Philippines in the 1990s was invested in stock and real-estate speculation rather than in agriculture or manufacturing. Thus, with the financial collapse of 1997, there was little of substance to fall back on. Even prior to the financial crisis, inflation had been above 8 percent per year and unemployment was nearing 9 percent. And one problem never seemed to go away: extreme social inequality. As in Malaysia, where ethnic Malays have constituted a seem-

ingly permanent class of poor peasants, Philippine society is fractured by distinct classes. Chinese and mestizos constitute the top of the hierarchy, while Muslims and most rural dwellers form the bottom. About half the Filipino population of 86 million make their living in agriculture and fishing; but even in Manila, where the economy is stronger than anywhere else, thousands of residents live in abject poverty as urban squatters. Officially, one third of Filipinos live below the poverty line. Disparities of wealth are striking. Worker discontent is so widespread that the Philippines often ranks near the top of Asian countries in days of work lost due to strikes and other protests. Economic leaders, once proud of the Philippines' economic strength, now see the country slipping further and further behind the rest of Asia. The per capita annual income is only US$4,600, ranking the Philippines 13th out of twenty-one Pacific Rim countries. Even China's per capita income exceeds the Philippines'.

Adding to the country's financial woes was the sudden loss of income from the six U.S. military bases that closed in 1991 and 1992. The government had wanted the United States to maintain a presence in the country, but in 1991, the Philippine Legislature, bowing to nationalist sentiment, refused to renew the land-lease agreements that had been in effect since 1947. Occupying many acres of valuable land and bringing as many as 40,000 Americans at one time into the Philippines, the bases had come to be seen as visible symbols of American colonialism or imperialism. But they had also been a huge boon to the economy. Subic Bay Naval Base alone had provided jobs for 32,000 Filipinos on base and, indirectly, to 200,000 more. Moreover, the United States paid nearly $390 million each year to lease the land and another $128 million for base-related expenses. Base-related monies entering the country amounted to 3 percent of the entire Philippines economy. After the base closures, the U.S. Congress cut other aid to the Philippines, from $200 million in 1992 to $48 million in 1993. To counterbalance the losses, the Philippines accepted a $60 million loan from Taiwan to develop 740 acres of the former Subic Bay Naval Base into an industrial park. The International Monetary Fund also loaned the country $683 million—funds that have been successfully used to transform the former military facilities into commercial zones.

Other efforts to revitalize the economy are also under way. In 1999, the Legislature passed a law allowing more foreign investment in the retail sector; and, in the most surprising development, the government approved resumption of large-scale military exercises with U.S. forces. Currently some 4,000 U.S. troops are operating in the Philippines.

Adding to the economic stress is the constant devastation from natural disasters. In recent years, Filipinos have suffered the massive explosion of the Mayon Volcano, which forced 66,000 people from their homes; Typhoon Nida which triggered landslides, destroyed houses, killed 19 people and left hundreds homeless; monsoon landslides and huge waves that killed 87 people in the eastern Philippines; a 6.4 magnitude earthquake in Manila; a storm and typhoon in the north that killed 650 people; and several other "acts of God." Some of the landslides appear to be the result of logging of the natural forests.

CULTURE

Philippine culture is a rich amalgam of Asian and European customs. Family life is valued, and few people have to spend their old age in nursing homes. Divorce is frowned upon. Women have traditionally involved themselves in the worlds of politics and business to a greater degree than have women in other Asian countries. Educational opportunities for women are about the same as those for men; adult literacy in the Philippines is estimated at 95 percent. Evidence of gender discrimination continues, however, in this Latin-based culture. For example, so many women are regularly groped by men in the tightly packed commuter rail cars in Manila that the rail line has established women-only coaches for their safety. Another serious problem is the inability of college-educated men and women to find employment befitting their skills. Discontent among these young workers continues to grow, as it does among the many rural and urban poor.

Nevertheless, many Filipinos take a rather relaxed attitude toward work and daily life. They enjoy hours of sports and folk dancing or spend their free time in conversation with neighbors and friends, with whom they construct patron/client relationships. In recent years, the growing nationalism has been expressed in the gradual replacement of the English language with Pilipino, a version of the Malay-based Tagalog language.

Singapore (Republic of Singapore)

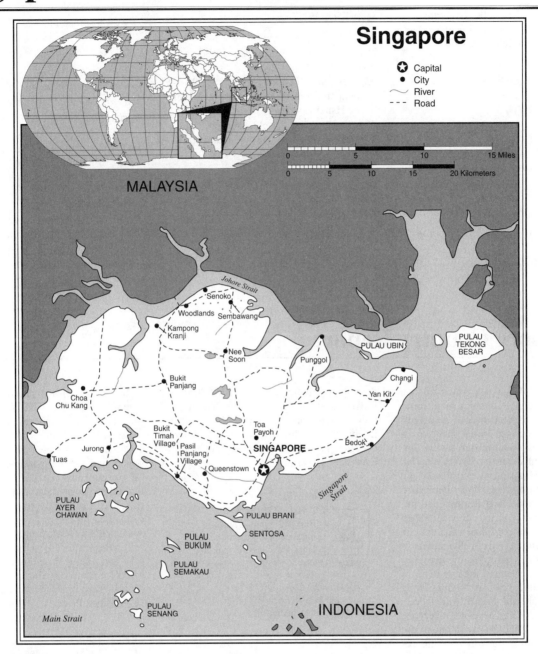

Singapore Statistics

GEOGRAPHY

Area in Square Miles (Kilometers): 250 (648) (about 3 1/2 times the size of Washington, D.C.)

Capital (Population): Singapore (4,108,000)

Environmental Concerns: air and industrial pollution; limited fresh water; waste-disposal problems

Geographical Features: lowlands; gently undulating central plateau; many small islands

Climate: tropical; hot, humid, rainy

PEOPLE

Population

Total: 4,353,893

Annual Growth Rate: 1.71%

Rural/Urban Population (Ratio): 0/100

Major Languages: Malay; Mandarin Chinese; Tamil; English

Ethnic Makeup: 77% Chinese; 14% Malay; 7% Indian; 2% others

Religions: 42% Buddhist and Taoist; 18% Christian; 16% Muslim; 5% Hindu; 19% others

Health

Life Expectancy at Birth: 78 years (male);
  84 years (female)
Infant Mortality: 3.6/1,000 live births
Physicians Available: 1/667 people
HIV/AIDS Rate in Adults: 0.19%

Education

Adult Literacy Rate: 93%

COMMUNICATION

Telephones: 1,950,000 main lines
Daily Newspaper Circulation: 340/1,000
  people
Televisions: 218/1,000 people
Internet Users: 2,310,000 (2002)

TRANSPORTATION

Highways in Miles (Kilometers): 1,936
  (3,122)
Railroads in Miles (Kilometers): 23 (38)

Usable Airfields: 9
Motor Vehicles in Use: 521,000

GOVERNMENT

Type: parliamentary republic
Independence Date: August 9, 1965
  (from Malaysia)
Head of State/Government: President
  Sellapan Rama Nathan; Prime Minister
  Lee Hsien Leong
Political Parties: People's Action Party;
  Workers' Party; Singapore Democratic
  Party; National Solidarity Party;
  Singapore People's Party
Suffrage: universal and compulsory at 21

MILITARY

Military Expenditures (% of GDP): 4.9%
Current Disputes: territorial and boundary
  disputes with Malaysia

ECONOMY

Currency ($ U.S. Equivalent): 1.73
  Singapore dollars = $1
Per Capita Income/GDP: $23,700/$109.4
  billion
GDP Growth Rate: 1.1%
Inflation Rate: 1.5%
Unemployment Rate: 4.8%
Natural Resources: fish; deepwater ports
Agriculture: rubber; copra; fruit;
  vegetables
Industry: petroleum refining; electronics;
  oil-drilling equipment; rubber processing
  and rubber products; processed food and
  beverages; ship repair; financial services;
  biotechnology
Exports: $142 billion (primary partners
  United States, Malaysia, Hong Kong)
Imports: $122 billion (primary partners
  Japan, United States, Malaysia)

Singapore Country Report

It is often said that North Americans are well off because they inhabit a huge continent that abounds with natural resources. How then can one explain the phenomenal prosperity of land- and resource-poor Singapore? The inhabitants of this tiny, flat, humid, tropical island, located near the equator off the tip of the Malay Peninsula, have so few resources that they must import even their own drinking water. With only 250 square miles of land (including 58 mostly uninhabited islets), Singapore is half the size of tiny Hong Kong, yet it has the 4th highest per capita income in all of the Pacific Rim; only Australia, Hong Kong, and Japan rank higher. Singapore's economic success has benefited everyone living there. For instance, with one of the highest population densities in the region, Singapore might be expected to have the horrific slums that characterize parts of other crowded urban areas. But almost 85 percent of its 4.4 million people live in spacious and well-equipped, albeit government-controlled, apartments.

Imperialism, geography, and racism help to explain Singapore's unique characteristics. For most of its recorded history, beginning in the thirteenth century A.D., Singapore was controlled variously by the rulers of Thailand, Java, Indonesia, and even India. In the early 1800s, the British were determined to wrest control of parts of Southeast Asia from the Dutch and expand their growing empire. Facilitating

their imperialistic aims was Sir Stamford Raffles, a Malay-speaking British administrator who not only helped defeat the Dutch in Java but also diminished the power of local elites in order to fortify his position as lieutenant-governor.

DEVELOPMENT

Development of the deepwater Port of Singapore has been so successful that at any single time, 400 ships are in port. Singapore has also become a base for fleets engaged in offshore oil exploration and a major financial center, the "Switzerland of Southeast Asia." Singapore has key attributes of a developed country.

Arriving in Singapore in 1819, Raffles found it to be a small, neglected settlement with an economy based on fishing. Yet he believed that the island's geographic location endowed it with great potential as a transshipment port. He established policies that facilitated its development just in time to benefit from the British exports of tin, rubber, and timber leaving Malaya. Perhaps most important was his declaration of Singapore as a free port. Skilled Chinese merchants and traders, escaping racist discrimination against them by Malays on the Malay Peninsula, flocked to Singapore, where they prospered in the free-trade atmosphere.

In 1924, the British began construction of a naval base on the island, the largest in Southeast Asia, which was nonetheless overcome by the Japanese in 1942. Returning in 1945, the British continued to build Singapore into a major maritime center. Today, oil supertankers from Saudi Arabia must exit the Indian Ocean through the Strait of Malacca and skirt Singapore to enter the South China Sea for deliveries to Japan and other Asian nations. Thus, Singapore has found itself in the enviable position of helping to refine and transship millions of barrels of Middle Eastern oil. Singapore's oil-refining capacities have been ranked the world's third largest since 1973.

FREEDOM

Under former prime minister Lee Kuan Yew, Singaporeans had to adjust to a strict regimen of behavior involving both political and personal freedoms. Citizens want more freedoms but also realize that law and order have helped produce their high quality of life. Political-opposition voices have largely been silenced since 1968, when the People's Action Party captured all the seats in the government.

Singapore is now the second-busiest port in the world (Rotterdam in the Netherlands is number one). It has become the largest shipbuilding and -repair port in the region, as well as a major shipping-related financial center. During the 1990s, Singapore's economy grew at the astounding rates of between 6 and 12 percent a year, making it one of the fastest-growing economies in the world. The Asian financial crisis of the late 1990s slashed growth to between 1.1 to 1.5 percent where it remained in 2004, despite government efforts to stimulate the economy through diversification.

HEALTH/WELFARE

About 85 percent of Singaporeans live in government-built dwellings. A government-created pension fund, the Central Provident Fund, takes up to one quarter of workers' paychecks; some of this goes into a compulsory savings account that can be used to finance the purchase of a residence. Other forms of social welfare are not condoned. Care of the elderly is the duty of the family, not the government.

In recent years, the government has aggressively sought out investment from non-shipping–related industries in order to diversify the economy. In 1992, Singapore hosted a summit of the Association of Southeast Asian Nations in which a decision was made to create a regional common market by the year 2008. In order to compete with the emerging European and North American regional trading blocs, it was decided that tariffs on products traded within the ASEAN region would be cut to 5 percent or less.

A UNIQUE CULTURE

Britain maintained an active interest in Singapore throughout its empire period. At its peak, some 100,000 British military men and their dependents were stationed on the island. The British military remained until 1971. (After the closure of U.S. military bases in the Philippines, logistics operations of the U.S. Navy's Seventh Fleet were transferred to Singapore, thereby increasing the number of U.S. military personnel in Singapore to about 300 persons.) Thus, British culture, from the architecture of the buildings, to the leisure of a cricket match, to the prevalence of the English language, is everywhere present in Singapore. Yet, because of the heterogeneity of the population (77 percent Chinese, 14 percent Malay, and 7 percent Indian), Singapore accommodates many philosophies and belief systems, including Confucianism, Buddhism, Islam, Hinduism, and Christianity. In recent years, the government has attempted to promote the Confucian ethic of hard work and respect for law, as well as the Mandarin Chinese language, in order to

develop a greater Asian consciousness among the people. But most Singaporeans seem content to avoid extreme ideology in favor of pragmatism; they prefer to believe in whatever approach works—that is, whatever allows them to make money and have a higher standard of living.

ACHIEVEMENTS

Housing remains a serious problem for many Asian countries, but virtually every Singaporean has access to adequate housing. Replacing swamplands with industrial parks has helped to lessen Singapore's reliance on its deepwater port. Singapore successfully overcame a Communist challenge in the 1950s to become a solid home for free enterprise in the region.

Their great material success has come with a price. The government keeps a firm hand on the people. For example, citizens can be fined as much as $250 for dropping a candy wrapper on the street or for driving without a seat belt. Worse offenses, such as importing or selling chewing gum, carry fines of $6000 and $1200 respectively. The chewing gum restriction—originally implemented because the former prime minister did not like discarded gum fouling sidewalks and train station platforms—was softened a little in 2004 when the government decided to allow gum to be imported for "medicinal" reasons. Thus, gum to help nicotine addicts break the smoking habit is now allowed, as are "dental" gum products. But there is still a catch: gum buyers must submit their names and government identification card numbers to the pharmacies (the only place gum can be sold) before they can buy a stick of gum, and if the pharmacists fail to record that information, they could be jailed for up to two years and fined nearly $3000. If gum buyers are treated with such Draconian measures, what about those who commit more serious crimes? Death by hanging is the punishment for murder, drug trafficking, and kidnapping; while lashing is inflicted on attempted murderers, robbers, rapists, and vandals. Being struck with a cane is the punishment for crimes such as malicious damage, as an American teenager, in a case that became a brief international cause célèbre in 1994, found out when he allegedly sprayed graffitti on cars in Singapore. Later that year, a Dutch businessperson was executed for alleged possession of heroin. The death penalty is required when one is convicted of using a gun in Singapore. The United Nations regularly cites Singapore for a variety of human-rights violations, and the world press frequently makes fun of the Singapore government

for such practices as giving prizes for the cleanest public toilet and for the incongruity of some laws. For example, a person has to register to buy a pack of nicotine gum but not to buy a pack of cigarettes or to visit a prostitute (which practice is legal in parts of Singapore).

If restrictions are the norm for ordinary citizens, they are even more severe for politicians, especially those who challenge the government. In recent years, opposition politicians have been fined and sent to jail for speaking in public without police permission, or for selling political pamphlets on the street without permission. Such actions can potentially bar would-be politicians from running for office for up to five years. Foreign media coverage of opposition parties is allowed only under strict guidelines, and in 1998, Parliament banned all political advertising on television. Government leaders argue that order and hard work are necessities

Timeline: PAST

A.D. 1200–1400
Singapore is controlled by several different nearby nations, including Thailand, Java, India, and Indonesia

1800s
British take control of the island

1942
The Japanese capture Singapore

1945
The British return to Singapore

1959
Full elections and self-government; Lee Kuan Yew comes to power

1963
Singapore, now unofficially independent of Britain, briefly joins the Malaysia Federation

1965
Singapore becomes an independent republic

1980s
Singapore becomes the second-busiest port in the world and achieves one of the highest per capita incomes in the Pacific Rim

1990s
The U.S. Navy moves some of its operations from the Philippines to Singapore; the Asian financial crisis briefly slows Singapore's economic growth

PRESENT

2000s
Singapore tries to position itself to become a regional financial center by liberalizing foreign investment in the banking sector

since, being a tiny island, Singapore could easily be overtaken by the envious and more politically unstable countries nearby; with few natural resources, Singapore must instead develop its people into disciplined, educated workers. Few deny that Singapore is an amazingly clean and efficient city-state; yet in recent years, younger residents have begun to clamor for more flexibility in their social lives—and their desires are producing changes in Singapore's image. Beginning in 2003, the Singapore government began allowing dance clubs to stay open all night, and young people now find they can get away with such things as body piercings, tattoos, and even tabletop dancing in clubs. Yet, government influence in people's private lives remains substantial; Singaporeans are regularly reminded via government-sponsored advertising campaigns to wash their hands, smile, speak properly, and never litter.

The law-and-order tone exists largely because after its separation from Malaysia in 1965, Singapore was controlled by one man and his personal hard-work ethic, Prime Minister Lee Kuan Yew, along with his Political Action Party (PAP). He remained in office for some 25 years, resigning in 1990, but his continuing role as "minister-mentor" gives him considerable clout. In 2000, for example, he was able to engineer the selection process for president in such a way that no election took place at all; his personal choice, S. R. Nathan, was simply appointed by the election board. Similarly, Lee saw to it that his personal preference for prime minister,

Goh Chok Tong, would, in fact, become his chosen successor. Goh had been the deputy prime minister and had been the designated successor-in-waiting since 1984. The transition was smooth, and the PAP's hold on the government remained intact. The strength of Lee's continuing influence on the government was revealed again when Prime Minister Goh announced that he would give up his seat to his deputy prime minister who was none other than Lee Hsin Loong, the son of Lee Kuan Yew. The transfer happened in 2004. Parliament is composed of 84 elected seats (and 9 appointed seats), and since 2001, the PAP has controlled all but 2 of them, with 75 per cent of the electors supporting the PAP. So strong is the party that its hold on the government was assured even before the election, because some 55 of its candidates ran unopposed!

The PAP originally came into prominence in 1959, when the issue of the day was whether Singapore should join the proposed Federation of Malaysia. Singapore joined Malaysia in 1963, but serious differences persuaded Singaporeans to declare independence two years later. Lee Kuan Yew, a Cambridge-educated, ardent anti-Communist with old roots in Singapore, gained such strong support as prime minister that not a single opposition-party member was elected for more than 20 years. Only two opposition seats exist today.

The two main goals of the administration have been to fully utilize Singapore's primary resource—its deepwater port—and to develop a strong Singaporean iden-

tity. The first goal has been achieved in a way that few would have thought possible; the question of national identity, however, continues to be problematic. Creating a Singaporean identity has been difficult because of the heterogeneity of the population, a situation that is only likely to increase as foreign workers are imported to fill gaps in the labor supply resulting from a very successful birth-control campaign started in the 1960s. Identity formation has also been difficult because of Singapore's seesaw history in modern times. First Singapore was a colony of Britain, then it became an outpost of the Japanese empire, followed by a return to Britain. Next Malaysia drew Singapore into its fold, and finally, in 1965, Singapore became independent. All these changes transpired within the lifetime of many contemporary Singaporeans, so their confusion regarding national identity is understandable.

Many still have a sense that their existence as a nation is tenuous, and they look for opportunities to build stabilizing networks. In 1996, Singapore reaffirmed its support for a five-nation defense agreement among itself, Australia, Malaysia, New Zealand, and Great Britain. It also strengthened its economic agreements with Australia. But the most controversial governmental decision was that of sending troops to support the U.S.-led invasion of Iraq. Singapore's most recent action in this regard was to send a troop landing ship with a crew of 180 to the Persian Gulf to conduct patrols and provide logistics support.

South Korea (Republic of Korea)

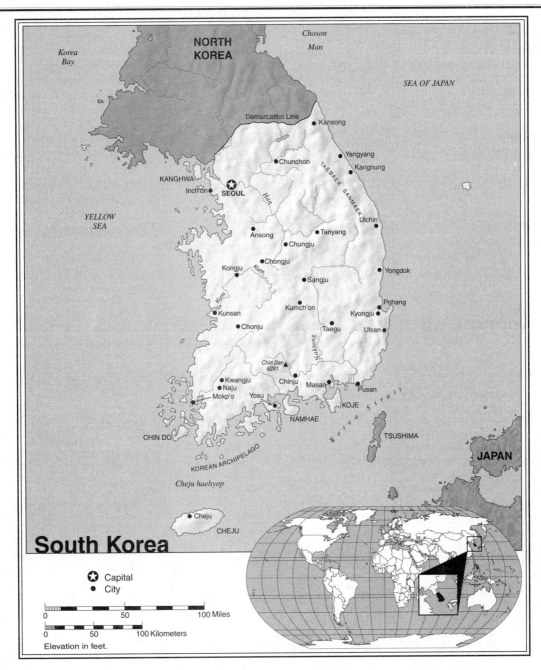

South Korea

○ Capital
● City

0   50   100 Miles
0   50   100 Kilometers
Elevation in feet.

South Korea Statistics

GEOGRAPHY

Area in Square Miles (Kilometers):
38,013 (98,480) (about the size of Indiana)
Capital (Population): Seoul (9,862,000)
Environmental Concerns: air and water pollution; overfishing
Geographical Features: mostly hills and mountains; wide coastal plains in west and south

Climate: temperate, with rainfall heaviest in summer

PEOPLE

Population

Total: 48,598,175
Annual Growth Rate: 0.62%
Rural/Urban Population (Ratio): 19/81
Major Language: Korean

Ethnic Makeup: homogeneous Korean
Religions: 49% Christian; 47% Buddhist; 3% Confucian; 1% Shamanist and Chondogyo

Health

Life Expectancy at Birth: 71 years (male); 79 years (female)
Infant Mortality: 7.18/1,000 live births
Physicians Available: 1/784 people
HIV/AIDS Rate in Adults: 0.01%

Education

Adult Literacy Rate: 98%
Compulsory (Ages): 6–12; free

COMMUNICATION

Telephones: 24,000,000 main lines
Daily Newspaper Circulation: 404/1,000 people
Televisions: 233/1,000 people
Internet Users: 25,600,000 (2002)

TRANSPORTATION

Highways in Miles (Kilometers): 53,940 (87,000)
Railroads in Miles (Kilometers): 1,874 (3,124)
Usable Airfields: 103
Motor Vehicles in Use: 10,420,000

GOVERNMENT

Type: republic

Independence Date: August 15, 1948 (from Japan)
Head of State/Government: President Roh Moo-hyun; Prime Minister Lee Hae-chan
Political Parties: Grand National Party; United Liberal Democratic Party; Millennium Democratic party
Suffrage: universal at 20

MILITARY

Military Expenditures (% of GDP): 2.7%
Current Disputes: Demarcation Line disputed with North Korea; Liancourt Rocks, claimed by Japan

ECONOMY

Currency ($ U.S. Equivalent): 1,191 won = $1
Per Capita Income/GDP: $17,800/$858 billion
GDP Growth Rate: 3.1%

Inflation Rate: 3.6%
Unemployment Rate: 3.4%
Labor Force by Occupation: 69% services; 22% industry; 10% agriculture
Natural Resources: coal; tungsten; graphite; molybdenum; lead; hydropower
Agriculture: rice; root crops; barley; vegetables; fruit; livestock; fish
Industry: textiles; clothing; footwear; food processing; chemicals; steel; electronics; automobile production; shipbuilding
Exports: $201.3 billion (primary partners United States, China, Japan)
Imports: $175.6 billion (primary partners Japan, United States, China)

SUGGESTED WEBSITES

http://memory.loc.gov/frd/cs/
   krtoc.html
http://www.skas.org

South Korea Country Report

EARLY HISTORY

Korea was inhabited for thousands of years by an early people who may or may not have been related to the Ainus of northern Japan, the inhabitants of Sakhalin Island, and the Siberian Eskimos. Distinct from this early civilization are today's Koreans whose ancestors migrated to the Korean Peninsula from Central Asia and, benefiting from close contact with the culture of China, established prosperous kingdoms as early as 1000 B.C. (legends put the date as early as 2333 B.C.). By A.D. 935, the Koryo Dynasty had established itself in Korea (the word *Korea* comes from the Koryo Dynasty). It was famous for, among other things, the invention of the world's first movable type printer.

DEVELOPMENT

The South Korean economy was so strong in the 1980s and early 1990s that many people thought Korea was going to be the next Japan of Asia. The standard of living was increasing for everyone until a major slowdown in 1997–1998. The resulting difficulties forced companies to abandon plans for wage increases or to decrease work hours. An industrial park to be built just across the border in North Korea will involve 1000 South Korean companies and draw billons of dollars in investment money.

The era of King Sejong (1418–1450) of the subsequent Choson or Yi Dynasty, is also notable for its many scientific and humanistic accomplishments. Ruling his subjects according to neo-Confucian thought, Sejong taught improved agricultural methods; published books on astronomy, history, religion, and medicine; and was instrumental in the invention of sundials, rain gauges to improve farming, and various musical instruments. Of singular importance was his invention of *hangul*, a simplified writing system that even uneducated peasants could easily learn. Before hangul, Koreans had to use the more complicated Chinese characters to represent sounds in the Korean language. Today, hangul is used in both North and South Korea, although some Chinese characters remain in use in newspapers and in family names.

REGIONAL RELATIONS

Except for the years 1905–1945, Korea has remained at least nominally independent of foreign powers. China, however, always wielded tremendous cultural influence and at times politically dominated the Korean Peninsula. Similarly, Japan often cast longing eyes toward Korea, controlling it directly for 40 years.

Korean influence on Japanese culture was pronounced in the 1400s and 1500s, when, through peaceful trade as well as forced labor, Korean artisans and technicians taught advanced skills in ceramics, textiles, painting, and other arts to the Japanese. (Historically, the Japanese received most of their cultural influence from China via Korea.)

FREEDOM

Suppression of political dissent, manipulation of the electoral process, and restrictions on labor union activity have been features of almost every South Korean government since 1948. Martial law has been frequently invoked, and governments have been overthrown by mass uprisings of the people. Reforms have been enacted under Presidents Roh Tae-woo, Kim Young-sam, and Kim Dae Jung, and are likely to continue under Roh Moo-hyun.

In this century, the direction of influence reversed—to the current Japan-to-Korea flow—with the result that the two cultures share numerous qualities. Ironically, cultural closeness has not eradicated Emotional distance: Some Japanese people today discriminate against Koreans who reside in Japan, while Japanese brutality during the four decades of Japan's occupation of Korea remains a frequent topic of conversation among Koreans. The Japanese emperor appears to want warmers ties between the two nations, for he recently announced that one of his royal ancestors was, in fact, a Korean.

Japan achieved its desire to rule Korea in 1905, when Russia's military, along with its imperialistic designs on Korea, was soundly defeated by the Japanese in the Russo–Japanese War; Korea was granted to Japan as part of the peace settlement. Unlike other expansionist nations, Japan was not content to rule Korea as a colony but, rather, attempted the complete cultural and political annexation of the Korean Peninsula. Koreans had to adopt Japanese names, serve in the Japanese Army, and pay homage to the Japanese emperor. Some 1.3 million Koreans were forcibly sent to Japan to work in coal mines or to serve in the military. The Korean language ceased to be taught in school, and more than 200,000 books on Korean history and culture were burned.

Many Koreans joined clandestine resistance organizations. In 1919, a "Declaration of Korean Independence" was announced in Seoul by resistance leaders, but the brutally efficient Japanese police and military crushed the movement. They killed thousands of demonstrators, tortured and executed the leaders, and set fire to the homes of those suspected of cooperating with the movement. Despite suppression of this kind throughout the 40 years of Japanese colonial rule, a provisional government was established by the resistance in 1919, with branches in Korea, China, and Russia. However, a very large police force—one Japanese for every 40 Koreans—kept active resistance inside Korea in check.

HEALTH/WELFARE

Korean men usually marry at about age 27, women at about 24. In 1960, Korean women, on average, gave birth to 6 children; in 1990, the expected births per woman were 1.6. The average South Korean baby born today can expect to live well into its 70s. More than 90% of all infants are immunized.

One resistance leader, Syngman Rhee, vigorously promoted the cause of Korean independence to government leaders in the United States and Europe. Rhee, supported by the United States, became the president of South Korea after the defeat of the Japanese in 1945.

Upon the surrender of Japan, the victorious Allied nations decided to divide Korea into two zones of temporary occupation, for the purposes of overseeing the orderly dismantling of Japanese rule and establishing a new Korean government. The United States was to occupy all of Korea south of the 38th Parallel of latitude (a demarcation running east and west across the peninsula, north of the capital city of Seoul), while the Soviet Union was to occupy Korea north of that line. The United States was uneasy about permitting the Soviets to move troops into Korea, as

the Soviet Union had entered the war against Japan just eight days before Japan surrendered, and its commitment to the democratic intentions of the Allies was questionable. Nevertheless, it was granted occupation rights.

Later, the United Nations attempted to enter the zone occupied by the Soviet Union in order to oversee democratic elections for all of Korea. Denied entry, UN advisers proceeded with elections in the South, which brought Syngman Rhee to the presidency. The North created its own government, with Kim Il-Song at the head. Tensions between the two governments resulted in the elimination of trade and other contacts across the new border. This was difficult for each side, because the Japanese had developed industries in the North while the South had remained primarily agricultural. Each side needed the other's resources; in their absence, considerable civil unrest occurred. Rhee's government responded by suppressing dissent, rigging elections, and using strong-arm tactics on critics. Rhee was forced to resign in 1960. Autocratic rule, not unlike that imposed on Koreans by the colonial Japanese, remained the norm in South Korea for several decades after the Japanese withdrew. Citizens, particularly university students, have frequently taken to the streets—often risking their lives and safety—to protest human rights violations by the various South Korean governments. Equally stern measures were instituted by the Communist government in the North, so that despite a half-century of rule as independent nations, the repressive legacy of the Japanese police state remains firmly in place in North Korea and has only just begun to be replaced by solid democratic practices in the South.

AN ECONOMIC POWERHOUSE

Upon the establishment of two separate political entities in Korea, the North pursued a Communist model of economic restructuring. South Korea, bolstered by massive infusions of economic and military aid from the United States, pursued a decidedly capitalist strategy. The result of these choices is dramatic, with South Korea having emerged as a powerhouse economy with a rapidly modernizing lifestyle, and the North stagnating under the weight of starvation and despair. Before the Asian financial crisis of the late 1990s, it had been predicted that South Korea's per capita income would rival that in European countries by the year 2010. Predictions have since been revised downward, but South Korea's success in improving the living standards of its people has nonetheless been phenomenal. South Korea is currently

the eighth strongest economy in Asia, based on annual per capita GDP of US$17,800. About 81 percent of South Korean people live in urban centers, where they have access to good education and jobs. Manufacturing accounts for about 30 percent of the gross domestic product. Economic success and recent improvements in the political climate seem to be slowing the rate of outward migration. In recent years, some Koreans have even returned home after spending years abroad.

North Koreans, on the other hand, are finding life unbearable. Hundreds have defected, some via a "safe house" system through China (similar to the famous Underground Railroad of U.S. slavery days). Some military pilots have flown their jets across the border to South Korea. Food shortages are increasingly evident, and some reports indicate that as many as 2,000 hungry North Koreans attempt to flee into China each month.

Imbued with the Confucian work ethic, and following the Japanese business model, South Korean businesspeople work long hours (a 6-day work week was the law until 2003, when it was reduced to 5-days) and have it as their goal to capture market share rather than to gain immediate profit—that is, they are willing to sell their products at or below cost for several years in order to gain the confidence of consumers, even if they make no profit. Once a sizable proportion of consumers buy their product and trust its reliability, the price is raised to a profitable level.

During the 1980s and much of the 1990s, South Korean businesses began investing in other countries, and South Korea became a creditor rather than a debtor member of the Asian Development Bank, putting it in a position to loan money to other countries. There was even talk that Japan (which is separated from Korea by only 150 miles of ocean) was worried that the two Koreas would soon unify and thus present an even more formidable challenge to its own economy—a situation not unlike some Europeans' concern about the economic strength of a reunified Germany.

The magic ended, however, in late 1997, when the world financial community would no longer provide money to Korean banks. This happened because Korean banks had been making questionable loans to Korean *chaebol*, or business conglomerates, for so long, that the banks' creditworthiness came into question. With companies unable to get loans, with stocks at an 11-year low, and with workers eager to take to the streets in mass demonstrations against industry cutbacks, many businesses went under. One of the more well-known firms that went into receivership in 1998 was Kia Motors Corporation. In 2000, the French automaker Renault bought the failed Samsung Motors company. With unemployment reaching 7 percent, the government applied for a financial bailout—with all of its restrictions and forced closures of unprofitable businesses—from the International Monetary Fund. Workers deeply resented the belt-tightening required by the IMF (shouting "No to layoffs!" thousands of them threw rocks at police, who responded with tear gas and arrests), but IMF funding probably prevented the entire economy from collapsing.

Without a doubt, the most exciting economic news is the creation of a mammoth industrial park just across the border in Kaesong, North Korea. The plan is to build manufacturing and assembly plants for over 1000 South Korean companies who will, in turn, employ some 730,000 North Koreans, taking advantage of the cheaper labor costs across the border. Not only will this plan help the economies of both countries, but it also represents a significant breakthrough in cooperation between North and South Korea.

ACHIEVEMENTS

In 1992, Korean students placed first in international math and science tests. South Korea achieved self-sufficiency in agricultural fertilizers in the 1970s and continues to show growth in the production of grains and vegetables. The formerly weak industrial sector is now a strong component of the economy.

SOCIAL PROBLEMS

Despite some recent setbacks, the South Korean economic recovery is well on its way, and South Korea will continue to be an impressive showcase for the fruits of capitalism. Politically, however, the country has been wracked with problems. Under Presidents Syngman Rhee (1948–1960), Park Chung Hee (1964–1979), and Chun Doo Hwan (1981–1987), South Korean government was so centralized as to constitute a virtual dictatorship. Human-rights violations, suppression of workers, and other acts incompatible with the tenets of democracy were frequent occurrences. Student uprisings, military revolutions, and political assassinations became more influential than the ballot box in forcing a change of government. President Roh Tae-woo came to power in 1987, in the wake of a mass protest against civil-rights abuses and other excesses of the previous government. Students began mass protests against various candidates long before the 1992 elections that brought to office the first civilian president in more than 30 years, Kim

Young-sam. Kim was once a dissident himself and was victimized by government policies against free speech; once elected, he promised to make major democratic reforms. The reforms, however, were not good enough for thousands of striking subway workers, farmers, or students whose demonstrations against low pay, foreign rice imports, or the deployment of U.S. Patriot missiles in South Korea sometimes had to be broken up by riot police.

Replacing Kim Young-sam as president in 1998 was opposition leader Kim Dae Jung. Like his predecessor, Kim had once been a political prisoner. Convicted of sedition by a corrupt government and sentenced to die, Kim had spent 13 years in prison or house arrest, and then, like Nelson Mandela of South Africa, rose to defeat the system that had abused him. That the Korean people were ready for real democratic reform was also revealed in the 1996 trial of former president Chun Doo Hwan, who was sentenced to death (later commuted) for his role in a 1979 coup. In 2003, a former human-rights lawyer, Roh Moo-hyun, occupied the Blue House (the presidential residence). Promising to continue Kim's "sushine policy" toward North Korea while establishing a more equal relationship with the United States, Roh won the support of many younger voters.

But Roh, a liberal with a labor-friendly platform, found himself outnumbered in the National Assembly by conservatives who tried to have him impeached over election law violations and who successfully overrode his veto of legislation. So intense was the acrimony that one Roh supporter set himself afire with gasoline just outside the Assembly building. Legislators got into fistfights on the floor of the Assembly during the debate over impeachment. The impeachment was dismissed after Roh's Uri Party was able to gain a slight majority in the one-chamber National Assembly in the 2004 elections, with many of those having voted for his impeachment losing their seats. Now, for the first time since 1987, the same party controls both the presidency and the Assembly.

A primary focus of the South Korean government's attention at the moment is the several U.S. military bases in South Korea, currently home to approximately 40,000 U.S. troops. The government (and apparently most of the 48 million South Korean people), although not always happy with the military presence, believes that the U.S. forces are useful in deterring possible aggression from North Korea, which, despite an enfeebled economy, still invests massive amounts of its budget in its military. Many university students, however, are offended by the presence of these troops. They claim

that the Americans have suppressed the growth of democracy by propping up authoritarian regimes—a claim readily admitted by the United States, which believed during the Cold War era that the containment of communism was a higher priority. Strong feelings against U.S. involvement in South Korean affairs have precipitated hundreds of violent demonstrations, sometimes involving as many as 100,000 protesters. The United States' refusal to withdraw its forces from South Korea leaves the impression with many Koreans that Americans are hard-line, Cold War ideologues who are unwilling to bend in the face of changing international alignments. This image has been bolstered by U.S. president George W. Bush's harsh rhetoric toward North Korea and his decision to invade Iraq (which the South Korean government supported by sending several thousand troops, but which many Korean's strongly opposed). After denouncing North Korea for developing nuclear capabilities, the U.S. government was chagrined to learn that South Korea had done the same—but without the tongue lashing the North had received from President Bush.

In 1990, U.S. officials announced that in an effort to reduce its military costs in the post–Cold War era, the United States would pull out several thousand of its troops from South Korea and close three of its five air bases. When the Iraq War was launched, the redeployment of troops from South Korea to Iraq was speeded up. The United States also declared that it expected South Korea to pay more of the cost of the U.S. military presence, in part as a way to reduce the unfavorable trade balance between the two countries. The South Korean government agreed to build a new U.S. military base about 50 miles south of the capital city of Seoul, where current operations would be relocated. South Korea would pay all construction costs—estimated at about $1 billion—and the United States would be able to reduce its presence within the Seoul metropolitan area, where many of the anti-U.S. demonstrations take place. Despite the effort to minimize the "footprint" of the U.S. military in Seoul, anti-American sentiment seems to be on the rise. When U.S. military courts acquitted American soldiers who had hit and killed two Korean teenagers in 2002, protests erupted again. Students broke into military installations, a white-robed Protestant minister led a march to the U.S. Embassy holding a wooden cross, and some 20 Catholic priests launched a hunger strike. As is often the case, the most strident protests come from those members of South Korean society who are the most Westernized—Christians and students educated overseas.

The issue of the South's relationship with North Korea has occupied the attention of every government since the 1950s. The division of the Korean Peninsula left many families unable to visit or even communicate with relatives on the opposite side. Moreover, the threat of military incursion from the North forced South Korea to spend huge sums on defense. Both sides engaged in spying, counter-spying, and other forms of subversive activities. Most worrisome of all was that the two antagonists would not sign a peace treaty, meaning that they were technically still at war—since the 1950s.

In 2000, just at the moment when the North was once again engaged in saber-rattling, an amazing breakthrough occurred: The two sides agreed to hold a summit in P'yongyang, the North Korean capital. President Kim, who had vigorously pushed his sunshine policy and had already achieved some improvements in communications between the two halves of the peninsula, was greeted with cheering crowds and feted at state dinners in the North. His counterpart, the reclusive Kim Jong-Il, seemed ready to make substantial concessions, including opening the border for family visits, halting the nonstop broadcasting of anti-South propaganda, and seeking a solution for long-term peace and reunification. The impetus behind the dramatic about-face appears to be North Korea's dire economic situation and its loss of solid diplomatic and economic partners, now that the Communist bloc of nations no longer exists. Reunification will, of course, take many years to realize, and some estimate that it would cost close to a trillion dollars. Still, it is a goal that virtually every Korean wants. South Korea made the first installment toward this goal in 2002 when it agreed to give the North $25 million to rebuild rail and road links between the two countries—links that have been severed for more than 50 years.

South Korean government leaders have to face a very active, vocal, and even violent populace when they initiate controversial policies. Among the more vocal groups for democracy and human rights are the various Christian congregations and their Westernized clergy. Other vocal groups include the college students who hold rallies annually to elect student protest leaders and to plan antigovernment demonstrations. In addition to the military-bases question, student protesters are angry at the South Korean government's willingness to open more Korean markets to U.S. products. The students want the United States to apologize for its alleged assistance to the South Korean government in violently suppressing an antigovernment demonstration in 1981 in Kwangju, a southern city that is a frequent locus of antigovernment as well as labor-related demonstrations and strikes. Protesters were particularly angered by then-president Roh Tae-woo's silencing of part of the opposition by convincing two opposition parties to merge with his own to form a large Democratic Liberal Party, not unlike that of the Liberal Democratic Party that governed Japan almost continuously for more than 40 years.

Ironically, demands for changes have increased at precisely the moment that the government has been instituting changes designed to strengthen both the economy and civil rights. Under Roh's administration, for example, trade and diplomatic initiatives were launched with Eastern/Central European nations and with China and the former Soviet Union. Under Kim Young-sam's administration, 41,000 prisoners, including some political prisoners, were granted amnesty, and the powerful chaebol business conglomerates were brought under a tighter rein. Similarly, relaxation of the tight controls on labor-union activity gave workers more leverage in negotiating with management. Unfortunately, union activity, exploding after decades of suppression, has produced crippling industrial strikes—as many as 2,400 a year—and the police have been called out to restore order. In fact, since 1980, riot police have fired an average of more than 500 tear-gas shells a day, at a cost to the South Korean government of tens of millions of dollars.

Domestic unrest notwithstanding, the South Korean people seemed to have put aside the era of dictatorial government and are moving proactively to solidify their role on the world stage. South Korea recently established unofficial diplomatic ties with Taiwan in order to facilitate freer trade. It then signed an industrial pact with China to merge South Korea's technological know-how with China's inexpensive labor force. Relations with Japan have also improved since Japan's apology for atrocities during World War II.

KOREAN CULTURE

Koreans occupy a beautiful peninsula, and they have created a vibrant culture to match it. Mountains make up about 70 percent of the Korean peninsula, and some 3,500 mountaintops poke out of the surrounding Yellow Sea and Sea of Japan, making Korea also a land of small islands. Monsoon rains bring water to the rice paddies and to the barley, potatoes, and soybean fields; Siberian winds bring snow to the mountains. Families are central to Korean life, and many rural families consist of 2 or even 3 generations. Households of city dwellers, who often live in modern high-rise apartment buildings, are usually smaller. Although Western foods such as hamburgers and fries are available everywhere in Seoul, Koreans are famous for the hot and spicy foods of their own making: *kimchi*, made from pickled cabbage, garlic, onions, oysters, and other ingredients, and *bulgogi*, a dish in which thin strips of meat are soaked in a mixture of sesame oil, garlic, onions, and black pepper, dipped in soy sauce and eaten using chopsticks. Some Korean cultural forms, such as *taekwondo*, have been exported all over the world.

Taiwan

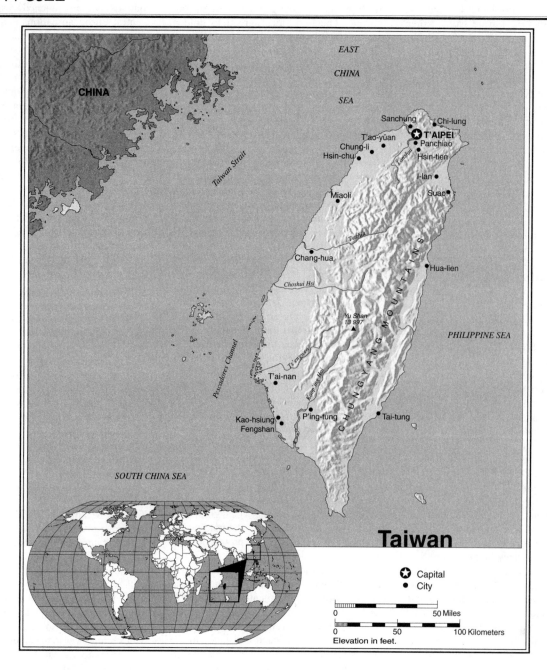

Taiwan Statistics

GEOGRAPHY

Area in Square Miles (Kilometers):
22,320 (36,002) (about the size of
Maryland and Delaware combined)
Capital (Population): Taipei (2,596,000)
Environmental Concerns: water and air
pollution; poaching; contamination of
drinking water; radioactive waste; trade
in endangered species

Geographical Features: mostly rugged
mountains in east; flat to gently rolling
plains in west
Climate: tropical; marine

PEOPLE

Population

Total: 22,749,838

Annual Growth Rate: 0.64%
Rural/Urban Population Ratio: 25/75
Major Languages: Mandarin Chinese;
Taiwanese and Hakka dialects also used
Ethnic Makeup: 84% Taiwanese; 14%
Mainlander Chinese; 2% aborigine
Religions: 93% mixture of Buddhism,
Confucianism, and Taoism; 4.5%
Christian; 2.5% others

Health

Life Expectancy at Birth: 74 years (male); 80 years (female)
Infant Mortality: 6.5/1,000 live births
Physicians Available: 1/867 people
HIV/AIDS Rate in Adults: na

Education

Adult Literacy Rate: 96%
Compulsory (Ages): 6–15; free

COMMUNICATION

Telephones: 12,500,000 main lines
Televisions: 327/1,000 people
Internet Users: 11,600,000 (2001)

TRANSPORTATION

Highways in Miles (Kilometers): 20,940 (34,901)
Railroads in Miles (Kilometers): 665 (1,108)
Usable Airfields: 39

Motor Vehicles in Use: 5,300,000

GOVERNMENT

Type: multiparty democratic regime
Head of State/Government: President Chen Shui-bian; Premier Yu Shyi-kun
Political Parties: Nationalist Party (Kuomintang); Democratic Progressive Party; Labour Party; People's First Party; others
Suffrage: universal at 20

MILITARY

Military Expenditures (% of GDP): 2.7%
Current Disputes: disputes with various countries over islands

ECONOMY

Currency ($ U.S. Equivalent): 34.55 New Taiwan dollars = $1
Per Capita Income/GDP: $23,400/$528 billion

GDP Growth Rate: 3.2%
Inflation Rate: 0.5%
Unemployment Rate: 4.5%
Labor Force by Occupation: 56% services; 36% industry; 8% agriculture
Population Below Poverty Line: 1%
Natural Resources: coal; natural gas; limestone; marble; asbestos
Agriculture: rice; tea; fruit; vegetables; corn; livestock; fish
Industry: steel; iron; chemicals; electronics; cement; textiles; food processing; petroleum refining
Exports: $143 billion (primary partners China, United States, Japan)
Imports: $119 billion (primary partners Japan, United States, China)

SUGGESTED WEBSITE

http://www.cia.gov/cia/
publications/factbook/geos/
tw.html

Taiwan Country Report

It has been called "beautiful island," "treasure island," and "terraced bay island," but to the people who have settled there, Taiwan (formerly known as Formosa) has come to mean "refuge island."

Typical of the earliest refugees of the island were the Hakka peoples of China, who, tired of persecution on the mainland, fled to Taiwan (and to Borneo) before A.D. 1000. In the seventeenth century, tens of thousands of Ming Chinese soldiers, defeated at the hands of the expanding Manchu Army, sought sanctuary in Taiwan. In 1949, a third major wave of immigration to Taiwan brought thousands of Chinese Nationalists, retreating in the face of the victorious Red Chinese armies. Hosting all these newcomers were the original inhabitants of the islands, various Malay-Polynesian-speaking tribes. Their descendants live today in mountain villages throughout the island, where some have been since 3000 B.C. These original inhabitants, generally grouped into about ten tribes, often feel neglected or abused by modern Taiwanese society, and about 2000 of them, dressed in traditional clothing, protested in Taipei in 2004 when the vice-president made statements about them they felt were untrue. They often resent the taking over of their island by so many Chinese people.

Since 1544, other outsiders have shown interest in Taiwan, too: Portugal, Spain, the Netherlands, Britain, and France have all

DEVELOPMENT

Taiwan has vigorously promoted export-oriented production, particularly of electronic equipment. In the 1980s, manufacturing became a leading sector of the economy, employing more than one third of the workforce. Virtually all Taiwanese households own color televisions, and other signs of affluence are abundant. Taiwan became the 144th member of the WTO.

either settled colonies or engaged in trade along the coasts. But the non-Chinese power that has had the most influence is Japan. Japan treated parts of Taiwan as its own for 400 years before it officially acquired the entire island in 1895, at the end of the Sino–Japanese War. From then until 1945, the Japanese ruled Taiwan with the intent of fully integrating it into Japanese culture. The Japanese language was taught in schools, college students were sent to Japan for their education, and the Japanese style of government was implemented. Many Taiwanese resented the harsh discipline imposed, but they also recognized that the Japanese were building a modern, productive society. Indeed, the basic infrastructure of contemporary Taiwan—roads, railways, schools, and so on—was constructed during the Japanese colonial era (1895–1945). Japan still lays claim to the Senkaku Islands, a chain of uninhabited is-

lands near Taiwan that the Taiwanese say belong to Taiwan.

In 1949, after Communist leader Mao Zedong defeated his rivals on the mainland, Taiwan became the island of refuge of the anti-Communist leader Chiang Kaishek and his 3 million Kuomintang (KMT, or Nationalist Party) followers, many of whom had been prosperous and well-educated businesspeople and intellectuals in China. These Mandarin-speaking mainland Chinese, called Mainlanders, now constitute about 14 percent of Taiwan's people.

During the 1950s, Mao Zedong, the leader of the People's Republic of China, planned an invasion of Taiwan. However, Taiwan's leaders succeeded in obtaining military support from the United States to prevent the attack. They also convinced the United States to provide substantial amounts of foreign aid to Taiwan (the U.S. government saw the funds as a way to contain communism) as well as to grant it diplomatic recognition as the only legitimate government for all of China. For many years Taiwan maintained offices in its government for every province in China, even though no one could ever go there.

China was denied membership in the United Nations for more than 20 years because Taiwan held the "China seat." World opinion on the "two Chinas" issue began to change in the early 1970s. Many countries

(UN photo by Chen Jr.)

Taiwan has one of the highest population densities in the world, but it has been able to expand its agricultural output rapidly and efficiently through utilization of a number of practices. By terracing, using high-yield seeds, and supplying adequate irrigation, Taiwanese can grow a succession of crops on the same piece of land throughout the year.

believed that a nation as large and powerful as the People's Republic of China should not be kept out of the United Nations, nor out of the mainstream of world trade, in favor of the much smaller Taiwan. In 1971, the United Nations withdrew its China seat from Taiwan and gave it to the P.R.C. Taiwan has consistently re-applied for membership, arguing that there is nothing wrong with there being either two China seats or one China seat and one Taiwan seat; its requests have been denied each time for some 12 years, most recently in 2004, when no country voted for its admission.

FREEDOM

For nearly 4 decades, Taiwan was under martial law. Opposition parties were not tolerated, and individual liberties were limited. A liberalization of this pattern began in 1986. Taiwan now seems to be on a path toward greater democratization. In 1991, 5,574 prisoners, including many political prisoners, were released in a general amnesty.

The United States and many other countries wished to establish diplomatic relations with China but could not get China to cooperate as long as they recognized the

sovereignty of Taiwan. In 1979, desiring access to China's huge market, the United States, preceded by many other nations, switched its diplomatic recognition from Taiwan to China. Foreign-trade offices in Taiwan remained unchanged but embassies were renamed; the U.S. Embassy, for example, was called the American Institute in Taiwan. As far as official diplomacy with the United States was concerned, Taiwan became a non-nation, but that did not stop the two countries from engaging in very profitable trade, including a controversial U.S. agreement in 1992 to sell $4 billion to $6 billion worth of F-16 fighter jets to Taiwan. Similarly, Taiwan has refused to establish diplomatic ties, yet continues to trade, with nations that recognize the mainland Chinese authorities as a legitimate government. In 1992, for instance, when South Korea established ties with mainland China, Taiwan immediately broke off formal relations with South Korea and suspended direct airline flights. However, Taiwan continued to permit trade in many commodities. Recognizing a potentially strong market in Vietnam, Taiwan also established air links with Vietnam in 1992, links that had been broken since the end of the Vietnam War. In 1993, a Taiwanese company collaborated with the Vietnamese government to construct a

$242 million highway in Ho Chi Minh City (formerly Saigon). Just over 30 states formally recognize Taiwan today, but Taiwan nevertheless maintains close economic ties with more than 140 countries. Of special interest to Taiwan are those countries with large populations of "overseas Chinese," or people of Chinese background who permanently live abroad. This community is estimated to number in excess of 33 million people, of whom 27 million live in the countries of Southeast Asia and 3 million live in the United States. Given that many of these overseas Chinese have become quite successful in business, Taiwan, which established a "Go South Policy" toward Southeast Asia, sees them as a source of potential investment as well as a way to bypass the diplomatic restrictions under which it suffers due to the United Nations' rejection of Taiwan as the singular, legitimate government of China. On this point, the rivalry between the mainland and Taiwan remains quite intense.

AN ECONOMIC POWERHOUSE

Diplomatic maneuvering has not affected Taiwan's stunning postwar economic growth. Like Japan, Taiwan has been described as an economic miracle. In the past two decades, Taiwan has enjoyed more years of double-digit economic growth than any other nation. With electronics leading the pack of exports, a substantial portion of Taiwan's gross domestic product comes from manufacturing. Taiwan has been open to foreign investment and, of course, to foreign trade. However, for many years, Taiwan insisted on a policy of no contact and no communication with mainland China. Private enterprises eventually were allowed to trade with China—as long as the products were transshipped through a third country, usually Hong Kong. In 1993, government-owned enterprises such as steel and fertilizer plants were allowed to trade with China, on the same condition. In 2000, the Legislature lifted the ban on direct trade, transportation, and communications with China from some of the Taiwanese islands nearest to the mainland. Little by little, the door toward China is opening. Taiwanese business leaders want direct shipping, aviation, and communication links with all of China. By the early 1990s, Taiwanese trade with China had exceeded $13 billion a year and China had become Taiwan's seventh-largest trading partner. The liberalization of trade between China (especially its southern and coastal provinces), Taiwan, and Hong Kong has made the region, now known as Greater China, an economic dynamo. Economists predict that Greater China will someday pass the United States to become the largest economy in the world.

HEALTH/WELFARE

Taiwan has one of the highest population densities in the world. Education is free and compulsory to age 15, and the country boasts more than 100 institutions of higher learning. Social programs, however, are less developed than those in Singapore, Japan, and some other Pacific Rim countries.

As one of the newly industrializing countries of Asia, Taiwan certainly no longer fits the label "underdeveloped." Taiwan holds large stocks of foreign reserves and carries a trade surplus with the United States (in Taiwan's favor) far greater than Japan's, when counted on a per capita basis. The Taipei stock market has been so successful—sometimes outperforming both Japan and the United States—that a number of workers reportedly have quit their jobs to play the market, thereby exacerbating Taiwan's already serious labor shortage. (This shortage has led to an influx of foreign workers, both legal and illegal.)

Successful Taiwanese companies have begun to invest heavily in other countries where land and labor are plentiful and less expensive. In 1993, the Philippines accepted a $60 million loan from Taiwan to build an industrial park and commercial port at Subic Bay, the former U.S. naval base; and Thailand, Australia, and the United States have also seen inflows of Taiwanese investment monies. By the early 1990s, some 200 Taiwanese companies had invested $1.3 billion in Malaysia alone (Taiwan supplanted Japan as the largest outside investor in Malaysia). Taiwanese investment in mainland China has also increased.

Taiwan's economic success is attributable in part to its educated population, many of whom constituted the cultural and economic elite of China before the Communist revolution. Despite resentment of the mainland immigrants by native-born Taiwanese, everyone, including the lower classes of Taiwan, has benefited from this infusion of talent and capital. Yet the Taiwanese people are beginning to pay a price for their sudden affluence. Taipei, the capital, is awash not only in money but also in air pollution and traffic congestion. The traffic in Taipei is rated near the worst in the world. Concrete high-rises have displaced the lush greenery of the mountains. Many residents spend their earnings on luxury foreign cars and on cigarettes and alcohol, the consumption rate of which has been increasing by about 10 percent a year. Many Chinese traditions—for instance, the roadside restaurant serving noodle soup—are giving way to 7-Elevens selling Coca-Cola and ice cream.

Some Taiwanese despair of ever turning back from the growing materialism; they wish for the revival of traditional Chinese (that is, mostly Confucian) ethics. They doubt that it will happen. Still, the government, which has been dominated almost continuously since 1949 by the conservative Mandarin migrants from the mainland, sees to it that Confucian ethics are vigorously taught in school. And there remains in Taiwan more of traditional China than in China itself, because, unlike the Chinese Communists, the Taiwanese authorities have had no reason to attempt an eradication of the values of Buddhism, Taoism, or Confucianism. Nor has grinding poverty—often the most serious threat to the cultural arts—negatively affected literature and the fine arts, as it has in China. Parents, with incense sticks burning before small religious altars, still emphasize respect for authority, the benefits of harmonious cooperative effort, and the inestimable value of education. Traditional festivals dot each year's calendar, among the most spectacular of which is Taiwan's National Day parade. Marching bands, traditional dancers, and a huge dragon carried by more than 50 young men please the crowds lining the streets of Taipei. Temples are filled with worshipers praying for health and good luck.

But the Taiwanese will need more than luck if they are to escape the consequences of their intensely rapid drive for material comfort. Some people contend that the island of refuge is being destroyed by success. Violent crime, for instance, once hardly known in Taiwan, is now commonplace. Six thousand violent crimes, including rapes, robberies, kidnappings, and murder, were reported in 1989—a 22 percent increase over the previous year, and the upward trend has continued since then. Extortion against wealthy companies and abductions of the children of successful families have created a wave of fear among the rich. Travelers at Chiang Kai Shek international airport find they are targets of pickpockets, who rob about 10 victims per day, on average.

ACHIEVEMENTS

From a largely agrarian economic base, Taiwan has been able to transform its economy into an export-based dynamo with international influence. Today, only about 10% of the population work in agriculture, and Taiwan ranks among the top 20 exporters in the world.

Like other Asian countries, Taiwan was affected by the Asian financial crisis of the late 1990s. Labor shortages have forced some companies to operate at only 60 percent of capacity, and low-interest loans are hard to get because the government fears that too many people will simply invest in get-rich stocks instead of in new businesses. In 2001, Taiwan recorded its first-ever year of negative growth, but slowly, the recession is fading, and Taiwan expects annual GDP growth to be about 4 percent for 2004 and beyond.

POLITICAL LIBERALIZATION

In recent years, the Taiwanese people have had much to be grateful for in the political sphere. Until 1986, the government, dominated by the influence of the Chiangs, had permitted only one political party, the Nationalists, and had kept Taiwan under martial law for nearly 4 decades. A marked political liberalization began near the time of Chiang Ching-Kuo's death, in 1987. The first opposition party, the Democratic Progressive Party, was formed; martial law (officially, the "Emergency Decree") was lifted; and the first two-party elections were held, in 1986. In 1988, for the first time a native-born Taiwanese, Lee Teng-hui, was elected to the presidency. He was reelected in 1996 in the first truly democratic, direct presidential election ever held in Taiwan. Although Lee never promoted the independence of Taiwan, his high-visibility campaign raised the ire of China, which attempted to intimidate the Taiwanese electorate into voting for a more pro-China candidate by conducting military exercises and firing missiles just 20 miles off the coast of Taiwan. As expected, the intimidation backfired, and Lee soundly defeated his opponents.

Under the Nationalists, it was against the law for any group or person to advocate publicly the independence of Taiwan—that is, to advocate international acceptance of Taiwan as a sovereign state, separate and apart from China. When the opposition Democratic Progressive Party (DPP) resolved in 1990 that Taiwan should become an independent country, the ruling Nationalist government immediately outlawed the DPP platform.

But times are changing. In 2000, after 50 years of Nationalist Party rule in Taiwan, the DPP was able to gain control of the presidency. Chen Shui-bian, a native-born Taiwanese, was elected. Both Chen and his running mate, Annette Lu, who were barely re-elected in a hotly contested 2004 election after an assassination attempt on their lives, had spent time in prison for activities that had angered the ruling Nationalists. Although his party had openly sought independence, Chen toned down that rhetoric and instead invited the mainland to begin talks for reconciliation. As usual, China had threatened armed intervention if Taiwan declared independence; but after the vote for Chen, the mainland seemed to moderate its position. The bilateral talks that were initiated in 1998 continue to hold promise. Some believe that such talks will eventually result in Taiwan being annexed by China, just as in the cases of Hong Kong and Macau (although from a strictly legalistic viewpoint, Taiwan has just as much right to annex China). Others believe that dialogue will eventually diminish animosity, allowing Taiwan to move toward independence without China's opposition. China's position on Taiwan ("It belongs to China and we are going to get it back") seems clearer to the world than does Taiwan's position ("We sort of want independence, but we won't say so, and we would sure like to

have close business ties with China"). Taiwan has recently tried to increase the pressure on China to accept reality. In a video conference aired in Japan, President Chen came as close to declaring independence as has ever been done, stating that in reality separate countries already exist on both sides of the 100-mile Taiwan Strait. His government has also upped the pressure on China's dismal human-rights record, issuing a critical report on rights abuses in the mainland and warning that relations between the two sides will not improve until human rights are protected on the mainland. He also proposed a national referendum that would have demanded of China that it redeploy the 500 missiles it has aimed at Taiwan (although he later softened the proposed wording, and also backpedaled on a proposed new constitution that would have taken effect in 2008 and probably would have included language suggesting the near independence of Taiwan. Still, the mainland seems to have the upper hand on most matters. For instance, the World Trade Organization succumbed to Chinese pressure not to allow Taiwan into the organization until China itself had been admitted, even though Taiwan had qualified years earlier. Thus, China became the 143rd member of the WTO, and Taiwan (technically "Chinese Taipei") became the 144th. Moreover, China continues to block Taiwan's membership in many other international organizations.

Opinion on the independence issue is clearly divided. Even some members of the anti-independence Nationalist Party have bolted and formed a new party (the New KMT Alliance, or the New Party) to promote closer ties with China. As opposition parties proliferate, the independence issue could become a more urgent topic of political debate. In the meantime, contacts with the P.R.C. increase daily; Taiwanese students are now being admitted to China's universities, and Taiwanese residents by the thousands now visit relatives on the mainland. After Chen's re-election in 2004, he even proposed a formal visit to China by himself and other leaders to discuss cooperation on such things as emergency military hotlines. Despite complaints from China Taiwanese government leaders have been courting their counterparts in the Philippines, Thailand, Indonesia, and South Korea. Despite a 2002 invitation from China for high-level visits from Taiwanese leaders, China continues to hold firm to its vow

to invade Taiwan if it should ever declare independence. Under these circumstances, many—probably most—Taiwanese will likely remain content to let the rhetoric of reunification continue while enjoying the reality of de facto independence.

Timeline: PAST

A.D. 1544
Portuguese sailors are the first Europeans to visit Taiwan

1700s
Taiwan becomes part of the Chinese Empire

1895
The Sino-Japanese War ends; China cedes Taiwan to Japan

1945
Taiwan achieves independence from Japan

1947–49
Nationalists, under Chiang Kai-shek, retreat to Taiwan

1950s
A de facto separation of Taiwan from China; Chinese aggression is deterred with U.S. assistance

1971
China replaces Taiwan in the United Nations

1975
Chiang Kai-shek dies and is succeeded by his son, Chiang Ching-Kuo

1980s
The first two-party elections in Taiwan's history are held; 38 years of martial law end

1990s
Relations with China improve; the United States sells F-16 jets to Taiwan

China conducts military exercises to intimidate Taiwanese voters

PRESENT

2000s

Trade and communication with China continue to expand

The opposition Democratic Progessive Party wins the presidency

Pro-independence president Chen Shubian is re-elected in 2004, but anti-independence candidates gain enough strength to deny his party a majority in the legislature.

Thailand (Kingdom of Thailand)

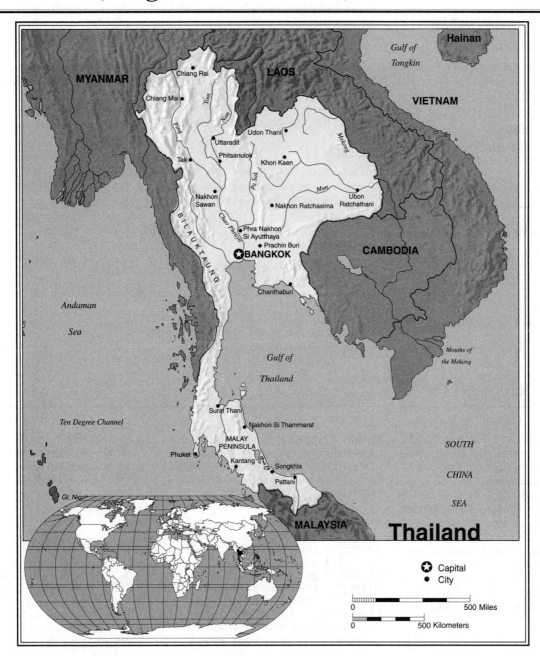

Thailand

GEOGRAPHY

Area in Square Miles (Kilometers):
198,404 (514,000) (about twice the size
of Wyoming)
Capital (Population): Bangkok
(7,527,000)
Environmental Concerns: air and water
pollution; poaching; deforestation; soil
erosion

Geographical Features: central plain;
Khorat Plateau in the east; mountains
elsewhere
Climate: tropical monsoon

PEOPLE

Population

Total: 64,865,523

Annual Growth Rate: 0.91%

Rural/Urban Population Ratio: 79/21

Major Languages: Thai; English; various
dialects

Ethnic Makeup: 75% Thai; 14% Chinese;
11% Malay and others

Religions: 95% Buddhist; 4% Muslim; 1%
others

Health

Life Expectancy at Birth: 66 years (male);
73 years (female)
Infant Mortality: 21.1/1,000 live births
Physicians Available: 1/4,165 people
HIV/AIDS Rate in Adults: 1.5%

Education

Adult Literacy Rate: 92%
Compulsory (Ages): 6–15

COMMUNICATION

Telephones: 5,600,000 main lines
Daily Newspaper Circulation: 47/1,000
people
Televisions: 56/1,000 people
Internet Users: 1,200,000 (2001)

TRANSPORTATION

Highways in Miles (Kilometers): 38,760
(64,600)
Railroads in Miles (Kilometers): 2,364
(3,940)
Usable Airfields: 110
Motor Vehicles in Use: 5,650,000

GOVERNMENT

Type: constitutional monarchy
Independence Date: founding date 1238;
never colonized
Head of State/Government: King
Phumiphon Adunyadet; Prime Minister
Thaksin Chinnawat
Political Parties: Thai Nation Party;
Democratic Party; National
Development Party; many others
Suffrage: universal and compulsory at 18

MILITARY

Military Expenditures (% of GDP): 1.8%
Current Disputes: boundary disputes with
Laos, Cambodia, and Myanmar; refugee
disputes

ECONOMY

Currency ($ U.S. Equivalent): 42.0 baht =
$1
Per Capita Income/GDP: $7,400/$477
billion
GDP Growth Rate: 6.7%

Inflation Rate: 1.8%
Unemployment Rate: 2.2%
Labor Force by Occupation: 54%
agriculture; 31% services; 15% industry
Population Below Poverty Line: 13%
Natural Resources: tin; rubber; natural
gas; tungsten; tantalum; timber; lead;
fish; gypsum; lignite; fluorite; arable
land
Agriculture: rice; cassava; rubber; corn;
sugarcane; coconuts; soybeans
Industry: tourism; textiles and garments;
agricultural processing; beverages;
tobacco; cement; electric appliances and
components; electronics; furniture;
plastics
Exports: $75.9 billion (primary partners
United States, Japan, China)
Imports: $65.3 billion (primary partners
Japan, United States, China)

SUGGESTED WEBSITE

http://www.mahidol.ac.th/
Thailand/Thailand-main.html

Thailand

One afternoon in late 2004, more than 50 Thai warplanes roared over the southern Thailand provinces of Yala, Pattani, and Narathiwat, where violence by Islamic extremists had taken the lives of over 500 people in several months of unrest. Instead of bombs, however, the air force dropped 120 million paper cranes, folded for the people of southern Thailand by their countrymen in the north. Like symbolic white doves of peace in Western cultures, cranes are Asian symbols of peace and reconciliation, and the airdrop was intended to show that, despite religious and other differences, Thai people care for one another. The prime minister, who promised to restore a mosque damaged in the violence, reported that the airdrop had an "enormous positive psychological effect," and that once people knew what was happening, crowds started lining the roads to catch the paper birds. Such is the kind of leadership that has made Thailand unique among Southeast Asian countries.

The roots of Thai culture extend into the distant past. People were living in Thailand at least as early as the Bronze Age. By the time Thai people from China (some scholars think from as far away as Mongolia) had established the first Thai dynasty in the Chao Phya Valley, in A.D. 1238, some communities, invariably with a Buddhist

temple or monastery at their centers, had been thriving in the area for 600 years. Early Thai culture was greatly influenced by Buddhist monks and traders from India and Sri Lanka (Ceylon).

DEVELOPMENT

Many Thais are small-plot or tenant farmers, but the government has energetically promoted economic diversification. Despite high taxes, Thailand has a reputation as a good place for foreign investment. Electronics and other high-tech industries from Japan, the United States, and other countries have been very successful in Thailand.

By the seventeenth century, Thailand's ancient capital, Ayutthaya, boasted a larger population than that of London. Ayutthaya was known around the world for its wealth and its architecture, particularly its religious edifices. Attempts by European nations to obtain a share of the wealth were so inordinate that in 1688, the king expelled all foreigners from the country. Later, warfare with Cambodia, Laos, and Malaya yielded tremendous gains in power and territory for Thailand, but it was periodically afflicted by Burma (present-day Myan-

mar), which briefly conquered Thailand in the 1760s (as it had done in the 1560s). The Burmese were finally defeated in 1780, but the destruction of the capital required the construction of a new city, near what, today, is Bangkok.

FREEDOM

Since 1932, when the absolute monarchy was abolished, Thailand has endured numerous military coups and countercoups, most recently in February 1991. Combined with the threat of Communist insurgencies, these have resulted in numerous declarations of martial law, press censorship, and suspensions of civil liberties. Thai censors banned the movie *Anna and the King* in 1999, just as they had done *The King and I* in 1956.

Generally speaking, the Thai people have been blessed over the centuries with benevolent kings, many of whom have been open to new ideas from Europe and North America. Gathering around them advisers from many nations, they improved transportation systems, education, and farming while maintaining the central place of Buddhism in Thai society. Occasionally royal support for religion overtook

Buddhism is an integral part of the Thai culture. Six hundred years ago, Buddhist monks traveled from India and Ceylon (present-day Sri Lanka) and built temples and monasteries throughout Thailand. These newly ordained monks are meditating in the courtyard of a temple in Bangkok.

other societal needs, at the expense of the power of the government.

The gravest threat to Thailand came during the era of European colonial expansion, but Thailand—whose name means "Free Land"—was never completely conquered by European powers. Today, the country occupies a land area about the size of France.

MODERN POLITICS

Since 1932, when a constitutional monarchy replaced the absolute monarchy, Thailand (formerly known as Siam) has weathered 17 attempted or successful military or political coups d'état, most recently in 1991. The Constitution has been revoked and replaced numerous times; governments have fallen under votes of no-confidence; students have mounted violent demonstrations against the government; and the military has, at various times, imposed martial law or otherwise curtailed civil liberties.

Clearly, Thai politics are far from stable. Nevertheless, there is a sense of stability in Thailand. Miraculously, its people were spared the direct ravages of the Vietnam War, which raged nearby for 20 years. Despite all the political upheavals, the same royal family has been in control of the Thai throne for nine generations, although its power has been severely delimited for some 60 years. Furthermore, before the first Constitution was enacted in 1932, the country had been ruled continuously, for more than 700 years, by often brilliant and progressive kings. At the height of Western imperialism, when France, Britain, the Netherlands, and Portugal were in control of every country on or near Thailand's borders, Thailand remained free of Western domination, although it was forced—sometimes at gunpoint—to relinquish sizable chunks of its holdings in Cambodia and Laos to France, and holdings in Malaya to Britain. The reasons for this singular state of independence were the diplomatic skill of Thai leaders, Thai willingness to Westernize the government,

and the desire of Britain and France to let Thailand remain interposed as a neutral buffer zone between their respective armies in Burma and Indochina.

HEALTH/WELFARE

About 2,000 Thais out of every 100,000 inhabitants attend college (as compared to only 200 per 100,000 Vietnamese). Thailand has devoted substantial sums to the care of refugees from Cambodia and Vietnam. The rate of nonimmigrant population growth has dropped substantially since World War II. HIV/AIDS is a significant problem in Thailand, as is bird flu, which has infected thousands of chickens, ducks, Bengal tigers, and some humans.

The current king, Phumiphon Adunyadet, born in the United States and educated in Switzerland, is highly respected as head of state. The king is also the nominal head of the armed forces, and his support is critical to any Thai government.

(Photo by Lisa Clyde Nielsen)

Bangkok is one of the largest cities in the world. The city is interlaced with canals, and the population crowds along river banks. With the enormous influx of people who are lured by industrialization and economic opportunity, the environment has been strained to the limit.

Despite Thailand's structures of democratic government, any administration that has not also received the approval of the military elites, many of whom hold seats in the Senate, has not prevailed for long. The military has been a rightist force in Thai politics, resisting reforms from the left that might have produced a stronger labor union movement, more freedom of expression (many television and radio stations in Thailand are controlled directly by the military), and less economic distance between the social classes. Military involvement in government increased substantially during the 1960s and 1970s, when a Communist insurgency threatened the government from within and the Vietnam War destabilized the external environment.

Until the February 1991 coup, there had been signs that the military was slowly withdrawing from direct meddling in the government. This may have been because the necessity for a strong military appeared

to have lessened with the end of the Cold War. In late 1989, for example, the Thai government signed a peace agreement with the Communist Party of Malaya, which had been harassing villagers along the Thai border for more than 40 years. Despite these political/military improvements, Commander Suchinda Kraprayoon led an army coup against the legally elected government in 1991 and, notwithstanding promises to the contrary, promptly had himself named prime minister. Immediately, Thai citizens, tired of the constant instability in government occasioned by military meddling, began staging mass demonstrations against Suchinda. The protesters were largely middle-class office workers who used their cellular telephones to communicate from one protest site to another. The demonstrations were the largest in 20 years, and the military responded with violence; nearly 50 people were killed and more than 600 injured. The public out-

cry was such that Suchinda was forced to appear on television being lectured by the king; he subsequently resigned. An interim premier dismissed several top military commanders and removed military personnel from the many government departments over which they had come to preside. Elections followed in 1992, and Thailand returned to civilian rule, with the military's influence greatly diminished.

The events of this latest coup show that the increasingly educated and affluent citizens of Thailand wish their country to be a true democracy. Still, unlike some democratic governments that have one dominant political party and one or two smaller opposition parties, party politics in Thailand is characterized by diversity. Indeed, so many parties compete for power that no single party is able to govern without forming coalitions with others. Parties are often founded on the strength of a single charismatic leader rather than on a distinct political philosophy, a circumstance that makes the entire political setting rather volatile. The Communist Party remains banned. Campaigns to elect the 360-seat Parliament often turn violent; in recent elections, 10 candidates were killed when their homes were bombed or sprayed with rifle fire, and nearly 50 gunmen-for-hire were arrested or killed by police, who were attempting to protect the candidates of the 11 political parties vying for office.

In 1999, corruption charges by the New Aspiration Party against the ruling Democrat Party failed to topple the government. Many citizens seemed to credit the government with bringing Thailand out of economic recession (but the International Monetary Fund loan of $17.2 billion probably helped more). Still, corruption was everywhere evident in 2000, when massive fraud was uncovered in Senate elections. It was the first time that senators were being directly elected instead of appointed, and many of them resorted to wholesale vote-buying and ballot-tampering. Out of 200 senators, 78 were disqualified as a result of election fraud, and the elections had to be held again. Even the former prime minister, Thaksin Chinnawat, was charged, but eventually cleared, with concealment of assets. The current prime minister, elected in January 2001 and up for re-election in 2005, is reputed to be the richest man in Thailand. He holds a doctorate from Sam Houston University in the United States. He too has been charged, but not convicted, of asset concealment and conflicts of interest involving his successful mobile phone business.

FOREIGN RELATIONS

Thailand is a member of the United Nations, the Association of Southeast Asian Nations, and many other regional and international organizations. Throughout most of its modern history, Thailand has maintained a pro-Western political position. During World War I, Thailand joined with the Allies; and during the Vietnam War, it allowed the United States to stage air attacks on North Vietnam from within its borders, and it served as a major rest and relaxation center for American soldiers. During World War II, Thailand briefly allied itself with Japan but made decided efforts after the war to reestablish its former Western ties. When the U.S. launched the war against Iraq, Thailand responded by sending 450 medical engineers to the war zone.

Thailand's international positions have seemingly been motivated more by practical need than by ideology. During the colonial era, Thailand linked itself with Britain because it needed to offset the influence of

France; during World War II, it joined with Japan in an apparent effort to prevent its country from being devastated by Japanese troops; during the Vietnam War, it supported the United States because the United States seemed to offer Thailand its only hope of not being directly engaged in military conflict in the region.

Thailand now seems to be tilting away from its close ties with the United States and toward a closer relationship with Japan. In the late 1980s, disputes with the United States over import tariffs and international copyright matters cooled the countries' warm relationship (the United States accused Thailand of allowing the manufacture of counterfeit brand-name watches, clothes, computer software, and many other items, including medicines). Moreover, Thailand found in Japan a more ready, willing, and cooperative economic partner than the United States. Many Thais also find Japanese culture to be interesting, and sign up for Japanese etiquette classes to improve their abilities in the business world.

During the Cold War and especially during the Vietnam War era, the Thai military strenuously resisted the growth of Communist ideology inside Thailand, and the Thai government refused to engage in normal diplomatic relations with the Communist regimes on its borders. Because of military pressure, elected officials refrained from advocating improved relations with the Communist governments. However, in 1988, Prime Minister Prem Tinslanond, a former general in the army who had been in control of the government for eight years, stepped down from office, and opposition to normalization of relations seemed to mellow. The subsequent prime minister, Chatichai Choonhavan, who was ousted in the 1991 military coup, invited Cambodian leader Hun Sen to visit Thailand; he also made overtures to Vietnam and Laos. Chatichai's goal was to open the way for trade in the region by helping to settle the agonizing Cambodian conflict. He also hoped to bring stability to

the region so that the huge refugee camps in Thailand, the largest in the world, could be dismantled and the refugees repatriated. Managing regional relations will continue to be difficult: Thailand fought a brief border war with Communist Laos in 1988. The influx of refugees from the civil wars in adjacent Cambodia and Myanmar continues to strain relations. Currently some 100,000 Karen refugees live precariously in 20 camps in Thailand along the border with Myanmar. The Karens, many of whom practice Christianity and are the second-largest ethnic group in Myanmar, have fought the various governments in their home country for years in an attempt to create an independent Karen state. Despite the patrol efforts of Thai troops, Myanmar soldiers frequently cross into Thailand at night to raid, rape, and kill the Karens. Thailand has tolerated the massive influx of war refugees from Myanmar, but its patience seems to have been wearing thin in recent years.

THE ECONOMY

Part of the thrust behind Thailand's diplomatic initiatives is the changing needs of its economy. For decades, Thailand saw itself as an agricultural country; indeed, more than half of the laborforce work in agriculture today, with rice as the primary commodity. Rice is Thailand's single most important export and a major source of government revenue. Every morning, Thai families sit on the floor of their homes around bowls of hot and spicy *tom yam goong* soup and a large bowl of rice; holidays and festivals are scheduled to coincide with the various stages of planting and harvesting rice; and, in rural areas, students are dismissed at harvest time so that all members of a family can help in the fields. So central is rice to the diet and the economy of the country that the Thai verb equivalent of "to eat," translated literally, means "to eat rice." Thailand is the fifth-largest exporter of rice in the world.

Unfortunately, Thailand's dependence on rice subjects its economy to the cyclical fluctuations of weather (sometimes the monsoons bring too little moisture) and market demand. Thus, in recent years, the government has invested millions of dollars in economic diversification. Not only have farmers been encouraged to grow a wider variety of crops, but tin, lumber, and offshore oil and gas production have also been promoted. Thailand is the world's largest rubber-producing country, but with prices at 30-year lows, that industry is struggling to survive. Foreign investment in export-oriented manufacturing has been warmly welcomed. Japan in particular benefits from trading with Thailand in food and other commodities, and it sees Thailand as one of the more promising places to relocate smokestack industries. For its part, Thailand seems to prefer Japanese investment over that from the United States, because the Japanese seem more willing to engage in joint ventures and to show patience while enterprises become profitable. Indeed, economic ties with Japan are very strong. For instance, in recent years Japan has been the largest single investor in Thailand and has accounted for more than 40 percent of foreign direct investment (Taiwan, Hong Kong, and the United States each have accounted for about 10 percent). About 24 percent of Thai imports come from Japan, while 14 percent of its exports go to Japan.

Thailand's shift to an export-oriented economy paid off until 1997, when pressures on its currency, the baht, required the government to allow it to float instead of having it pegged to the U.S. dollar. That action triggered the Southeast Asian financial crisis. Until that time, Thailand's gross domestic product growth rate had averaged about 10 percent a year—one of the highest in the world, and as high, or higher than, all the newly industrializing countries of Asia (Hong Kong, South Korea, Singapore, Taiwan, and China). Furthermore, unlike the Philippines and Indonesia, Thailand was able to achieve this incredible growth without very high inflation. The 1997–1998 financial crisis hit Thailand very hard, but with the government welcoming foreign investment and pushing to expand the economy, the annual rate of economic growth is now in excess of 6 percent.

SOCIAL PROBLEMS

Industrialization in Thailand, as everywhere, draws people to the cities. Bangkok is one of the largest cities in the world. Numerous problems, particularly air pollution, traffic congestion, and overcrowding, complicate life for Bangkok residents. An international airport that opened near Bangkok in 1987 was so overcrowded just four years later that a new one had to be planned, and new harbors had to be constructed south of the city to alleviate congestion in the main port. Demographic projections indicate that there will be a decline in population growth in the future as the birth rate drops and the average Thai household shrinks from the six people it was in 1970 to only three people by 2015. This will alter the social structure of urban families, especially as increased life expectancy adds older people to the population and forces the country to provide more services for the elderly.

Today, however, many Thai people still make their living on farms, where they grow rice, rubber, and corn, or tend chickens and cattle, including the ever-present water buffalo. Thus, it is in the countryside (or "up-country," as everywhere but Bangkok is called in Thailand) that the traditional Thai culture may be found. There, one still finds villages of typically fewer than 1,000 inhabitants, with houses built on wooden stilts alongside a canal or around a Buddhist monastery. One also finds, however, unsanitary conditions, higher rates of illiteracy, and lack of access to potable water. Of increasing concern is deforestation, as Thailand's growing population continues to use wood as its primary fuel for cooking and heat. The provision of social services does not meet demand even in the cities, but rural residents are particularly deprived.

Relative tolerance has mitigated ethnic conflict among Thailand's numerous minority groups. The Chinese, for instance, who are often disliked in other Asian countries because of their dominance of the business sectors, are able to live with little or no discrimination in Thailand; indeed, they constitute the backbone of Thailand's new industrial thrust.

In addition to the Chinese, Thailand is home to many other ethnic groups: the Lisu, the Mien, the Hmong, the Akha, the Karen (a largely Christian group), and the Lahu, to name a few. As many as 550,000 semi-nomadic tribal peoples live in Thailand's northern mountains, where they have little in the way of modern conveniences, and live more or less as they were living hundreds of years ago. In fact, many of them have little or emotional link with the nation of Thailand, preferring to crisscross the borders between Tibet, Myanmar, Laos, and China as they make their subsistence living without electricity or running water. The 59,000 members of the Lahu tribe, for example, entered Thailand from Tibet about 200 years ago. They live in the midst of a lush jungle, where they build their bamboo huts on stilts and wear homespun clothing. With no phones, no television, and no cars, the Lahu use wooden plows pulled by oxen to prepare their fields for rice planting.

Culturally, Thai people are known for their willingness to tolerate (although not necessarily to assimilate) diverse lifestyles and opinions. Buddhist monks, who shave their heads and make a vow of celibacy, do not find it incongruous to beg for rice in districts of Bangkok known for prostitution and wild nightlife. And worshippers seldom object when a noisy, congested highway is built alongside the serenity of an ancient Buddhist temple. (However, the mammoth scale of the proposed $3.2 billion, four-level road-and-railway system in the city and its likely effect on cultural and religious sites prompted the Thai cabinet to order the construction underground; but the cabinet had to recant when the Hong Kong firm designing the project announced that it was technically impossible to build it underground.)

Timeline: PAST

A.D. 1200s
The formal beginning of Thailand as a nation

1782
King Rama I ascends the throne, beginning a nine-generation dynasty

1932
Coup; constitutional monarchy

1939
The country's name is changed from Siam to Thailand

1942
Thailand joins Japan and declares war on the United States and Britain

1946
Thailand resumes its historical pro-Western stance

1960s–1970s
Communist insurgency threatens Thailand's stability

1973
Student protests usher in democratic reforms

1990s
Currency decisions in Thailand precipitate the Southeast Asian financial crisis; mass demonstrations force a return to civilian rule

PRESENT

2000s
Thailand rebounds from the economic crisis of the late 1990s

Bangkok looks for ways to reduce its terrible pollution and traffic congestion

The 2004 Indian Ocean earthquake and tsunami took the lives of up to 11,000 people, many of them European tourists vacationing on the powdery resort beaches of Phuket Island.

Despite their generally tolerate attitudes, Thais have been sore pressed to handle Muslim unrest in the provinces near Malaysia, where residents feel neglected and persecuted by the government. In 2004, violence by Muslims provoked a strong reaction by Thai police, resulting in the shooting deaths of some who were rioting and the deaths by suffocation by many more detainees as they were being transported in overcrowded military vehicles. Militants have attacked schools and police stations and raided military depots. With Muslims preachers using increasingly volatile speech during worship services and in schools, it is not likely that the unrest—already close to a full-blown separatist movement—will be calmed any time soon.

The Thai government seems to be making headway on at least two serious social problems, HIV/AIDS and smoking. Thailand has been known for years as a hotbed for the transmission of the virus that causes AIDS. Education programs and tighter regulation of the sex industry have produced dramatic results, with the number of new HIV infections on the decline. In 2002, the government passed a strict antismoking law. Smoking is now outlawed in virtually all indoor spaces, including Buddhist temples. Not only can smokers be fined if they light up indoors, but so can the establishment that allows it. According to the World Health Organization, Thailand now joins Hong Kong and Singapore as the Asian countries with the strongest antismoking laws. Thailand is known as a laissez-faire culture where "never mind" is the solution to vexing problems; thus the implementation of these strict policies demonstrates how seriously the Thai government regards the health of its citizens.

Vietnam (Socialist Republic of Vietnam)

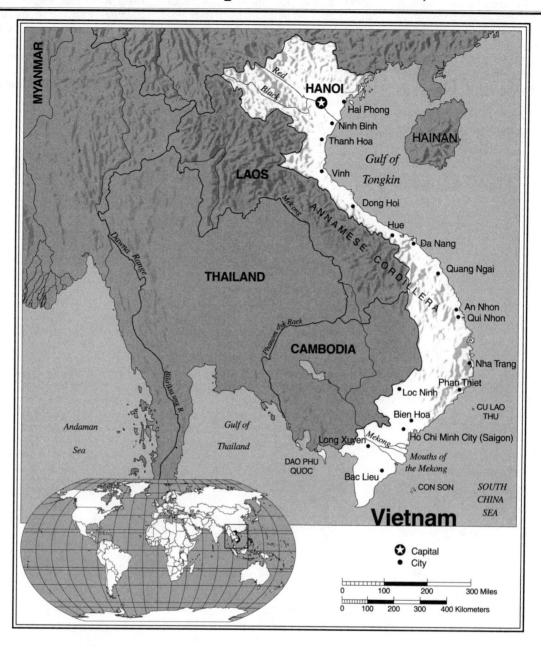

Vietnam Statistics

GEOGRAPHY

Area in Square Miles (Kilometers):
121,278 (329,560) (about the size of
New Mexico)
Capital (Population): Hanoi (3,822,000)
Environmental Concerns: deforestation;
soil degradation; overfishing; water and
air pollution; groundwater contamination
Geographical Features: low, flat delta in
south and north; central highlands; hilly
and mountainous in far north and
northwest

Climate: tropical in south; monsoonal in
north

PEOPLE

Population

Total: 82,689,518
Annual Growth Rate: 1.3%
Rural/Urban Population Ratio: 80/20
Major Languages: Vietnamese; French;
Chinese; English; Khmer; tribal
languages

Ethnic Makeup: 90% Vietnamese; 7%
Muong, Thai, Meo, and other mountain
tribes; 3% Chinese
Religions: Buddhists, Confucians, and
Taoists most numerous; Roman
Catholics; Cao Dai; animists; Muslims;
Protestants

Health

Life Expectancy at Birth: 67 years (male);
73 years (female)
Infant Mortality: 29.8/1,000 live births

Physicians Available: 1/2,444 people
HIV/AIDS Rate in Adults: 0.4%

Education

Adult Literacy Rate: 90.3%
Compulsory (Ages): 6–11; free

COMMUNICATION

Telephones: 2,600,000 main lines
Daily Newspaper Circulation: 8/1,000
  people
Televisions: 43/1,000 people
Internet Users: 400,000 (2002)

TRANSPORTATION

Highways in Miles (Kilometers): 63,629
  (106,048)
Railroads in Miles (Kilometers): 1,701
  (2,835)
Usable Airfields: 34
Motor Vehicles in Use: 177,000

GOVERNMENT

Type: Communist state
Independence Date: September 2, 1945
  (from France)
Head of State/Government: President Tran
  Duc Luong; Prime Minister Phan Van
  Khai
Political Party: Communist Party of
  Vietnam
Suffrage: universal at 18

MILITARY

Military Expenditures (% of GDP): 2.5%
Current Disputes: boundary/border
  disputes with Cambodia and other
  countries

ECONOMY

Currency ($ U.S. Equivalent): 15,462
  dong = $1

Per Capita Income/GDP: $2,500/$203
  billion
GDP Growth Rate: 7.2%
Inflation Rate: 3.1%
Unemployment Rate: 6.1%
Population Below Poverty Line: 37%
Natural Resources: phosphates; coal;
  manganese; bauxite; chromate; oil and
  gas deposits; forests; hydropower
Agriculture: rice; corn; potatoes; rubber;
  soybeans; coffee; tea; animal products;
  fish
Industry: food processing; textiles;
  machine building; mining; cement;
  chemical fertilizer; glass; tires;
  petroleum; fishing; shoes; steel; coal;
  paper
Exports: $19.8 billion (primary partners
  Japan, China, Australia)
Imports: $22.5 billion (primary partners
  Singapore, Japan, Taiwan)

Vietnam Country Report

Foreign powers have tried to control Vietnam for 2,000 years. Most of that time it has been the Chinese who have had their eye on control—specifically of the food and timber resources of the Red River Valley in northern Vietnam.

Most of the northern Vietnamese were originally ethnic Chinese themselves; but over the years, they forged a separate identity for themselves and came to resent Chinese rule. Vietnam was conquered by China as early as 214 B.C. and again in 111 B.C., when the Han Chinese emperor Wu Ti established firm control. For about 1,000 years (until A.D. 939, and sporadically thereafter by the Mongols and other Chinese), the Chinese so thoroughly dominated the region that the Vietnamese people spoke and wrote in Chinese, built their homes like those of the Chinese, and organized their society according to Confucian values. In fact, Vietnam (*viet* means "people" and *nam* is Chinese for "south") is distinct among Southeast Asian nations because it is the only one whose early culture—in the north, at least—was influenced more by China than by India.

The Chinese did not, however, directly control all of what constitutes modern Vietnam. Until the late 1400s, the southern half of the country was a separate kingdom known as Champa. It was inhabited by the Chams, who originally came from Indonesia. For a time Champa was annexed by the

north. However, between the northern region called Tonkin and the southern Chams-dominated region was a narrow strip of land occupied by Annamese peoples (a mixture of Chinese, Indonesian, and Indian ethnic groups), who eventually overthrew the Cham rulers and came to dominate the entire southern half of the country. In the 1500s, the northern Tonkin region and the southern Annamese region were ruled separately by two Vietnamese family dynasties. In the 1700s, military generals took power, unifying the two regions and attempting to annex or control parts of Cambodia and Laos as well.

DEVELOPMENT

The government is reluctant to allow a capitalist economy to take hold, so per capita gross domestic product remains at only $2500 per year, and $2 billion of the economy comes in each year from remittances sent from overseas Vietnamese, mostly in the United States. Still, the country has resumed exports of rice, cement, fertilizer, steel, and coal.

In 1787, Nguyen-Anh, a general with imperial ambitions, signed a military-aid treaty with France. The French had already established Roman Catholic missions in the south, were providing mercenary soldiers for the Vietnamese generals, and were interested in opening up trade along the Red

River. The Vietnamese eventually came to resent the increasingly active French involvement in their internal affairs and took steps to curtail French influence. The French, however, impressed by the resources of the Red River Valley in the north and the Mekong River Delta in the south, were in no mood to pull out. Vietnam's geography contains rich tropical rain forests in the south, valuable mineral deposits in the north, and oil deposits offshore.

FREEDOM

Vietnam is nominally governed by an elected National Assembly. Real power, however, resides in the Communist Party and with those military leaders who helped defeat the U.S. and South Vietnamese armies. Civil rights, such as the right of free speech, are curtailed. Private-property rights are limited. In 1995, Vietnam adopted its first civil code providing property and inheritance rights for citizens.

War broke out in 1858, and by 1863 the French had won control of many parts of the country, particularly in the south around the city of Saigon (now known as Ho Chi Minh City). Between 1884 and 1893, France solidified its gains in Southeast Asia by taking the northern city of Hanoi and the surrounding Tonkin region and by putting Cambodia, Laos, and Vietnam under one administrative unit, which it named *Indochina.*

Ruling Indochina was not easy for the French. For one thing, the region comprised hundreds of different ethnic groups, many of whom had been traditional enemies long before the French arrived. Within the borders of Vietnam proper lived Thais, Laotians, Khmers, northern and southern Vietnamese, and mountain peoples whom the French called Montagnards. Most of the people could not read or write—and those who could wrote in Chinese, because the Vietnamese language did not have a writing system until the French created it. Most people were Buddhists and Taoists, but many also followed animist beliefs.

In addition to the social complexity, the French had to contend with a rugged and inhospitable land filled with high mountains and plateaus as well as lowland swamps kept damp by yearly monsoon rains. The French were eager to obtain the abundant rice, rubber, tea, coffee, and minerals of Vietnam, but they found that transporting these commodities to the coast for shipping was extremely difficult.

HEALTH/WELFARE

Health care has been nationalized and the government operates a social-security system, but the chronically stagnant economy has meant that few Vietnamese receive sufficient health care or have an adequate nutritional intake. The World Health Organization has been involved in disease-abatement programs since reunification of the country in 1975.

VIETNAMESE RESISTANCE

France's biggest problem, however, was local resistance. Anti-French sentiment began to solidify in the 1920s; by the 1930s, Vietnamese youths were beginning to engage in open resistance. Prominent among these was Nguyen ai Quoc, who founded the Indochinese Communist Party in 1930 as a way of encouraging the Vietnamese people to overthrow the French. He is better known to the world today as *Ho Chi Minh*, meaning "He Who Shines."

Probably none of the resisters would have succeeded in evicting the French had it not been for Nazi Germany's overrunning of France in 1940 and Japan's subsequent military occupation of Vietnam. These events convinced many Vietnamese that French power was no longer a threat to independence; the French remained nominally in control of Vietnam, but everyone knew that the Japanese had the real power. In 1941, Ho

Chi Minh, having been trained in China by Maoist leaders, organized the League for the Independence of Vietnam, or Viet Minh. Upon the defeat of Japan in 1945, the Viet Minh assumed that they would take control of the government. France, however, insisted on reestablishing a French government. Within a year, the French and the Viet Minh were engaged in intense warfare, which lasted for eight years.

The Viet Minh initially fought the French with weapons supplied by the United States when that country was helping local peoples to resist the Japanese. Communist China later became the main supplier of assistance to the Viet Minh. This development convinced U.S. leaders that Vietnam under the Viet Minh would very likely become another Communist state. To prevent this occurrence, U.S. president Harry S. Truman decided to back France's efforts to recontrol Indochina (although the United States had originally opposed France's desire to regain its colonial holdings). In 1950, the United States gave $10 million in military aid to the French—an act that began a long, costly, and painful U.S. involvement in Vietnam.

In 1954, the French lost a major battle in the north of Vietnam, at Dien Bien Phu, after which they agreed to a settlement with the Viet Minh. The country was to be temporarily divided at the 17th Parallel (a latitude above which the Communist Viet Minh held sway and below which non-Communist Vietnamese had the upper hand), and country-wide elections were to be held in 1956. The elections were never held, however; and under Ho Chi Minh, Hanoi became the capital of North Vietnam, while Ngo Dinh Diem became president of South Vietnam, with its capital in Saigon.

THE UNITED STATES GOES TO WAR

Ho Chi Minh viewed the United States as yet another foreign power trying to control the Vietnamese people through its backing of the government in the South. The United States, concerned about the continuing attacks on the south by northern Communists and by southern Communist sympathizers known as Viet Cong, increased funding and sent military advisers to help prop up the increasingly fragile southern government. Unlike the French who had wanted to control Vietnam for its resources, the United States wanted to control the spread of communism. In 1955, The U.S. had helped airlift a million Vietnamese Catholics who were fleeing communism in the north, and U.S. President Eisenhower had been urged, at that early date, to either send in troops or to even drop two small nuclear

bombs (!) on the communist north. He did neither, but the fear of a "domino effect" in which it was believed that one country after another would fall to communism, propelled U.S. President John F. Kennedy, in 1963, to send 12,000 military "advisors" to Vietnam. In 1964, an American destroyer ship was attacked in the Gulf of Tonkin by North Vietnam (or, as some people believe, the attack was a hoax, intentionally fabricated to justify direct U.S. military involvement in Vietnam). The U.S. Congress responded by giving then-president Lyndon Johnson a free hand in ordering U.S. military action against the north; before this time, U.S. troops had not been involved in direct combat.

ACHIEVEMENTS

Vietnam provides free and compulsory schooling for all children. The curricular content has been changed in an attempt to eliminate Western influences. New Economic Zones have been created in rural areas to try to lure people away from the major cities of Hanoi, Hue, and Ho Chi Minh City (formerly Saigon).

By 1969, some 542,000 American soldiers and nearly 66,000 soldiers from 40 other countries were in battle against North Vietnamese and Viet Cong troops. Despite unprecedented levels of bombing and use of sophisticated electronic weaponry and powerful chemicals such as Agent Orange to strip the leaves off of jungle trees to make it easier to find the enemy. U.S. and South Vietnamese forces continued to lose ground to the Communists, who used guerrilla tactics and built their successes on latent antiforeign sentiment among the masses as well as extensive Soviet military aid. At the height of the war, as many as 300 U.S. soldiers were being killed every week.

Watching the war on the evening television news, many Americans began to feel that the war was a mistake. Anti-Vietnam rallies became a daily occurrence on American university campuses, and many people began finding ways to protest U.S. involvement: dodging the military draft by fleeing to Canada, burning down ROTC buildings, and publicly challenging the U.S. government to withdraw. President Richard Nixon had once declared that he was not going to be the first president to lose a war; but after his expansion of the bombing into Cambodia to destroy Communist supply lines and after significant battlefield losses, domestic resistance became so great that an American withdrawal seemed inevitable. The U.S. attempt to "Vietnamize" the war by training South

Timeline: PAST

214 B.C.
China begins 1,000 years of control or influence over the northern part of Vietnam

A.D. 1500s
Northern and southern Vietnam are ruled separately by two Vietnamese families

1700s
Military generals overthrow the ruling families and unite the country

1787
General Nguyen-Anh signs a military-aid treaty with France

1863
After 5 years of war, France acquires its first holdings in Vietnam

1893
France establishes the colony of Indochina

1930
Ho Chi Minh founds the Indochinese Communist Party

1940s
The Japanese control Vietnam

post-1945
France attempts to regain control

1950s
The United States begins to aid France to contain the spread of communism

1954
Geneva agreements end 8 years of warfare with the French; Vietnam is divided into North and South

1961
South Vietnam's regime is overthrown by a military coup

1965
The United States begins bombing North Vietnam

1973
The United States withdraws its troops and signs a cease-fire

1975
North Vietnamese troops capture Saigon and reunite the country; a U.S. embargo begins

1979
Vietnamese troops capture Cambodia; China invades Vietnam

1980s
Communist Vietnam begins liberalization of the economy; naval clashes with China over the Spratly Islands

1990s
The U.S. economic embargo of Vietnam is lifted; the United States establishes full diplomatic relations

PRESENT

2000s
Relations with China improve with the resolution of a border dispute

A bilateral trade agreement is signed with the United States

Vietnam and the United States sign an agreement allowing for direct air flights from the two countries for the first time in three decades

Vietnamese troops and supplying them with advanced weapons did little to change South Vietnam's sense of having been sold out by the Americans.

Secretary of State Henry Kissinger negotiated a cease-fire settlement with the North in 1973; but most people believed that as soon as the Americans left, the North would resume fighting and would probably take control of the entire country. This indeed happened, and in April 1975, under imminent attack by victorious North Vietnamese soldiers, the last Americans lifted off in helicopters from the roof of the U.S. Embassy in Saigon. The South Vietnamese government subsequently surrendered, and South Vietnam ceased to exist. The country had, at last, been reunited, but in a way that no one would have predicted.

The war wreaked devastation on Vietnam. It has been estimated that nearly 2 million people were killed during just the American phase of the war; another 2.5 million were killed during the French era. In addition, 4.5 million people were wounded, and nearly 9 million lost their homes. U.S. casualties included more than 58,000 soldiers killed and 300,000 wounded. In the end, the Vietnamese got what they had wanted all along: the right to run their own country without outside interference.

A CULTURE, NOT JUST A BATTLEFIELD

Because of the Vietnam War (which the Vietnamese call the American War), many people think of Vietnam as if it were just a battlefield. But Vietnam is much more than that. It is a rich culture made up of peoples representing diverse aspects of Asian life. In good times, Vietnam's dinner tables are supplied with dozens of varieties of fish and shrimp, and the ever-present bowls of rice and beef soup, or *pho*. *Pho* is a mixture of beef soup, rice noodles, and meat spiced with cloves, Asian basil, white onions, and star anise. It is enjoyed at both breakfast and dinner. Sugarcane and bananas are also favorites. Because about 80 percent of the people live in the countryside, the population as a whole possesses a living library of practical know-how about farming, livestock raising, fishing, and home manufacture. Today, only about 200 out of every 100,000 Vietnamese people attend college, but most children attend elementary school and nearly 94 percent of the adult population can read and write.

Literacy was not always so high; much of the credit is due the Communist government, which, for political-education reasons, has promoted schooling throughout the country. Another government initiative—unifying the two halves of the country—has not been

an easy task, for, upon the division of the country in 1954, the North followed a socialist route of economic development, while in the South, capitalism became the norm.

Religious belief in Vietnam is an eclectic affair and reflects the history of the nation; on top of a Confucian and Taoist foundation, created during the centuries of Chinese rule, rests Buddhism (a modern version of which is called Hoa Hao and claims 1 million believers); French Catholicism, which claims about 15 percent of the population; and a syncretist faith called Cao Dai, which claims about 2 million followers. Cao Dai models itself after Catholicism in terms of hierarchy and religious architecture, but it differs in that it accepts many Gods—Jesus, Buddha, Mohammed, Lao-Tse, and others—as part of its pantheon. Many Vietnamese pray to their ancestors and ask for blessings at small shrines located inside their homes. Animism, the worship of spirits believed to live in nature, is also practiced by many of the Montagnards. These mountain peoples, most of whom are Protestants, actively resisted the Communists during and after the Vietnam War. The government has now officially labeled them a separatist movement and has banned their demonstrations for religious freedom and private land rights. Many of them fled to Cambodia in 2001, and nearly 1,000 await resettlement in the United States. Religious tension remains high in Vietnam. In 2004, over a thousand, mostly Protestant, ethnic Christians rioted over the Easter weekend in Vietnam's Central Highland city of Daklak. What started out as a peaceful demonstration against religious repression and land confiscation turned violent, and government troops ended up arresting scores of protestors and sealing off the entire region. Vietnam remains on the U.S. State Department's list of countries where full religious freedom is not allowed.

Vietnamese citizens are permitted to believe in God and to participate in religious rituals, but Western religions are regarded warily, and those who practice them are often harassed, especially if, along with Christianity, they begin demanding democratic reforms. Periodically, the government launches a wave of arrests against those advocating democratic reforms. When one Vietnamese actor, Don Duong, appeared in Holywood movies about the Vietnam War that the government did not like, he was labeled a "national traitor" and told to prepare for deportation. These, and many other, examples confirm that the communist government intends to remain in control and to prevent a "gradual evolution" toward capitalism and Western-style democracy—a formidable task, given that half the country already was just such a place before the North won the war.

THE ECONOMY

When the Communists won the war in 1975 and brought the capitalist South under its jurisdiction, the United States imposed an economic embargo on Vietnam, which most other nations honored and which remained in effect for 19 years, until President Bill Clinton ended it in 1994. As a consequence of war damage as well as the embargo and the continuing military involvement of Vietnam in the Cambodian War and against the Chinese along their mutual borders, the first decade after the end of the Vietnam War saw the entire nation fall into a severe economic slump. Whereas Vietnam had once been an exporter of rice, it now had to import rice from abroad. Inflation raged at 250 percent a year, and the government was hard-pressed to cover its debts. Many South Vietnamese were, of course, opposed to Communist rule and attempted to flee on boats—some 100,000 of them—but, contrary to popular opinion, most refugees left Vietnam because they could not get enough to eat, not because they were being persecuted.

Beginning in the mid-1980s, the Vietnamese government began to liberalize the economy. Under a restructuring plan called *doi moi* (similar in meaning to Soviet *perestroika*), the government began to introduce elements of free enterprise into the economy. Moreover, despite the Communist victory, the South remained largely capitalist. For a while after the U.S. trade embargo was lifted in the mid-1990s, outside investors poured money into the Vietnamese economy. Australia, Japan, and France invested $3 billion in a single year; and in 1994, more than 540 firms from Singapore, Hong Kong, Japan, and France were doing business with Vietnam. Furthermore, the World Bank, the Asian Development Bank, and the International Monetary Fund were providing development loans.

Unfortunately, the financial crises in nearby Malaysia, Indonesia, and other countries caused Vietnam to devalue its currency in 1998. Combined with El Niño-caused drought, forest fires (some 900 in 1998), and floods (1999 floods killed more than 700 people), the currency devaluation slowed economic growth and gave government leaders an excuse to trumpet the failures of capitalism. But the worst shock was the withdrawal of many of the companies that had wanted to invest in Vietnam. They withdrew, reluctantly, because of crippling government red tape, unfair business poli-

cies, and obvious corruption (a situation often heard in China today as well).

Much to the worry of government traditionalists, the Vietnamese people seem fascinated with foreign products. They want to move ahead and put the decades of warfare behind them. Western travelers in Vietnam are treated warmly, and the Vietnamese government has cooperated with the U.S. government's demands for more information about missing U.S. soldiers—the remains of five more of whom were located as recently as 2004, thirty years after the war's end. In 1994, after a 40-year absence, the United States opened up a diplomatic mission in Hanoi as a first step toward full diplomatic recognition. So eager are the Vietnamese to reestablish economic ties with the West that the Communist authorities have even offered to allow the U.S. Navy to lease its former port at Cam Ranh Bay (the offer has not yet been accepted). Diplomatic bridge-building between the United States and Vietnam increased in the 1990s, when a desire to end the agony of the Cambodian conflict created opportunities for the two sides to talk together. Telecommunications were established in 1992, and in the same year, the United States gave $1 million in aid to assist handicapped Vietnamese war Veterans. In 2004, Vietnam was the only Asian country included in the U.S. government's $15 billion AIDS program. Two decades after the end of the Vietnam War, the United States established full diplomatic relations with Vietnam. Then, in 2001, after difficult and lengthy negotiations, the two countries signed a trade pact. With tariffs reduced or eliminated, exports to the United States doubled, and cotton, fertilizer, seafood, shoes, and other products began being shipped across the Pacific. The agreement was difficult for many of the hard-line Communist leaders; but after a 1997 farmers' riot and other evidences of citizen unrest, they recognized that they had to do something to salvage the struggling economy. Used wisely, the agreement could be a first step toward Vietnamese membership in the World Trade Organization.

Despite this gradual warming of relations, however, anti-Western sentiment remains strong in some parts of the population, particularly the military. As recently as 1996, police were still tearing down or covering up signs advertising Western products, and anti-open-door policy editorials were still appearing in official newspapers. The situation was not helped by the visit, in 2000, of U.S. senator John McCain, a former prisoner of war in Vietnam, during which he publicly declared that the wrong side had won the war. Still, the Vietnamese gave a formal state

welcome to U.S. defense secretary William Cohen, who discussed military and other topics, especially the missing-in-action issue, with Vietnamese leaders.

HEARTS AND MINDS

As one might expect, resistance to the current Vietnamese government comes largely from the South Vietnamese, who, under both French and American tutelage, adopted Western values of capitalism and consumerism. Many South Vietnamese had feared that after the North's victory, South Vietnamese soldiers would be mercilessly killed by the victors; thousands were, in fact, killed, but many—up to 400,000—former government leaders and military officers were instead sent to "reeducation camps," where, combined with hard labor, they were taught the values of socialist thinking. Several hundred such internees still remain incarcerated. Many of the well-known leaders of the South fled the country when the Communists arrived and have long since made new lives for themselves in the United States, Canada, Australia, and other Western countries. Those who have remained—for example, Vietnamese members of the Roman Catholic Church—have occasionally resisted the Communists openly, but their protests have been silenced. Hanoi continues to insist on policies that remove the rights to which the South Vietnamese had become accustomed. For instance, the regime has halted publication and dissemination of books that it judges to have "harmful contents." There is not much that average Vietnamese can do to change these policies except passive obstruction, which many are doing even though it damages the efficiency of the economy.

Vietnam has made progress with some of its neighbors in recent years. In 1999, it settled a long-standing, 740-mile border dispute with China; and in 2000, it accepted the last repatriated Vietnamese from refugee camps in Hong Kong. Also in 2000, the government announced that it would build a two-lane highway along the route of the famous Ho Chi Minh Trail. It will border Cambodia and Laos, at a cost of $375 million.

Of growing concern is the presence of Bird Flu, a devastating avian illness that has infected over 600,000 chickens in Vietnam and thousands more in Hong Kong, China, and other countries. World Health Organization doctors believe that as many as 20 Vietnamese children and adults have also died as a result of the flu. If not contained, the disease could result in a world epidemic.

Meeting the Needs of the Developing World

Japanese ODA has been building bridges of cooperation for 50 years. What has been accomplished during the half-century, and what changes are under way?

What exactly is offical development assistance, or ODA? According to the generally accepted international definition, ODA is the provision of funds or technical cooperation by an industrialized country's government or government entity in order to promote economic development or imporve the quality of like in a developing country.

Japan's ODA began when Japan joined the Colombo Plan in 1954. An international organization created after the end of World War II, the Colombo Plan was established to provide technical cooperation aimed at promoting the socioeconomic development of countries in the Asia-Pacific region. Around the time Japan joined the Colombo Plan, it began reparations to Asian countries devastated during World War II. In 1958, Japan began providing loan assistance, also known as "yen loans," with a loan to India. The loan was made on the condition that it be used to purchase materials from Japan. Yen loan ODA conveniently incorporated the additional policy aim of promoting exports from Japan.

During the 1960s there was an increase in the overall amount of ODA delivered, and at the same time the system for delivering ODA underwent development. Agencies for carrying out ODA programs were established, starting with the Overseas Economic Cooperation Fund of Japan (OECF) in 1961, and followed by the Japan International Cooperation Agency (JICA) in 1974. Japan also joined the Organization for Economic cooperation and Development (OECD) in 1964. In this way, Japan's system for providing aid within an international framework gradually took shape.

Japanese war reparations were completed during the latter half of the 1970s. During that same period, the Japanese public grew increasingly supportive of a greater role for Japan in the international community. ODA disbursements began to expand dramatically as Japan matured into an economic powerhouse and its trade surplus ballooned—there had been calls for ODA that would share these trade profits with developing countries. The ODA budget continued its rapid growth as the Japanese economy experienced the economic bubble of the 1980s. In 1989, Japan surpassed the United States to become the world's biggest donor nation. With the exception of 1990, Japan remained the largest donor of aid in the world for the 10 years leading up to 2000. In the 50 years since the inception of Japanese ODA, a total of US $221 billion has been disbursed in 185 countries and regions.

The international environment changed significantly following the end of the Cold War, and in 1992 the Japanese cabinet drew up and passed the ODA Charter. The principles outlined in the

Charter were fleshed out in 1999 with the development of a Mid-Term Policy on ODA; this policy clarified Japan's aid-related policy and principles for the benefit of both the Japanese public and the international community.

Just as the international environment has changed, Japanese domestic circumstances have also evolved. The Japanese economy has been struggling since the 1990s, and the Japanese public has criticized ODA-related inefficiencies. Against this domestic background, the ODA budget has undergone a series of cutbacks. As a result, Japanese ODA in 2003 was approximately US$8.88 billion, 4.3% drop from the previous year. Even so, Japan is currently the second largest donor of ODA among the member countries of the OECD's Development Assistance Committee (DAC), after the United States.

What are the current international trends for development assistance? Most programs since the beginning of the 1990s have focused on the development of human resources, aiming to help the world's poorest people. A report adopted by DAC member countries in 1996 set forth strategic orientations for development cooperation into the 21st century. The report set a goal of reducing by half the proportion of people living in extreme poverty by 2015. This goal was again adopted at the Millennium Summit in

2000. During the 1990s, a downturn in the financial conditions in the industrialized countries contributed to a sense of "aid fatigue," but the 9/11 terrorist attacks on the United States brought a renewed realization that poverty can give rise to terrorism. The past few years have thus seen an international trend toward more active implementation of ODA.

With evolution of the international environment continuing apace, Japan has revised its ODA charter, striving to enhance the transparency and efficiency of its aid programs. The stated objective of Japan's revised ODA Charter is "to contribute to the peace and development of the international community, and thereby to help ensure Japan's own security and prosperity." In order to achieve this objective, the Charter places a dual focus on peace-building and human security. In focusing on human security, the revised approach alters course from the conventional method of supplying aid on a country bais. Rather, the approach aims to protect the dignity of individuals, which may at times be threatened by a government or state. This shift to a person-based approach is perhaps the most significant change in the revised ODA Charter.

The Charter also clarifies Japan's ODA policy stance as based on careful selection of the recipient and concentration of resources, and it specifies Asia as a priority region. Careful research into the conditions of each recipient country in order to provide efficient aid that meets the recipient country's needs is noted as another important principle. In the past, Japan's ODA was generally aid-on-demand cooperation provided in response to a request from a recipient country. Now, in addition to that approach, aid-related organizations including the Japanese embassy, JICA and the Japan bank for International Cooperation (JBIC) set up an ODA task force to meet representatives of a recipient country to discuss assistance needs.

Japan's ODA has come to a turning point. Today, assistance programs and institutions are incorporating a variety of reforms, including the revised ODA Charter. The future will likely demand that Japan's ODA continue to respond flexibly to changing conditions both at home and abroad.

United East Asia

A trend towards integration is slowly starting to gather momentum across East Asia, Japan, China, the Republic of Korea, and the countries of ASEAN are strengthening their economic ties through free trade agreements (FTAs) and economic partnership agreements (EPAs); these agreements could be the beginnings of a regional economic community along the lines of the EU of NAFTA.

Fukunari Kimura

We live in a time of ever-increasing economic globalization, but there are also regionally-based movements aiming to build regional economic zones.

Such regional zones are being built through two types of agreement between different countries: the free trade agreement (FTA) and the economic partnership agreement (EPA). FTAs aim to reduce or eliminate tariffs and other barriers to trade between countries; an FTA can be concluded between two countries (a bilateral FTA) or among three or more countries (a multilateral FTA). The Japanese government sees the EPA as taking the FTA concept one step further. Under an EPA, participant countries agree to allow the free movement of persons, goods, and money within the trading zone.

FTAs and EPAs unify multiple countries and regions into a single market, stimulating competition and invigorating the economy. FTAs and EPAs form the nucleus of regional economic integration. As of 2003, there were a total of 189 FTAs worldwide (according to a Ministry of Economy, Trade and Industry study based on notifications to the World Trade Organization). Considering that in 1990 there were only 31 FTAs, and in 1970 a mere six, the trend toward the creation of regional economic zones through the use of FTAs has positively mushroomed over the past 10 years or so. Perhaps the best example of regionally-based integration is the European Union (EU). The forerunner of the EU was the European Economic Community (EEC), a customs union established in 1958 to promote freer inter-zone trade and to apply the same tariffs to all external trading partners. The EU expanded eastwards following the end of the Cold War in 1989, and has evolved into a powerful regional economic zone that boasts 25 member countries with a total population of 455 million.

In North America, the United States, Canada, and Mexico make up the North American Free Trade Agreement (NAFTA), which was signed in 1992. NAFTA is another major regional economic zone, claiming a total population of 411 million and a GDP totaling 11.5 trillion dollars. South America has also seen increasing economic integration through the establishment of the Southern Common Market (MERCOSUR) by member states Argentina, Brazil, Paraguay, and Uruguay in 1991. Details of an even larger free trade zone are currently being negotiated among 34 countries in North and South America and the Caribbean (with the exception of Cuba). Known as the Free Trade Area of the Americas (FTAA), this economic alliance is due to take effect in December 2005.

Asia makes up the third major economic zone in today's world economy, yet economic integration in Asia has made a slow start. The biggest reason for this delay is that Japan—the biggest economy and long the only "advanced" country in the region—has generally been reluctant to sign on to FTAs and EPAs. After World War II Japan built up its economy by manufacturing inexpensive, high-quality products and selling them overseas. In order to pursue this trade policy, there had to be overseas markets open to Japanese products; the government of post-war Japan has thus always based its trade policy around maintaining and strengthening the GATT (General Agreement on Tariffs and Trade) and WTO multilateral free trade framework. This policy has certainly been no mistake; Japan has reaped the greatest benefits of the free trade system in the post-war world, and consequently has experienced great economic growth. However, it cannot be denied that Japan has not kept up with the new movement toward regional economic cooperation that started in the 1990s.

During the mid-1980s Japanese manufacturing and other companies began to transfer production facilities to other countries in East Asia. Now almost 20 years later, an open and extremely sophisticated international production and distribution network has been created among those Japanese corporations and local companies in the destination countries. However, systematic support for this interaction through FTAs lags behind. There are business costs such as tariffs even on simple intra-region movement of product components. Average customs tariffs in Asia—25.6% in the Philippines, 25.8% in Thailand, and 37.1% in Indonesia, for example—are set at much higher levels than the 3.6% of the United States or the 4.1% of the EU, creating a huge barrier for companies trying to do business in the region.

Fukunari Kimura of Keio University is a leading proponent of economic integration. He explains its importance in East Asia.

Creation of an FTA that includes all the countries of East Asia would allow goods to move freely throughout the region, resulting in better operations for Japanese corporations and for local companies, too. In turn, this would contribute greatly toward improving international competitiveness across the region.

When considering economic cooperation, I think that for Asian countries liberalization of investment is even more important than liberalization of trade. In contrast to the industrialization strategies pursued by Japan and the Republic of Korea during the 1950s and 1960s, some of the members of ASEAN (the Association of Southeast Asian Nations, made up of Brunei Darussalam, Cambodia, Indonesia, Laos, Malaysia, Myanmar, the Philippines, Singapore, Thailand and Vietnam) and especially China chose to seek investment from foreign-owned corporations to fund industrialization. In other words, the economic prosperity in East Asia today is largely thanks to investment and technical assistance from overseas. In or-

der to maintain and further develop this prosperity, the countries of East Asia need to work hard to ensure that they present an attractive investment opportunity for funding from external sources. Drafting of an EPA is probably the most effective means of creating that sort of investment environment.

Furthermore, economic cooperation will help free the movement of people—liberalizing labor markets would foster greater societal internationalization. This could be expected to bring even greater development and interchange of human resources in the economic and technical sectors throughout the region.

In January 2002, Japan and Singapore concluded the Japan-Singapore Economic Agreement for a New Age Partnership (JSEPA); this agreement was Japan's first EPA. Importantly, it is also an impressively open EPA that fulfills the stipulations made in GATT Article 24 that all the trade between member countries be substantially barrier-free and that no additional barriers to non-member countries be erected.

As economic integration moves forward among the ASEAN+3 group—the countries of ASEAN together with Japan, China and the Republic of Korea—the tendency will be to adopt overall the highest level of cooperation already present between two or more of the countries. For that reason, I think Japan should wholeheartedly commit to a grand design for a future East Asian Community through the conclusion of open FTAs and EPAs on a par with the JSEPA. I believe that this commitment to the future is Japan's obligation to the region as an advanced country.(Compiled from an interview)

FUKUNARI KIMURA is a professor at the Department of Economics, Keio University, and a specialist in international trade and developing economies.

From *Asia Pacific Perspective*, vol. 2, no.7, November 2004, pp. 8-9. Copyright © 2004 by Jijigaho.

CHINA GOES SHOPPING

Lenovo-IBM is only the highest-profile deal yet in a wave of Western acquisitions that is certain to build

By Dexter Roberts, in Beijing

GOT A TIRED BRAND that needs a shot of adrenalin? The answer to your troubles may reside in China. Billions of dollars, euros, and yen have been invested to build up companies on the mainland in the last decade. Now Chinese companies, flush with cash and in command of the world's lowest-cost manufacturing plants, are doing some foreign investing of their own. First it was Huizhou-based TCL Corp. merging its television business with France's Thomson to create the world's biggest TV manufacturer—one of whose brands is venerable RCA. In September, Shanghai Automotive Industry Corp. announced it was buying a 48.9% stake in Korean truckmaker Ssangyong for close to $500 million, and there are rumors that SAIC's next target is Britain's troubled MG Rover Group Ltd. Then, on Dec. 7, Beijing-based Lenovo Group Ltd., China's biggest computer maker, agreed to buy a controlling stake in IBM's PC operations for $1.75 billion. "We saw a unique opportunity in front of us," says Mary Ma, Lenovo's chief financial officer. "It is an opportunity to differentiate our products and our technology."

GOVERNMENT PUSH

PLENTY OF OTHER Chinese execs will be looking for similar opportunities. China's Commerce Ministry figures the country's corporations spent $2.85 billion buying foreign companies and other assets in 2003. Straszheim Global Advisors, a Los Angeles-based research outfit that specializes in Chinese equities, says total outlays by Chinese companies on foreign deals may total as much as $7 billion in 2004 and could reach $14 billion in 2005.

Don't expect a bid for Exxon Mobil Corp. or Dell Inc. anytime soon, though—and don't expect the process to be easy. The Chinese have plenty of issues to sort out before they completely master the art of the deal, from smoothing over cultural differences to building up brands to running global supply networks.

But the momentum that's driving China's acquisitive streak is bound to grow. For starters, Beijing wants to create global champions, such as Lenovo in computers or TCL in TVs and mobile phones. "Go global" has become a catchphrase in China's official media. Beijing has just loosened the rules for companies investing abroad. "This is part of the more developed stage of China's opening to the world," says Gu Kejian, a professor in the Business School at People's University in Beijing.

BEIJING Lenovo, No. 1 in desktop computers, is buying IBM's PC business

What's more, though the amounts involved are small now, allowing further overseas investment could eventually help relieve pressure on China's overvalued currency. Investing yuan outside the mainland will balance the inward flow of money lured by a possible revaluation. And when China lets the yuan appreciate, the buying power of Chinese companies will increase.

SHANGHAI Lenovo faces pressure from Western brands

Finally, the Chinese need to go abroad for resources to feed their industrial machine and to scare up the talent and research they can't find at home. China National Petroleum Corp., parent of New York Stock Exchange-listed PetroChina; Sinopec, an oil-and-gas company; and CNOOC, an offshore driller, have each invested billions in oil-and-gas projects in Africa, Southeast Asia, and Latin America. More recently talk has been rife that Beijing-based China Minmetals Corp. will spend up to $5.5 billion to buy Canada's biggest mining company, Noranda Inc.

The next big wave of Chinese overseas investment will come from manufacturers such as Lenovo. China will buy into the industries in which it already competes heavily, particularly electronics, auto parts, appliances, textiles, and apparel. These are industries where acquiring global supply chains can confer big advantages, even for low-cost players like the Chinese. "The Chinese understand that global scale gives them lower-cost components, and that R&D and branding are now done globally," says Paul DiPaola, a managing partner at Bain & Co. China. "Chinese companies in industries where the economics are global will have no choice but to make these deals."

COMMUNICATION GLITCHES

GLOBAL SCALE HAS motivated Hangzhou-based auto-parts maker Wanxiang Group to buy 10 overseas companies in the last few years, including NASDAQ-listed Universal Automotive Industries Inc. "We combine international and domestic resources to speed up development," says Chairman Lu Guanqiu. The Chinese certainly have to do something. Their cell-phone, washing-machine, and electronics makers face overcapacity and razor-thin margins in their home market, where they already often compete with the world's best brands. Lenovo, for example, is No. 1 in desktop computers in China, but it has been under pressure from Dell and other non-Chinese brands. And as they enter foreign markets, Chinese execs realize they lack essential skills. "China needs brand names, reach, logos, marketing, distribution—and the management that attends to all of those," says Donald Straszheim, head of Straszheim Global Advisors.

The pressing question is whether the Chinese can get what they're looking for. Asians have a long history of overpaying for foreign assets: Remember the Japanese, with their ill-fated U.S. real estate and entertainment-industry deals? Ana-

---

CHINA ABROAD

Big Chinese companies have recently made or are considering lots of foreign acquisitions. Some deals:

TCL-THOMSON The Huizhou electronics company merged its TV unit with that of France's Thomson, acquiring the RCA brand in the U.S. Revenues this year: $4 billion.

SHANGHAI AUTOMOTIVE-SSANGYONG The Chinese auto maker paid $500 million for almost 50% of the Korean truckmaker. It may also buy control of Britain's Rover.

WANXIANG-UNIVERSAL AUTOMOTIVE The auto-parts company has bought 10 U.S. companies, including a stake in Nasdaq-listed Universal Automotive Industries.

UTSTARCOM-AUDIOVOX The U.S. gearmaker, which has most of its sales and workers in China, bought the cell-phone business of Long Island-based Audiovox for $165 million.

CHINA MINMETALS-NORANDA China's mining giant is considering offering as much as $5.5 billion for the Canadian nickel-and-copper miner.

---

lysts say China's oil companies have already spent too much on overseas reserves.

The Chinese have internal problems as well. Too many of their best-known companies, such as appliance maker Haier Group Co., rely on the vision of a charismatic founder and have little management depth. A merger can struggle with cultural differences, including management styles and big pay gaps between Chinese and Western executives.

Thomson and TCL, for example, have a lot to thrash out. In China, says one TCL official, "if the leader says something is right, even if he is wrong, employees will agree with him. But in foreign companies, they will not agree with him. We have two different cultures." Alar E. Arras, president of TCL-Thomson Electronics, says the venture is working through pay issues and other problems, but he concedes the hurdles: Try solving a technical glitch through translators, for example. "The communication challenge is one of the unique items we face," he says.

Given the problems, the Chinese may try various tactics in their forays abroad. Edward King, executive director for mergers and acquisitions at Morgan Stanley in Hong Kong, figures they will use joint ventures more than outright acquisitions to spread responsibility for making a deal work. But one way or another, more of those deals are on the way.

By Dexter Roberts in Beijing, with Frederik Balfour in Hong Kong; Pete Engardio in New York and Joseph Weber in Chicago.

Animé—Japan's Animated Pop Culture

The first Japanese animated film was made about 90 years ago, and Japan is now the anime capital of the world. How did animé develop into an industry and a pop culture, and why have people around the world become animé fans? Here we examine the evolution of animated films in Japan, and discover the secret of their success.

The Worldwide Phenomenon of Animé: Past and Present

Written by Yonezawa Yoshihiro

Japanese Animated Films Sweep the World

After *Spirited Away* won an Oscar for best animated feature film at the 75th Academy Awards in 2003, Japanese animated films (animé) achieved instant fame worldwide. Of course, even before then, animé were being broadcast and enjoyed by children everywhere, although most did not realize they were watching a Japanese production. Animé have attracted fans all over the world since the early 1990s. In the U.S., the heros in early TV animé like *Astro Boy* and *Speed Racer* have become icons. *Heidi, Girl of the Alps* and *CandyCandy* were televised in Europe. It has even been reported that almost 90% of Spain's population was in the habit of watching *Mazinger Z*. Some professional soccer players say they started playing avidly after watching *Captain Tsubasa*. The *Sailor Moon* boom in Germany a few years ago created tremendous interest in other animé as well, in that country. In Asia, *Doraemon* and *Dragon Ball* are apparently better known than Disney's animated films.

Cyberpunk animé that explore the near future—including Otomo Katsuhiro's *Akira*, Shiro Masamune's *Ghost in the Shell: Stand Alone Complex,* and *Neon Genesis Evangelion*—created a model that influenced films like *The Matrix*. The movie *Pocket Monsters*, a spin-off from the *Pokemon* video games, made waves in the U.S, and has become a classic. Around 50 or 60 new animé are produced every week in Japan, and a number of these are being exported.

Animé and Manga:
An Interdependent Relationship

Why are animé so popular in Japan, and why are they now beginning to attract so much attention overseas? When answering these questions, we cannot ignore the strong appeal of manga, the ancestor of animé. Other possible reasons: many people

enjoy the animated film genre, and the animé format makes it easy to change elements to fit the sensitivities of the audience in different countries. As a result, the fantasy world and characters of Japanese manga-animé pop culture are now widely known.

Actually, animated films were being produced in Japan before World War II by artists such as Masaoka Kenzo and Seo Taro. After the war, animated feature films for cinema, beginning with Toei Animation's *White Colored Snake*, followed in the footsteps of Disney masterpieces—and did it so well that some were released worldwide, surpassing Japanese expectations. And yet it was *Astro Boy,* a TV show quite different from the Disney model, that set the ground rule for modern Japanese animé—emphasis on character and story line.

Tezuka Osamu was a great admirer and follower of Disney animated productions. When working on *Astro Boy,* he decided to cut the number of cell pictures and use the same actions often. This approach, while reducing movement and aesthetic appeal, made it easier to concentrate on the story and dramatic moment. Tezuka tried various ways to reduce production costs and keep turning out new work. He developed a unique art form by taking immobile pictures and making them appear to move, and stacking short shots together.

Astro Boy was a tremendous hit and launched a new era of science fiction animé, including titles such as *Iron Man 28* and *8th Man*. It also set the stage for the creation of a new TV animé business model, which was the development of animé character products and tie-ups with sponsors.

More than half of all animé produced in the 1960s were inspired by manga stories. After the science fiction boom, the TV animé that were based on manga stories went off in different directions, with such hits as *Sally the Witch,* for young girls, *Q-taro the Ghost,* featuring a timid ghost, and *Star of the Giants*, about a young baseball player who seeks the truth. Popular manga stories were transformed into TV animé, and established themselves in pop culture. Paradoxically, by attracting a large following among the very young and the not-so-young, animé ended up promoting greater interest in manga, boosting comic book sales.

Bigger Audiences, More Varied Productions

Nagai Go created a hit with his *Mazinger Z* in the early 1970s, and this pushed the world of animé in a new direction—giant robots. Toy manufacturers then began making toys based on these mechanical creatures. Rather than depending on manga for inspiration, the new animé were inspiring spin-offs of their own. Science fiction thrillers with giant, fighting robots were churned out one after the other in the mid-1970s, the most notable being *Space Cruiser Yamato*, which spawned an animé generation. This film, so closely followed by teenagers, launched another science fiction boom, opened the door to the publication of animé magazines, and attracted the attention of older people, as well. One science fiction blockbuster riding this trend was *Mobile Suit Gundam*.

Animé began portraying teenagers and young adults 10 years after manga had done so, introducing complicated stories, youthful emotions, and philosophical themes. Animé such as *Legendary Giant God Ideon, Botomusu,* and *Daguramu,* all from the Gundam series, continued to focus on this age group into the 1980s. But the golden age of youth-oriented works that were not inspired by manga ended soon after 1980, and once again popular manga formed the basis for animated blockbusters. These included *Those Obnoxious Aliens, Touch,* and *Kimagure Orange Road.* Another big hit, *Dr. Slump,* forced animators to take a second look at manga. The manga-based trend was seen again in *Doraemon,* a feature animated film released in 1980.

Some stories in *Shonen Jump,* a very popular comic magazine even then, became animé hits. Three of these were *Captain Tsubasa, Knights of the Zodiac,* and *Muscleman.* On the other hand, films like Otomo Katsuhiro's *Akira* sprang from a new style of manga. Then, in the mid-1980s, both manga and animé took on new forms, helping them broaden their audience base by including other age groups. The various possibilities being explored within this pop culture attracted children, teenage boys and girls, and even adults.

Manga, which had inspired animé in the first place, began diversifying even further and underwent various transformations in style.

Most non-Japanese animated films were wholesome and suitable for children—especially those made by Disney—or else they were experimental art. On the other hand, Japanese animated films presented story lines suitable for teenagers and young adults, attracting a different type of audience. Japanese animated dramas shied away from uplifting themes leading to moral or poetic justice. Instead, they encouraged their audience to discover new possibilities and be swept up in some magic spell. High-quality works, like those of Otomo Katsuhiro and Shiro Masamune, became accepted as cyberpunk art.

Miyazaki Hayao, who had already made waves with *Lupin III: The Castle of Cagliostro,* was one animator who brought animé to the level of a cult with his *Nausica of the Valley of Wind.* Oshii Mamoru was another who achieved the same result with his *Those Obnoxious Aliens 2: Beautiful Dreamer.* As a result, the viewing public began seeing animators as celebrities. The entertainment and artistic aspects of animé began attracting serious attention in the 1990s.

- Number of animators in Japan 3,567
- Number of Japanese companies producing animé 247
- Revenues from Japanese animé in the United States US$4,359,110,000 (approx. 520 billion yen)

  Breakdown
  Licensing revenue (including fees for TV broadcast rights): US$3,937,000,000
  Video and DVD sales: US$414,000,000
  Revenue from cinema showings: US$8,110,000

- Toei Animation revenue from animé sales in Japan and abroad
Unit: million yen

Year	Overseas	Japan
1999	1,561	8,233
2000	5,452	9,393
2001	7,154	8,876
2002	6,253	11,442

The Evolution Continues

In Japan, one direction taken was a focus on mature themes. These animé were difficult to export. The sexual overtones and violent scenes were considered unsuitable for audiences in countries where animated films were made for children. Some works were edited, with objectionable scenes cut so they could be broadcast in countries where strict criteria were enforced. Many others were seen only in Japan. Even today, the animé-for-teenagers model is still not fully understood overseas, and films that fit this model are viewed only by enthusiasts.

Animé offer many possibilities, because they speak in an international language and express a story in a way understood universally. Many works have not yet been exported, but they are waiting to get out there. Animé have evolved to a high level because anything can happen in them, and everything can be expressed freely. They introduce fascinating characters, complex inner workings, and stories with deep meaning, all combined with rapid action, expressive power and exceptional visual effects. These superlative aspects explain why they are so popular in Japan. But continued success can only come if they can keep satisfying the fans, who always want to experience new and more exciting possibilities.

Japan's Aging Society

Fewer children, more elderly people—Japan's demographics are changing rapidly. The aging society is a major issue for the 21st century.

—Tadahiro Ohkoshi & Masaki Yamada

According to the national census taken in 2000, Japan's total population is 126.9 million. Children (ages 0 to 14) make up only 14.6% of that number, at just over 18.5 million. No other industrialized country has such a low proportion of children. People aged 65 and older make up 17.3% of the total population, numbering just over 22 million. That means that one in every 5.7 people is a senior citizen.

The main reason for this aging of society is a major increase in life expectancy. Economic growth following the end of World War II brought about vastly improved living conditions, and life expectancy grew dramatically. In 1947 Japanese men could expect to live for 50.1 years, and women for 54.0 years. By 2000, according to a 2002 study by the Ministry of Health, Labour and Welfare those numbers had increased to 78.3 years for men and 85.2 years for women; these are the highest life expectancies in the world.

The birthrate has also declined significantly. A 2003 study by the Ministry of Public Management, Home Affairs, Posts and Telecommunications showed that 1,153,866 babies were born in Japan in 2002, and the average woman had 1.32 babies during her lifetime. These numbers are historic lows. Until recently, the main cause of the falling birthrate was thought to be women's increased workforce participation, which resulted in both fewer and later marriages, and in later childbirth. However, in recent years another factor has emerged, accelerating the decline: more couples are choosing not to have children, even after marriage. A study conducted in June 2002 by the National Institute of Population and Social Security Research looked at the average number of children married couples planned to have. The overall average was 2.13 children, but, notably, the average for couples married for less than five years was 1.99, the first time this indicator had dropped below two.

What are the implications for Japan's society and economy if the current trends towards increased longevity and decreased birthrate continue?

The first major problem is pensions and social security programs. According to predictions laid out in an April 2002 report by the National Institute of Population and Social Security Research, the proportion of Japan's population aged 65 or older will grow from 17.4% in 2000 to 22.5% in 2010. It will continue to climb, and is expected to reach 35.7% in 2050. This means one in 2.8 people will be 65 or older, and pension and social security programs will have approximately two people to support each elderly recipient. Moreover, since this 'approximately two people' includes children, students, and others who are not working, the burden on people who are working will be greater still.

There are also economic repercussions. Fumikatsu Kimura, head of policy research with Mitsubishi Research Institute, Inc., explains; "If the current trends continue, the younger segment of the population will shrink, but this is the segment which comprises the labor force. And with a smaller population, demand for consumables will also shrink. That will result in industrial decline and economic stagnation. We will likely see either zero or negative growth in GDP."

There are other issues associated with a declining birthrate and an aging population: these include economic failure of universities and other educational institutions, depopulation of residential areas due to a population exodus, and possibly even labor shortages in the caregiving service industry.

Of course, the aging society does not only mean doom and gloom. A decrease in the population will reduce the burden on the environment, and alleviate some of the housing and land-use problems facing urban areas. Fewer children may encourage smaller pupil-to-teacher ratios, and the pressure on children of school entrance examinations may lighten. However, the reality is that the issues requiring immediate attention far outweigh the positive aspects.

Japan is not alone in facing the challenges of an aging society. China, South Korea, Thailand and other Asian countries are also experiencing similar population trends. By the 2050s, one person in five worldwide is expected to be age 65 or older (according to a 2002 United Nations study). Japan is joined by the entire world in facing a declining birthrate and an aging population as important issues for the 21st century.

From , vol. 1, no. 6, October 2003. Copyright © 2003 by Jijigaho.

As Japan's Women Move Up, Many Are Moving Out

By HOWARD W. FRENCH

TOKYO—Life was happy during the first 10 years of Tomoko Masunaga's married life. At the very least, as a middle-class housewife with two small children to raise, she was far too busy to focus on the nettles, Ms. Masunaga said.

Serious problems in her marriage began to surface, though, as her children grew older, and Ms. Masunaga began doing things outside the home, first busying herself with the local parent-teacher association, and eventually writing articles for the teachers' union and environmental groups.

"He had promised he would support me if I decided to work someday, and then he betrayed me," Ms. Masunaga said of her husband, an executive. "What's worse, he got old very quickly.

"For the first 10 years at least he made an effort at conversation," she said. "But the company was everything for him, and after awhile, he would just come home tired and sit silently watching TV, drinking his beer."

Finally, after more than two decades of marriage, Ms. Masunaga moved boldly to cast off her unhappiness and, taking a step that stunned her husband, got a divorce against his wishes.

Ms. Masunaga, now a vivacious 60-year-old who went on to write a popular book about her experience, teaches English and has resumed a practice abandoned since her college days: dating.

Little more than a decade ago, middle-aged divorces like these were almost unheard of in Japan, even while the divorce rate among younger couples was steadily creeping upward to levels comparable with many European countries.

While the overall divorce rate in Japan still appears flat when compared with America and Europe, in the last few years divorces among older people have been skyrocketing, reflecting profound changes in a traditionally conservative society.

Novel concepts like individualism, materialism and personal happiness, experts say, are breaking down age-

old notions of the collective good, of harmony, and, above all, of "gaman," or self-denial.

"Middle-aged divorces in my practice have gone up 300 percent in the last 10 years," said Atsuko Okano, who runs a highly successful marriage counseling business, with offices around the country. "Gaman used to be considered Japan's greatest virtue, and for the wife, that meant responsibility toward the children, and knowing how to cope.

"There has been a real shift in the winds since Japan became rich, and more and more women started working outside the home," Ms. Okano added. "Nowadays, without a doubt, one's own life is the most important thing."

Japan's rate of divorce per 1,000 people has jumped in recent years.

Technically speaking, divorce has long been a simple matter in Japan, requiring little more than the signatures of both parties and a trip to city hall to file the papers. For women, however, a hundred more or less hidden constraints have long conspired to make legal separation morally or financially prohibitive, in all but the most abusive relationships.

The wife, in particular, traditionally faced opprobrium from family, which expected her to suffer in silent dignity. The whispers of neighbors in a shame-sensitive society were another powerful disincentive. If that were not enough, divorce courts were stingy with alimony, the job market all but closed to middle-aged women, and banks unwilling to issue loans, mortgages or even credit cards to the former wife, a situation that is only slowly changing.

"I knew that I couldn't get a credit card once I was divorced, so I established one in my name beforehand," said Keiko, 43, who divorced a retired diplomat two years ago and asked that only her first name be used.

"Nonetheless, the difficulties were greater than I had expected," she said. "I was a housewife—with no career, no qualifications and no skills—and it was difficult to find a job.

"Even for part-time jobs and contract workers, most companies only accept people under 35."

Sociologists often lump the significant shifts taking place in marriage with other profound social changes under way here: the demise of lifetime employment, the postponement of marriage, the collapsing birthrate and the dropping out of many younger adults, who drift between part-time jobs and live with their parents well into their 30's.

Taken together, many experts say, these changes are comparable to the seismic cultural shifts seen in the United States in the 1960's and 1970's.

"It is very hard to look at Japan today and imagine the way it was just a decade ago," said Sean Curtin, an expert in family studies at the Japan Red Cross University, on Hokkaido. "Usually when we think of changes in the family we think of younger people, but these things are sweeping through the entire society."

Mr. Curtin, who has tracked the evolution of television in Japan, says the changing attitudes surrounding middle-aged divorce are both reflected in and propelled by racy daytime dramas and boisterous talk programs, known as "wide shows," which have gone from schooling housewives in how to become the perfect spouse to tutoring women in issues like divorce, postmotherhood careers and sexual freedom.

"In the 1950's, TV advertisements typically showed smiling women serving coffee to their husbands, but by the 1960's people were breaking out of such stereotyped roles," he said. "That's what's taking place right now in Japan."

Other experts attribute the explosion in middle-aged divorce to a sort of trickle-up women's liberation, in which grown daughters, often still living at home, prod their mothers to stop putting up with emotionally barren or abusive relationships with their fathers.

"For a very long time I was unhappy, but when I consulted with my parents and friends, they all said to me that I was being too selfish," said Keiko Imaizumi, 56, who is in the midst of a divorce from an art gallery manager, after over 20 years of marriage. "For years, I resolved to simply be a good wife, but I was never happy. Finally, when I told my children I wanted a divorce, all three children said they supported me.

"One of them said, 'Mom, you've been cleaning up his messes all these years, at last you should be free to enjoy your own life.'"

Whither North Korea?

North Korea is often no more than a blip on the radar screens of international news agencies. However, over the last two years it has attracted more media coverage, as a perfectly manageable crisis over North Korea has been teetering out of control. Usually referred to as the nuclear crisis, and dated back to October 2002, this crisis is far more fundamental and comprehensive than the gradually increasing nuclear bravado of the North, and can be traced back to the coming to power of the Bush administration.

By Koen De Ceuster

Out of a growing concern about and frustration with the confrontational policy of the US government towards North Korea, scholars from the US and around the world united in March 2003 in an 'Alliance of Scholars Concerned about Korea.' Convinced that political problems 'can only be solved through dialogue, cooperation and active pursuit of peace,' the association is dedicated to 'the promotion of mutual understanding between the people of the United States and the people of Korea, both North and South.' By providing accurate, historically informed analyses, it seeks to help scholars, students, policy makers, and the general public to learn about Korea, and to contribute to the constructive and peaceful development of US-Korean relations.[1]

In a similar though unrelated initiative, a panoply of speakers from different national and disciplinary backgrounds, but all motivated by the same concern, gathered late last June in the once divided city of Berlin to ponder the future of North Korea.[2]

At the end of the day, the participants left the Berlin symposium with the bewildering feeling that all issues touched upon—famine relief and the humanitarian crisis, economic reform, inter-Korean cooperation and reunification policies, and the nuclear crisis—were conditional on the willingness of the US government

to engage North Korea. The key to unlock the gridlock in and over North Korea clearly lies in the White House. Coincidentally, this would have to be the same key that firmly locked the door to any meaningful détente when George W. Bush took over the American presidency in 2001.

A cold shower during Sunshine

Determined to prove himself the anti-Clinton in foreign policy, George W. Bush abruptly withdrew all contact with North Korea and ordered a policy review, not unlike the review Bill Clinton had ordered back in 1998. This felt like a cold shower in Korea, following the rapid improvement of inter-Korean relations since the historic June 2000 summit between South Korean president Kim Dae Jung and the North Korean leader Kim Jong Il. That meeting proved to be the start of a thawing period on the Korean peninsula. The North Korean regime inched forward in its engagement with the outside world, while many allies of South Korea, in line with Kim Dae Jung's Sunshine policy, established diplomatic relations with the North. October 2000 proved a watershed in US-DPRK (Democratic People's Republic of Korea) relations with the visit of

the first vice chairman of North Korea's National Defence Commission, Vice-Marshal Jo Myong-Rok, to Washington, followed later in that month by a return visit to Pyongyang by American Secretary of State Madeleine Albright. Short of formal diplomatic recognition, this was the closest the US ever came to acknowledging the DPRK. The swift progress in solving outstanding nuclear and missile proliferation issues was such that even a state visit to Pyongyang by outgoing President Bill Clinton was on the drawing board. The institutional crisis over the American presidential elections, and their eventual outcome, decided differently. The moment the Bush administration took over in Washington, a new chill came over US DPRK relations. All contacts were put on hold pending a review of the US government's North Korea policy. North Korean gestures of goodwill towards Washington: the prompt official condemnation of the WTC attacks of 11 September, its professed opposition to any form of terrorism, and the North's signing of two UN treaties against terrorism, all went unacknowledged.[3] The visceral dislike for the likes of Kim Jong Il in the White House made the Bush administration up the ante all the time. Any North Korean concession only led to stiffer demands from Washington. Pyongyang's hopes for improved relations with the US were finally dashed on 29 January 2002 when George W. Bush, in his State of the Union address, singled out North Korea as belonging to 'an Axis of Evil', thereby earmarking the North as a potential target for a preemptive strike.

Bluff and rebuff

In October 2002, nearly two years after Madeleine Albright's visit to Pyongyang, US Assistant Secretary of State for East Asian and Pacific Affairs James Kelly travelled to the North not so much to re-open a dialogue but to confront the North Koreans on their home turf with 'conclusive' evidence of Pyongyang's secret uranium enrichment programme. He brought the message that Washington would not talk to the North until it had totally and verifiably dismantled this secret programme. Through press leaks orchestrated from Washington indications first trickled through that Pyongyang had in fact been rebuffed. From the various versions of events now in circulation, it is obvious that the American visit was hardly an attempt at diplomacy. Also clear is that the North Korean delegates did not anticipate such high-handedness. As a (typical) response, they bluffed their way out of it by confirming on the sidelines of the meeting that indeed they had this secret programme going, adding in the same breath that they were willing to negotiate about its dismantling.

As 2002 ended, the situation was getting out of control. What followed was a sequel to the 1992–94 crisis which had ended in the 1994 Geneva Agreed Framework. Then as now, suspicions about the exact nature of North Korea's nuclear ambitions had led to a confrontation with the US which was only dispelled following the intervention of former American President Jimmy Carter. The agreement that was eventually brokered offered the North two less proliferation-prone 1,000 MW light water reactors in return for the internationally supervised mothballing of the Yongbyon nuclear complex. The Republican opposition in the American Congress cried appeasement and accession to nuclear blackmail, and tried to block its implementation. The Bush administration lost no time in using the disclosure of the uranium enrichment programme to once and for all derail the Agreed Framework. Despite the recognition of KEDO, the international consortium overseeing the implementation of the Agreed Frame-

work that the North had scrupulously lived up to the letter of the Agreement (though obviously not the spirit, given its secret uranium enrichment programme), the October disclosure offered the Bush administration the ammunition to blow the much maligned Agreed Framework irretrievably to pieces. Washington stopped the yearly delivery of 500,000 metric tons of heavy fuel under the Geneva Agreement, which in turn provoked the North into announcing it did not feel bound by the Agreement anymore. Pyongyang declared its immediate withdrawal from the Non-Proliferation Treaty, expelled the two IAEA (International Atomic Energy Agency) inspectors from the country, broke the seals of the Yongbyon complex and, in April 2003, following another failed attempt at renewing proper dialogue with the US, proclaimed its intention to begin the reprocessing of 8,000 spent fuel rods.

Cognitive dissonance

This crisis could have been avoided, and the threat of nuclear proliferation could have been contained. The secret uranium enrichment programme that started this renewed nuclear crisis over North Korea was to all accounts up to four years away from maturation. What is more, the North has time and again indicated that it was willing to find a negotiated solution with Washington.[4] American mismanagement of this crisis, provoked by Washington in the first place, led to the restarting of the Yongbyon nuclear complex and the very real possibility that the North is (capable of) producing nuclear warheads.

However, this is not how this crisis is usually reported. Media follow Washington's lead; news about North Korea is often filed from Washington, where State or Defence Department briefings set the tone. North Korea hardly has a voice, and the voice it has is distorted through a haze of cognitive dissonance. Flustered by the bombastic rhetoric of the North, and unwilling to question the motives behind the US government's policy, no effort is made to understand the intentions of the North. Instead, the media seem to take the image of an immovable, monolithic North Korea frozen in time for granted. Strangely enough, contrary to the customary image of an erratic North Korea, Washington's motives have become hard to gauge. With ongoing squabbles between 'hawks' and 'doves' in the Defence and State Depart-

ment, the US administration speaks with a split tongue.[5] While publicly paying lip service to South Korea's Sunshine policy of engagement and reapprochement, administration officials in Washington come out in support of an induced collapse of the North Korean regime. Even with the Bush administration currently shifting towards a more accommodating position, it is hard to believe that this is any more than window dressing. The US participates in the Six Party Talks in Beijing with the sole purpose of having the North unconditionally acquiesce to all American demands. While ruling out a military invasion of the North, Washington has made no secret of the fact that its 'Proliferation Security Initiative' is clearly aimed at North Korea, and is second best to an economic blockade, which it cannot enforce. By maintaining this policy confusion, the American government can rest assured that the North will stick to its provocative posturing. Unable to fathom the true intentions of the American administration, the North has no intention to let its guard down. Iraq was a clear reminder that concessions and cooperation with this administration can be counterproductive.

With all attention focused on the ongoing international stand-off over North Korea's nuclear ambitions, the Korean people are once again threatened by a renewed deterioration of the food situation. Not so much donor fatigue, as a (renewed) politicization of food aid is menacing the stability that had been reached. The nuclear crisis is also overshadowing the real efforts the North Korean regime is making to implement economic reforms. At the Berlin symposium, the question was 'whither North Korea'; the answer may have to be found in Washington.

Notes

1. See the mission statement of the ASCK, on: www.asck.org/statement.html
2. 'Wohin Steuert Nordkorea? Soziale Verhältnisse, Entwicklungstendenzen und Perspektiven', an international symposium organized by the Korea-Verband e.V. (in Asienhaus, Bullmannaue 11, 45327 Essen, Germany. www.koreaverband.de) on 25 June 2003 in the Centre Monbijou im Haus der Bank für Sozialwissenschaft, Berlin.
3. On 12 November 2001, the North Korean representative to the UN, Ri Hyong Chol, signed the 1999 International Convention for the Sup-

pression of the Financing of Terrorism, and the 1979 international convention against hostage taking. Hwang Jang-jin, 'N.K. Signs U.N. Convention on Anti-Terrorism', *The Korea Herald*, 29 November 2001.

4. North Korea's voice is seldom heard (undistorted). In Berlin, a very balanced justification of North Korea's right to have a deterrent was read out. See 'Die Berechtigung der DVRK zum Besitz militärischer Abschreckungskraft' (The DPRK's justification for the possession of a military deterrent), a collective document prepared by the Institute for the Reunification of the Fatherland for presentation at the Berlin symposium and included (in German translation) in the (unpublished) symposium materials.

5. At the Berlin Symposium, Bruce Cumings spoke on 'North Korea, the Sequel,' addressing the Washington wrangle over some form of North Korea policy. In his upcoming book, *North Korea: The Hermit Kingdom*, New York and London: The New Press, (2003), he paints a tantalizing portrait of North Korea.

Dr Koen De Ceuster is Assistant Professor at the Centre for Korean Studies, Leiden University. He is Korea editor of the IIAS Newsletter and a member of the Association of Scholars Concerned about Korea.k.de.ceuster@let.leidenuniv.nl

From *IIAS Newsletter*, November 2003. Copyright © 2003 by IIAS Newsletter. Reprinted by permission.

Parasites in Prêt-à-Porter

By Peggy Orenstein

Japan's young women are shunning marriage, spending big and still living with their parents. Are these 'Parasite Singles' the harbingers of a feminist revolution, or have they just gone *wagamama*?

On a Sunday afternoon, the Omotesando neighborhood in Tokyo swarms with women in their 20's and 30's. They spill out of stores with bags marked Gucci, Jil Sander, Issey Miyake, Comme des Garçons. They crowd cafes and snack on coffee and cake specials. They finger sleek tchotchkes on the shelves of home design stores, occasionally shouting, "*Kawaaaaaiiiii*!" the compliment supreme, which means, essentially, "cuuuuuuute!" As I cross the main thoroughfare, three women in their late 20's pass by, arms linked. They are actually singing "Girls Just Want to Have Fun."

Sumiko Arai waits for me at the Anniversaire cafe. Her hair is lightened to an of-the-moment reddish brown, cut in a layered bob. She carries a trendy canvas purse with sequined straps and is shod in high-heeled, animal- print mules with cripplingly pointed toes. Fish-net stockings? Of course. She even has the makeup right: neutral, with clear lip gloss and a bit of shimmer across the eyes. She waves when she spots me, a cell phone in her manicured hand, her right cheek dimpling in a grin.

Anniversaire, arguably Omotesando's most popular cafe, is attached to a department store specializing in weddings. Brides can shop for trousseau jewelry on the first floor, register for gifts on the second and third and buy frothy, Western-style dresses on the appointment-only upper levels. There is even a quaint wooden chapel where ceremonies are performed. Yet when a pair of cake-perfect newlyweds glide past the cafe, store clerks ringing hand bells behind them, the fashionable young women sipping cappuccino take little notice. Few of them are married or want to be any time soon. "If I were married," Arai says, "I'd have to worry about what my husband thinks, about cooking for him, cleaning. And if I had kids. . . . " She shrugs. "Child-rearing takes up so much time." Arai's cell phone chimes, a bell she always heeds. "Right now," she adds before answering it, "I don't have to worry about anything."

MORE THAN HALF OF JAPANESE WOMEN are still single by 30—compared with about 37 percent of American women—and nearly all of them live at home with Mom and Dad. Labeled "Parasite Singles" (after "Parasite Eve," a Japanese horror flick in which extraterrestrial hatchlings feed off unsuspecting human hosts before bursting, "Alien"-style, through their bellies), they pay no rent, do no housework and come and go freely. Although they earn, on average, just $27,000 a year, they are Japan's leading consumers, since their entire income is disposable. Despite Japan's continuing recession, they have created a boom in haute couture accessories by Louis Vuitton, Bulgari, Fendi and Prada, as well as in cell phones, minicars and other luxury goods. They travel more widely than their higher-earning male peers, dress more fashionably and are more sophisticated about food and culture.

While their spending sprees keep the Japanese economy afloat, their skittishness about traditional roles may soon threaten to capsize it. Japan's population is aging more rapidly than any on the planet—by 2015 one in four Japanese will be elderly. The birthrate has sunk to 1.34 per woman, well below replacement levels. (The birthrate in the United States, by contrast, is 2.08.) Last year, Japan dropped from the eighth-largest nation in the world to the ninth. The smallest class in recorded history just entered elementary school. Demographers predict that within two decades the shrinking labor force will make pension taxes and health care costs untenable, not to mention that there will not be enough workers to provide basic services for the elderly. There are whispers that to avoid ruin, Japan may have to do the unthinkable: encourage mass immigration, changing the very notion of what it means to be Japanese.

Politicians, economists and the media blame parasite women for the predicament. (Unmarried men can also be parasitic, but they have received far less scrutiny.) "They are like the ancient aristocrats of feudal times, but their parents play the role of servants," says Masahiro Yamada, a sociologist who coined the derogatory but instantly popular term "Parasite Single." (The clock on his 15 minutes of fame has been ticking ever since.) "Their lives are spoiled. The only thing that's important to them is seeking pleasure."

He may be right: parasite women may indeed be a sign of decadence, a hangover from the intoxicating materialism of the Bubble years of the 80's. But that conclusion, the most common one in the Japanese press, misses something more substantive: an unconscious protest against the rigidity of both traditional family roles and Japan's punishing professional system.

"Maybe they appear to be spoiled," says Yoko Kunihiro, a sociologist who studies dissatisfaction among women in their 30's, "but you could also perceive Parasite Singles as the embodiment of a criticism against society. Seen from the perspective of conventional values, even feminist values, they seem like a very negative force, but I see something positive in them."

THERE WAS A TIME WHEN A WOMAN SUMIKO Arai's age would have been dismissed as "Christmas cake": like a holiday pastry, her shelf life would have expired at 25. But sell-by dates have changed in Japan, along with male predilections: high-profile sports heroes like the Seattle Mariner Ichiro Suzuki and the sumo grand champion Takanohana are married to women several years their senior. Instead of calling her a stale sweet, Arai's parents, with more affection than disapproval, call her *para-chan* ("little parasite": "chan" is the diminutive in Japanese). "They tell me to get married and leave the house," she says. "But if they really thought that, they'd try to set me up on *omiai*"—matchmaking meetings. "My mom has said to me, 'Make sure you find the right guy.' I think she's speaking from personal experience. Maybe she feels she did find the right guy. But I think sometimes she's telling me because she wishes she'd chosen better herself."

Arai is in her late 20's, the younger of two children—her older brother is married but childless. Her mother is a housewife, her father a salaryman whom she speaks of fondly. That is somewhat unusual; most of the women I met felt little connection to their dads, whose careers took precedence over family life. "We don't have an emotional bond," one woman explained. "I try to be nice to him now that he's retired, but I hardly saw him growing up."

Arai is the publicist for Girlsgate.com, one of several Web sites for women that have sprung up over the last year in Japan. Girlsgate hopes to lure the discerning parasite with articles on the history of

Hermès fashions and tips on customizing designer shoes. The editor in chief, Yoshiko Izumi, 31, is a former Miss Fairlady, the generic term for a woman who stands smilingly next to new-model vehicles at car shows. Although not a parasite herself—Izumi has been married to her business partner for seven years—she says she understands the parasite psychology. "They're not dependent on men financially," she explains as we huddle around an i-Mac at Girlsgate headquarters, touring the site. "They're enjoying their lives. They don't want to give up that pleasure for marriage." She turns to Arai. "Isn't that right?"

> 'Maybe I am *wagamama*. I don't know. I do want to get married eventually, but I have to find the right guy. **Meanwhile, I want to look cool.**'

Like employees of dot-com start-ups anywhere, Arai frequently clocks 12-hour days, sometimes more, which separates her from the parasite pack. "I didn't expect that when I took the job," she confides later. Although she finds the work satisfying and is carefully politic in her enthusiasm for the company, the long hours are, quite simply, tedious. She never set out to be the woman in the gray flannel kimono. "You have to put in a ridiculous amount of time to succeed in Japanese society," she says. "But if you have a typical office lady job, you work regular hours. After work you can take lessons, spend time with friends, have dinner. That's why women in Japan have much better lives than men."

Girlsgate was launched last year on March 3, which in Japan is Girls' Day: families with daughters display dolls representing the ancient royal court, eat foods symbolizing purity and pray for their girls' happiness. Legend has it that if mothers don't pack the dolls away promptly after the holiday, their daughters will be slow to find a husband. Apparently, in Sumiko Arai's generation, there were a lot of lazy moms. "No one is actually rejecting marriage," Izumi says quickly. "Not even Sumiko. They all think they'll probably get married some day. It's just..." She breaks off and laughs. "Women today are *wagamama*."

Her word choice is significant. *Wagamama* means selfish, willful; in a culture where personal sacrifice is the highest virtue, the connotation is far harsher, especially for women. Yet, as the parasite trend has emerged, women like Arai have taken on the word *wagamama*, albeit slightly tongue in cheek, as a term of defiance—somewhat like the way American women use "girl" or African-Americans say "nigga"—transforming its meaning in the process to something closer to "choosy" or even "self-determining." Makiko Tanaka, Japan's first female foreign minister, who has been flayed in the press as "rude" and "incompetent," is *wagamama*. Princess Masako, whose sole purpose is to produce an heir to the throne, is not. Women's magazines have caught the trend, featuring headlines like "Restaurants for the Wagamama You." One afternoon, I even walked by an office building on which WAGAMAMA was painted in English letters 10 stories high. Although women would initially startle when I asked if they were wagamama, they would, with some self-mockery, accept the label: it had clearly become a kind of resistance against expectations.

"Maybe I am *wagamama*," Arai says. "I don't know. I do want to get married eventually, but I have to find the right guy." She grins mischievously. "Meanwhile, I want to look cool."

If you were a 30-ish parasite in the spring of 2001, you would mix your Uniqlo T-shirts (Uniqlo being the Japanese version of the Gap) with high-end designer accessories. You would be considering the purchase of a wide belt. You would take lessons in English or French. You would frequent galleries. You would be planning a vacation to Vietnam. You would be tiring of Italian restaurants and returning to Japanese, served in a Western setting. In a nation where the G.D.P. is driven by consumer spending, you would be part of an economic powerhouse. The entertainment, travel and fashion industries would cater to your slightest whim. Consider: Since the current recession began in 1994, the G.D.P. has dropped nearly 20 percent. Japanese sales of Louis Vuitton products, meanwhile, have soared from $36 million to $863 million annually, accounting for a full third of the company's worldwide sales.

"We're heading into a market in which mothers in their 50's and daughters in their 20's and 30's are the main consumers," explains Jun Aburatani, founder of the Tokyo marketing firm Gauss. He calls it the New 50 Pattern Society. "The mothers are beginning to live their lives after long years spent child-rearing. The daughters are liberated from social pressure to get married, so they, too, are beginning to live their lives. You can see it already: even during the recession we're in now, entertainment, designer products and healthy products are strong sellers: red wine, olive oil, vitamins, travel, performing arts, diet products. Young women are very positive about enjoying their lives. They go to hot-springs resorts. They buy clothes, shoes, purses, cosmetics. Men don't. And that reflects a difference in attitude between fathers and mothers. The women are much more vivacious." According to Gauss, men over 50 want to die before their wives—and their wives want them to as well. Widowhood, the women say, is the best time of their lives.

Switch on Japanese TV, and you'll see this new trend mirrored in advertising. In America, where there has been a second baby boom, children symbolize satisfaction and fulfillment. Soft-focus images of infants or families are used to hawk cars, insurance, coffee, prescription drugs. Japanese ads are comparatively baby-free. "We're becoming a society that excludes children," Aburatani says. "Whether you think it's good or bad, that's the way it is. Many women over 50 found marriage to be a disappointment and motherhood to be a burden. They tell that to their adult daughters, and that makes their daughters want to stay single. They doubt whether husbands and children are worth it."

As I leave Gauss, the woman who served the obligatory tea during my meeting with Aburatani stops me. "Would you like to ask me some questions?" she says. "I fit into your demographic—I'm 26 years old with no plans to marry." I agree, and we head downstairs to a Subway sandwich shop. Because she has her own apartment, Chiho Kashiwagi is not strictly speaking a Parasite Single, although she would be if her parents didn't live so far from Tokyo. She points out that those who condemn parasites tend to overlook the fact that women in Japan, as in most countries, earn less than men

and that Tokyo rents are prohibitively expensive. Not to mention that until 1986, many employers required single women to live with their parents, and some continue to look askance at those who don't.

Kashiwagi can't imagine modeling her life on her mother's, who is a housewife. "I see her dissatisfaction," she says. But would she like to live like her father? Kashiwagi laughs and tucks her hair behind her ears. "How can I answer that?" she says. "I'd like to be like him in the sense that he's independent, but if you mean working like a traditional Japanese man, no. He worked very long hours and devoted his life almost exclusively to his work. He has no hobbies, no outside interests. That conflicts with other things I want to do."

Kashiwagi admires Makiko Tanaka's political career, especially the very qualities that have drawn fire from the press: her outspokenness and common touch. Kashiwagi confides that her secret ambition is to pursue a career in politics, but since she'd also like to be a mother, she's not sure it's a realistic dream. Women like Tanaka—a prominent public figure who identifies herself as a "housewife"—are still the exception. For most women, it's still nearly impossible to have both a family and a career. "There's a term in Japan called the *ie*," Kashiwagi says. "It's like the household, but it means more than that." The word feels weighty, implying the family as institution. "Even now, Japanese marriage isn't between two individuals, but two ies. And that's the reason I'm single now. If I were married, I'd be influenced by the idea of the ie, by the expectations I'd feel."

> In Japan, where women for centuries were raised to be 'good wives and wise mothers,' Kashiwagi's simple assertion—
> **'I want to enjoy my life'**
> —is itself a radical act.

Kashiwagi wants to put off taking her place in the *ie* for as long as possible—perhaps until her mid-30's. Why would she be more willing to relinquish her professional goals then? "Well, I won't have much time left to have children, and I don't want to be a single mother," she says. We sit in silence a moment, then

she perks up. "For now, though, I want to live for myself and enjoy my life."

"Live for myself and enjoy my life." I heard that phrase, articulated precisely that way, from every single woman I interviewed. It was as if they had all read it in the same magazine (which is, in fact, possible). They said it with such finality, as if it explained something. When pressed, they would cite the freedom to go to nice restaurants, hang out with friends, buy clothes on a whim, do what they want. But listening to Kashiwagi, it dawns on me: for American women, self-determination may be the bedrock on which we build our dreams, but in Japan, where women for centuries were raised to be "good wives and wise mothers," simply asserting "I want to live for myself" or "I want to enjoy my life" is itself a radical act—even if it's unclear where it may lead.

I ask Kashiwagi, If I come back to visit her in 10 years, where will she be? "I wonder," she muses. "I suppose I could be a politician. But maybe I'll just be an ordinary housewife."

LONG BEFORE MAKIKO TANAKA THERE WAS Mariko Bando, one of Japan's best-known female politicians; she is currently director general of the Gender Equality Bureau Cabinet Office and a regular on the pundit circuit. Her bureau is charged with the amorphous task of "encouraging other ministries to look at gender issues" as well as strategizing on how to better support mothers in the workplace. So far, she has generated lots of paper with few results, which has been frustrating.

Rather than dismissing the parasites as merely spoiled children, overindulged by their parents, Bando says she believes there is also an economic explanation for the phenomenon. "We understand why Japanese women don't want to have children," she says. "Once they're married, they have to do all the housework. Japanese husbands may help some, but they won't share the burden. Also, if women work as hard as men they can be promoted—not always, but it's possible—but if they have children and stop working, it's virtually impossible to re-enter the work force. Many well-educated women quit and become housewives whether they want to or not. So instead, women are postponing marriage and/or childbirth."

The typical employment pattern for women in Japan is age-related, following an M curve: it peaks by 24, drops sharply, then spikes again in the early 50's (when former housewives take low-level part-time jobs) before falling off for good at 55. By comparison, American women's employment stays steady from their 20's until around 60. Some Japanese economists believe that boosting women's labor-force participation rate to U.S. levels during their 30's, 40's and 50's would lower inflation and raise the G.D.P. It would offset the labor shortage caused by the declining birthrate and revitalize Japan's economy. But that doesn't appear to be an argument either government or business is heeding; according to the Economic Planning Agency, working conditions for women have actually worsened in Japan since the 1980's. "They agree in principle," Bando says, "but a lot of men in the government are themselves uncomfortable with working mothers as colleagues."

Meanwhile, the psychological impact of the M curve, which Kashiwagi and Arai are preparing for, goes a long way toward explaining the parasite phenomenon. If, because of social pressures and discrimination—not to mention long, inflexible working hours, grueling commutes, lack of support from her husband and limited child care—a woman has to quit her job after having children, and never return, what is the motivation for someone who wants kids to push herself professionally? What is the motivation for an ambitious woman to contemplate motherhood? I recalled something that Yumi Matsushita, a 33-year-old interpreter, said to me: "You commute long hours in unfashionable trains and eat bad canteen food and for what? Do these men have good lives? And even if you get promoted, you have more drinking to do, more time at the office, more time away from family. And if you have a child, it will become even more difficult. So you have to wonder, what's the point in pushing harder?

"At the same time, once you become a mother, you're a mother. That's it. You're not a woman anymore. You can't work anymore. And the father's not involved. It's very confining. It limits your activities, your financial freedom. It's really not attractive."

So far, the government's main response to the baby bust has been to hike child allowances to about $2,400 a year per child for six years. Some conservative politicians would like to go further, increasing them tenfold. The idea is to offer an incentive for women to stay home,

making larger reforms unnecessary. Bando scoffs at that. "Women don't need a child allowance, they need services," she says, especially more day-care centers open longer hours. (Japanese nannies are virtually nonexistent, and hiring foreigners is illegal.) But creating more places to park the kids would not challenge the system a whole lot more than child subsidies do. Business meetings would still start at 8 p.m. Leaving work to tend to a sick child would still be considered a sign of disloyalty. "It's a workaholic culture," Bando agrees. "We have to change the structure of Japanese companies." And how will that be done? "This is the most difficult challenge," she says, shaking her head.

Bando is herself the mother of a 28-year-old Parasite Single who works in a pharmaceutical laboratory. Her dreams for her daughter often seem muddled. "Like many mothers, we think of our daughters as another version of ourselves," she says. "Most of my college friends quit their jobs to become housewives. They encouraged their daughters to have careers, and some do. My friend's daughter is a rising star in her profession, but she's single. And my friend says: 'I pity her. She can't marry and have children.' Even though this career is what my friend wanted for her. Maybe we can change it. I'm hoping it will change. But I'm not seeing it. Not yet."

I f you watch TV, it seems like American women feel a lot of pressure to marry or to be in couples," says Shuko Sadamoto, a single 34-year-old economist. "Do you think they just marry because it's *time*?"

Mihoka Iida, a 34-year-old magazine editor who lives with her parents, adds: "That obsession with having a boyfriend. . . . We just don't feel that paranoia. I mean, I enjoy 'Ally McBeal'"—which recently came to Japan as "Ally My Love"—but she seems so extreme. That dancing baby?" She rolls her eyes. "I think I'm doing Japan a favor by not having children. There are too many people in this country anyway." They both laugh.

Sadamoto and Iida have each traveled widely in Asia, Europe and North America. Sadamoto, who is shy and a little tomboyish, attended graduate school at Columbia University. Iida, tall and elegant in Comme des Garçons, spent her high-school years in Vancouver. They

say it's easier to be single in Japan, removed from America's pervasive "couples culture." "In the U.S. you're supposed to be together with your boyfriend or husband all the time," says Iida. "In Japan, women have their ways of having fun and men have their ways. You're not expected to bring a date everywhere, and you don't feel excluded if you're not involved with someone."

We are having drinks after work at Shunju, a restaurant at the top of a new skyscraper with a panoramic view of the city. Pinpoints of light beam down on our plates from artfully placed halogens, refracting onto our faces with a flattering glow. We are drinking wine and joking, but then Iida suddenly turns serious. "In a way, the men have to pay for what they've created in Japan," she says. "They hire all these educated, intelligent, even bilingual women in their companies, but then they don't utilize them the way they could. So that means the men have to work 24 hours a day, but the women don't. I suppose if this were America, women in that position would feel discriminated against, and they'd try to do something about it. We just react by going out and having fun, by not being part of it."

Of course women would like to have broader professional opportunities, they say, but not under the current conditions. "I don't know how men do it," Iida says. "When I first started working, I looked at the guys above me—and they were all guys—and I thought, I don't aspire to that. Living and working like men in Japan is not something to dream of. But then, I think that's related to the difficulty of having dreams in general."

Ever since the Bubble economy burst, Iida says, young people in Japan—both women and men—have lost hope. The Parasite Single phenomenon is merely a symptom of their pessimism. Women do tend to defer marriage and children during economic downturns: America's birthrate reached a historic low during the Great Depression. Perhaps in Japan it just looks different, with women deflecting despair by pursuing the perfect pair of shoes. "It's easy to blame women for the declining birthrate," Iida says. "To blame so-called Parasite Singles. But this isn't our making. We didn't create all these problems with the economy that have brought us to where we are now, where women have no hope for the future."

Sadamoto admits that she's not optimistic about the future. "Maybe if I were,

I'd have children," she says. "It's difficult to have dreams when the economy is so bad."

Iida says: "I don't really understand how Americans can be optimistic about marriage and children. It's possible, even probable, that it won't work out, but you do it anyway. I don't get it. We don't idealize family life. We never had a Japanese version of 'The Brady Bunch.'"

Sadamoto adds: "That's why being single doesn't feel like a sacrifice. It's the other way around—you have your own time and your own income. You weren't forced to get married to someone you didn't really like just because you were supposed to. That seems like a good thing."

THOSE WHO HAVE TRIED TO MAKE PARASITE singles the whipping girls for Japan's declining birthrate tend to believe that the solution is a return to the traditional family, in which men work and women stay home. In fact, the two factors that are keeping birthrates up in the United States are both distinctly nontraditional. One is single motherhood, which in America accounts for one-third of births. The sec-ond, according to World Bank data, is female employment. Women's earning power appears to raise confidence in the future: it gives young couples hope. Economic conditions in the West helped push women into the workplace; perhaps the situation in Japan will need to become significantly more dire before that solution is seriously considered. "In the United States and other countries, the economy went through gut-wrenching pain and got to a point where you couldn't afford one income to support a family," said Kathy Matsui, an analyst for Goldman Sachs in Tokyo. "That forced change."

Japan is nothing if not adaptable. It is a country with near-archaeological layers of tradition that can, nonetheless, change as quickly as Tokyo fashions. Consider: One day Japan operated on the feudal system. The next day it did not. One day the emperor was a god. The next day he was not. One day it was unthinkable for a woman to hold a cabinet post. Today Makiko Tanaka is the country's most popular politician. If such vast, deep transformation can happen so rapidly, there may yet be hope for break-ing the deadlock between Parasite Singles and a government that needs more bodies.

One spring afternoon, I visited Mitsuko Shimomura, a pioneering female journalist who, in her 60's, has taken over for her 90-year-old mother as administrator of her family's health clinic in Tokyo. She is also director of the Gender Equity Center in Fukushima prefecture, about an hour and a half outside Tokyo. "I don't regret the decline in the birthrate," Shimomura told me. "I think it's a good thing. The Parasites have unintentionally created an interesting movement. Politicians now have to beg women to have babies. Unless they create a society where women feel comfortable having children and working, Japan will be destroyed in a matter of 50 or 100 years. And child subsidies aren't going to do it. Only equality is."

Peggy Orenstein is a contributing writer for the [*New York Times* Magazine] and author of "Flux: Women on Sex, Work, Love, Kids and Life in a Half-Changed World."

Taiwan's democratic movement and push for independence

Taiwanese nationalism can be traced back to resistance against Japanese colonialism in the early 1920s. Upon Japan's defeat in 1945, Taiwan was returned to the 'motherland', the Republic of China. Taiwanese rebelled in 1947; the Guomindang's suppression of the uprising—the February 28 Incident—aliented the population and helped create the contemporary Taiwanese independence movement.

By Chang Mau-kuej

As is well known, there were two Chinas after 1949; following the Communist victory on the mainland, the island of Taiwan became the last holdout of Jiang Zieshi's Guomindang (GMD) regime. Until the mid 1980s, the GMD rules Taiwan with an iron fist; in the name of countering communist insurgency, the regime was inclined to punish all signs of political assertion from below. During this period, the independence movement was forced underground or into exile; it had little or no impact on cross-Straits relations or on Taiwan's domestic politics, though resentment against the ROC—the 'Chinese outsider regime'—remained.

Increased prosperity in the 1970s created a social base desiring political change. Opposition to the GMD grew, especially after 1978 when the US and the PRC established diplomatic ties. Diplomatically isolated and its legitimacy challenged, the GMD had to loosen its grip to include more Taiwanese in politics. This set the background for the political struggle during the process of democratisation between 1986 and 1995.

Indigenising Taiwanese politics

The opposition to the GMD regime called for democracy, social reform, and the asser-tion of Taiwanese identity and pride. The call to determine Taiwan's own future grew as control over the levers of political power and cultural domination shifted from Mainland Chinese to Taiwanese. Political indigenisation was prompted first and foremost by the GMD's desire to retain dominance; without its transformation, the GMD would likely have lost power much earlier. Institutionally, indigenisation included phasing out the National Assembly, which in theory still represented all of China, and revisions to the constitution to accommodate democratic politics and direct presidential elections.

From 1986 onwards, the GMD had to compete with the newly formed Democratic Progressive Party (DPP). In addition, the GMD had to face Taiwanese self-assertion from within the party—led by its own chairman Li Denghui. Li came to power in 1988, succeeding the last strongman of the ROC, Jiang Jingguo. Li's twelve-year rule—termed the 'silent revolution'—featured indigenisation as its basic philosophy in international relations and domestic politics, in trade and culture, the military and education.

Unsurprisingly, the programme provoked backlash. The power struggle within the GMD, the expulsion and marginalisation of mainlander elites from important positions, and the replacement of Chinese nationalism with Taiwanese consciousness resulted in the break-up of the GMD, first with the emergence of the New Party in 1993, and again in 2000 with the emergence of the people First Party. Feuds within the GMD benefited the DPP, allowing it to win key elections. The DPP not only sided with the GMD-promoted indigenisation campaign, but allied with Li in his intra-party fight, helping to split the GMD. Li led the GMD and the country until he was expelled in 2000 for 'destroying the party and selling the country'.

The 'silent revolution' encouraged citizens to cultivate their love and loyalty to Taiwan. Though the name and constitution of the ROC remain, people can now justifiably think of the ROC as equivalent to 'Taiwan', a source for new loyalty and pride.

The current dilemma

Beijing's influence on Taiwanese domestic politics has grown since the mid 1990s. This can be attributed to China's new economic weight, and the need felt by Taiwanese businesses to 'go west' across the Straits to compete successfully in the global economy. In 2003, China, including Hong Kong, accounted for about one fourth of Taiwan's trade surplus, and about one-half of Taiwan's foreign investment. To further complicate matters, an estimated one million Taiwanese live, study, do business or travel in China every day; others have chosen to live on the mainland more permanently. The number of cross-Straits marriages has also risen. As a result, China can now play Taiwan's domestic political game by manipulating Taiwan's vested and perceived interests. This has made Taiwanese party politics a nastier game, with both sides mobilizing appeals to national identity.

Li Denghui's visit to the US in June 1995, followed by his 'two countries', triggered the 1995-96 Taiwan Straits crisis which saw large-scale military exercises in southern china, and the shooting of ballistic missiles into Taiwanese waters. This was followed by Hong Kong's uncontested return to the mainland in 1997, which Taiwanese viewed with alarm. For the PRC, Taiwan remains the last lost territory, the final would caused by a century and a half of national humiliation. Many suspect that, if provoked, Beijing will use force to unify China; no one in power in Beijing can afford to appear soft on Taiwan.

Despite Beijing's repeated warnings, Taiwan held its first-ever referendum on March 20, 2004. Coinciding with presidential elections, the referendum was to call on China to remove its 500 mid-range missiles aimed at Taiwan. The proposed referendum invited citizens to vote on Beijing's stand—the legitimacy of its op-tion to use force to unify China. The referendum drew criticism from the US, where President Bush accused Taiwan's President Chen Shuibian of wanting to 'change the status quo unilaterally'.

Mounting pressure finally forced Chen to compromise. He replaced the original referendum question with two awkwardly worded queries that addressed funding for national defence, and the creation of a special department to promote peaceful relations with the PRC. The referendum failed to pass the threshold required by law (an absolute majority of eligible voters had to vote in favour). Only 45% of eligible voters participated, though 90% of them voted in favour of the two proposals. The referendum, however, demonstrated strong Taiwanese assertion in the face of pressure from both Beijing and Washington.

As the campaign ended, Chen regained the presidency by a margin of 0.2%. Protests questioning the legitimacy of Chen's victory plunged Taipei into chaos for weeks. Taiwan's voters are now divided into two camps. The first, Pan Green Camp, led by the current DPP government, sees the PRC as an immanent threat. While they may desire better relations with China, their main concern is Taiwan's hard-earned democracy, prosperity and pride. The DPP, under the pretext of improving government efficiency, wants to revise or draft a new Taiwanese constitution. As openly pushing for independence remains risky, the Pan Green Camp has chosen a defensive approach to the sovereignty issue: resistance to unification, and, as a last resort, insistence on the right to declare independence should Beijing invade.

The Pan Blue Camp is led by the GMD and other opposition parties. Viewing China as the land of economic opportunity, they want Taiwan to make use of its relative advantages before it is too late. They do not 'wish' for better relations with China; they demand the government improves relations immediately. Criticizing Taiwanese independence as parochial and risky, they present themselves as the true sons of the ROC.

Taiwan's domestic politics—the processes of indigenization, democratization and electoral competition—are driving the country's zigzagged route towards self-assertion. So far, the Taiwanese have been unable to establish a clear and sustainable consensus over their own future. The island is pulled by forces from different directions, and is plagued by internal divisions. The overall trend, however, is in favour of greater sovereignty. The dust is far from settled, and the trouble is likely to continue.

Chang Mau-kuei (IIAS fellow December 2003-June 2004) is a political sociologist at the Institute of Sociology, Academia Sinica, Taiwan. His research interests include Taiwanese social movements and the ethnography of the Dachen people, resettled from a Zhejiang Province island to Taiwan in 1954, and now dispersed around the world. etpower@gate.sinica.edu.tw

From *IIAS Newsletter*, #34, July 2004. Copyright © 2004 by IIAS Newsletter.

The Japan that can say yes

Japan under Koizumi has become a more assertive country in world affairs; a certain intransigence can be observed in its foreign policy. Contemporary Japanese self-assertion is driven by an internal logic set in motion by Japan's defeat in World War Two, given new scope for expression by changes in the international environment.

By Kazuhiko Togo

Japan had not been defeated and occupied by outside forces prior to 1945. Defeat in World War Two was nothing short of traumatic for the majority of the population. The Allied Occupation had as its initial goal the complete and permanent demilitarisation of Japan. Article 9 of the Constitution, promulgated in 1946, stated: 'the Japanese people forever renounce war as a sovereign right of the nation.... Land, sea, and air forces, as well as other war potential, will never be maintained.' Pacifist idealism, a major current running through post-war Japanese society, dates from this period.

The Cold War descended on East Asia in 1947. Under the US strategy of containing communism, Japan re-emerged as Western democracy's bulwark in East Asia. Economic recovery became the priority; with its newly established Self Defence Forces, Japan entered a security alliance with the US. Those prepared to face the reality which surrounded Japan welcomed this change. But their views clashed with pacifist idealism, which had established itself in the vacuum of defeat. Under the iron umbrella of US-Soviet rivalry, a deep rift descended on Japanese society. Pacifist idealism was supported by the Socialist and Communist Parties, labour unions, the media, influential intellectuals, and public opinion. The conservative parties, government agencies and minority of intellectuals espoused realism.

The end of thecold War transformed the context of Japanese foreign and security policy and brought the country into the arena of international politics. Japan's internal political power structure changed as well. In 1993, forty years of Liberal Democratic Party rule was brought to an

end by a reform-minded coalition government. In 1994, the LDP returned to power in a most unlikely coalition with the Socialist Party. Reversing its previous stance of unarmed neutrality, the Socialist Party acknowledged the legality of the Self Defence Forces and the security treaty with the United States. The largest political party carrying the banner of pacifism therey lost its *raison-d'être*; the newly formed Democratic Party, with a much more pro-active security policy, became the opposition in 1996.

Against the background of these internal and external changes, Japan moved towards a more realistic, pro-active and responsible stance in international affairs. Offended by the derision that met Japan's $14 billion contribution to the 1990-01 Gulf War, many Japanese became convinced that active participation in the international arena required political and military contributions. The Peace Keeping Operations (PKO) Law was passed in 1992, enabling troops to be sent to Cambodia, the golan Height and East Timor on UN peacekeeping operations. The 1993-94 North Korean nuclear crisis and the 1995-96 Taiwan Strait crisis resulted in the reaffirmation of the US-Japan security alliance in 1996 and 1997. North Korean encroachments by sea and sky in 1998-99 further enhanced awareness of Japan's own responsibilities for national self-defence.

9/11, North Korea, and Koizumi's security policy

Koizumi came to power in April 2001 and was immediately faced with the challenge of global terrorism. Declaring any

terrorist attack to be an attack on Japan's security, Koizumi ordered the Maritime Self Defence Forces to the Indian Ocean to offer logistical support to US, UK and other coalition forces. In October 2002, the North Korean nuclear crisis erupted, further heightening Japan's sense of vulnerability. Tokyo reacted by enacting new laws to respond to armed attack; an missle defence program was introduced in the 2004 budget.

Koizumi's decision to send troops to Iraq must be understood as part of Japan's readiness to bear greater responsibilities towards global security. While the government came under heavy criticism for following America's lead, its decision was based on a calculation of long-term Japanese strategic interests. Had its security policy been more mature, Japan, while still supporting the US, could have entered into dialogue with nations in the Middle East and Europe and pushed for a greater United Nations role.

Koizumi's foreign policy

Under Koisumi a new intransigence has appeared in important foreign policy arenas. As a result, Japan missed several opportunities to strengthen its foreign policy leverage. If the 1990s were a period of realist victory over pacifist idealism, the turn of the century witnessed the beginning of a new rift between realists and nationalists pursuing narrowly—and emotionally—defined national interests.

The first sign of intransigence appeared immediately after Koizumi took power, in his policy towards Russia. Japan and Russia had been working to settle the territorial dispute over four islands

'the Japanese people forever renounce war as a sovereign right of the nation... Land, sea, and air forces, as well as other war potential, will never be maintained' *–Japanese Constitution, 1946, Article IX*

northeast of Hokkaido since the late 1980s. Both sides had failed to grasp opportunities during the Gorbachev and Yeltsin presidencies. In 2000-01, Prime Minister Mori and President Putin came close to resolving the issue and signing a peace treaty. After Koizumi came to power, confusion reigned within the Japanese Ministry of Foreign Affairs, while raging public sentiment against compromise practically crushed the accumulated results of a decade of negotiations.

Intransigence was evident in Koizumi's policy towards North Korea. Public outrage following Koizumi's visit to Pyongyang—when it was revealed that eight of the thirteen Japanese abducted by North Korean agents were dead—was understandable, but the anger froze Japanese policy, preventing it from acknowledging Kim John Il's unexpected apology. Public pressure compelled the government to take a tough position; Japan thus lost important diplomatic leverage in the ongoing North Korean nuclear crisis.

Japanese policy towards China remains difficult. On the one hand, the policy of engagement has bveen consistent since the end of the 1970s. China's gowing military power, assertiveness in the South China Sea, and continuous pursuit of Japanese war guilt has, however, fuelled antipathy in Japan, particularly since the mid 1990s. Tokyo's policy of cintinuing Official Development Assistance while the economy was in the doldrums only added fuel to the fire.

The issue of Taiwan has only complicated matters. Following Japan's diplomatic recognition of the People's Republic of China in 1972, Japan served relations with the Republic of China, thought the island remained a mjor trading partner. Taiwan's democratisation, affirmation of national identity, and Li Deng-hui's praise of Japanese colonial governance appealed to certain politicians and intellectuals; anti-Chinese feelings became mixed up with pro-Taiwanese emotion.

Koizumi's China policy rests on engagement, but is on delicate ground. His repeated vists to Yasukuni Shrine to pay homage to the war dead keep the two leaderships from engaging in meaningful dialogue. Japan's China and Taiwan policies remain unclear at the outset of the twenty-first century.

Domestic terrain

Koizumi's policy of greater self-assertion cannot be understood without analysing the domestic context of its formulation. Since the end of the Cold War, the legitimacy of the iron triangle of politicians, bureaucrats and businessmen who governed post-war Japan has been shattered. The long time rule of the Liberal Democratic Party ended in 1993, bringing fluidity to politics. Bureaucratcies were brought down by a series of scandals, while the burst of the bubble economy shook financial institutions and small and middle-scale enterprises. The economic crisis, and the limitations it placed on Japan's once mighty checkbook diplomacy, made the country more sensitive to its political role.

Koizumi came to power in 2001 upon a wave of popular discontent and desire for a new reform-minded leader. Public opinion favours greater Japanese self-assertion; Koizumi's policy plays to this media-led self-assertive public opinion. Among his supporters are members of the older generation whose sense of national pride has long chafed under the post-war ascendancy of pacifism. The younger generation, too, is quite vocal in asserting Japan's need for participation in global issues of peace and security.

Ways ahead

The momentum towards greater self-assertion in security policy will probably continue for some time. A few years hence, the revision of the Constitution's Article 9 may appear on the agenda. Realists have a crucial role to play in con-

vincing neighbouring countries that the revision of Article 9 does not signal a Japanese return to militarism; rather, it reflects Japan's desire to become a more responsible and pro-active member of the international community. This task is important due to the legacy of war. Japan's quest for greater self-assertion has not yet found solid ground for true reconciliation with neighbouring countries.

True self-assertion can only be achieved through the understanding of the position of others. Self-assertion inevitably brings states into the international arena where the conflicting interests of other nations confront them. Without the peaceful resolution of these conflicting interests, few states will have their interests realised. Self-assertion that can understand the positions of other, however, can only be namifested when there is real national self-confidence. After the void that engulfed the country nearly 60 years ago, Japan's whole post-war history can be seen as a long painful process in trying to regain a true sense of self-confidence.

Japan's responsible and active participation in the cause of regional and global security was restricted for many decades after World War Two by the influence of pacifist idealism. One can only hope that Japan, by gaining a true sense of self-confindence, develops a wise and balanced policy, conductive to realising its true interests in harmony with its neighbours in East Asia and beyond.

Togo Kazuhiko *recently retired from the Japanese Foreign Ministry where he specialised in relations with the Soviet Union/Russia. A former Japanese ambassador to the Netherlands, he is currently a Canon Foundation-sponsored professorial fellow at IIAS. His research interests include foreign and security policy in East Asia. k.togo@let.leidenuniv.nl*

Hong Kong: Still "One Country, Two Systems"?

"After four years of Hong Kong self-rule, the overriding question remains whether the 'one country, two systems' experiment is working. So far the answer is yes—with some qualification."

CRAIG N. CANNING

The handover ceremonies on June 30, 1997, returning Hong Kong to Chinese sovereignty after 156 years as a British colony marked a significant turning point in the city's unique history and launched an unprecedented historical experiment: "one country, two systems." Hong Kong under British colonial rule had established a reputation as one of the world's centers of manufacturing and trade, with enhanced leadership potential and competitive drive provided by a massive influx of mainland China refugees after the 1949 Communist revolution. In the 1980s and 1990s, when Hong Kong shifted its manufacturing industry to south China and transformed itself into a global hub of trade, finance, communications, and other business services, the city strengthened its reputation as an economic dynamo.

By the time it reverted to Chinese rule in mid-1997, Hong Kong was widely recognized as a locus for international business opportunities, offering the advantages of low tax rates and a supportive legal infrastructure, a broad pool of managerial talent, a well-educated and efficient workforce, and a government based on the rule of law and administered by a competent and accountable civil service meritocracy. The question on everyone's mind was whether these favorable conditions would continue after responsibility for governing Hong Kong was placed in the hands of Hong Kong residents themselves under the "one country, two systems" formula. While the implementation of Hong Kong self-rule remains a work in progress, the four years since the handover provide some preliminary answers to this question.

DEFINING THE GROUND RULES

The "one country, two systems" model introduced by China's paramount leader, Deng Xiaoping, during the negotiations leading to the December 1984 Sino-British joint agreement for the return of Hong Kong to China was conceptually simple: for 50 years after Hong Kong's reversion in 1997, Hong Kong could retain its capitalist system and its own government, administered by the people of Hong Kong, while the Chinese central government in Beijing would assume responsibility for Hong Kong's foreign affairs and national defense. Details of this plan were painstakingly hammered out between 1985 and 1990 by a joint committee and enshrined in the Basic Law (1990), Hong Kong's miniconstitution. According to the Basic Law, the Hong Kong special administrative region (SAR) would be governed by a chief executive elected to a five-year renewable term by a 400-member selection committee; an Executive Council (Exco) selected by the chief executive to serve as special advisers; a civil service bureaucracy under the supervision of a chief secretary for administration; a 60-member Legislative Council (Legco) elected according to a detailed formula involving functional constituencies and geographic representation; and an independent judiciary.[1]

To understand Hong Kong's governance as an SAR since mid-1997, it is important to keep in mind the historical background under which the Basic Law was conceived. Neither the British colonial government nor the People's Republic of China was a democratic institution. The late 1980s, when China was undergoing broad-based domestic reforms and rapidly opening to the world, also saw several popular demonstrations perceived by the government as potential challenges to Beijing's authority, including the large prodemocracy movement in spring 1989 that was suppressed in the Tiananmen Square massacre on June 4. In this context the Basic Law was formulated to emphasize stability—not democratic rule—by concentrating power in the hands of the chief executive and civil service bureaucracy, circumscribing and diffusing the authority of Legco, and allowing no more than incremental democratic reforms. For example, until 2007—10 years after the establishment of the Hong Kong SAR—the Basic Law permits only minor adjustments to the procedures for electing the chief executive and Legco representatives.

IMPLEMENTING "ONE COUNTRY, TWO SYSTEMS"

Elected in December 1996, Tung Chee-hwa became the first chief executive of the Hong Kong SAR when the "one country, two systems" experiment began on July 1, 1997. A conservative Hong Kong businessman born in Shanghai and educated in China, Hong Kong, and Britain, Tung also had extensive experience living and working in the United States. Anson Chan—who in November 1993, under British colonial Governor Chris Patten, became the first female and first Chinese chief secretary—stayed on to head the civil service.

The first post-1997 Legislative Council election was held in May 1998, returning 60 Legco members. While the colonial administration under Governor Patten (1992–1997) had introduced a series of reforms intended to increase the level of democratic participation in Hong Kong government, Beijing charged that Patten and the British had violated the Sino-British Joint Declaration, the Basic Law, and supplemental understandings on Hong Kong's reversion in attempting to change the ground rules at the eleventh hour. Working through the Preparatory Committee for the Hong Kong SAR, Beijing established a Provisional Legislative Council, which took power on July 1, 1997, in effect expelling Legco members elected in 1995 under British rule who had hoped to ride a "through train" into Hong Kong's new era.

The first and most formidable challenge facing the new Hong Kong government was launching self-rule—a difficult process made more complicated by bickering between Beijing and Patten's colonial government. In addition, officials in the new government faced two immediate problems—a deadly "chicken flu" virus and the Asian financial crisis. The H5N1 bird flu virus, which first infected and killed over 6,000 chickens in March 1997, was identified by August as a cause of human infection. The government ordered the slaughter of 1.4 million chickens in late 1997 and early 1998, but before the disease was brought under control, 18 people had been infected, including 6 who died. The Asian financial crisis, triggered by currency devaluation in Thailand in mid-1997 and spreading rapidly to other countries in Southeast and East Asia, struck Hong Kong in the fall of 1997 and contributed to the city's first recession, record-breaking unemployment, a plunge in the stock market, and a precipitous drop in highly inflated property values. Hong Kong citizens and the media complained vociferously as confidence in the new government slipped.

Often ranked as the world's ninth- or tenth-largest trading economy, Hong Kong had produced consistent annual GDP growth rates of around 5 percent for nearly two decades. Consequently, the Asian financial crisis and ensuing recession of 1997–1998 introduced a sense of economic insecurity to Hong Kong that has yet to be completely dispelled. Many Hong Kong citizens who had purchased apartments and other property as values rose in the mid-1990s found themselves holding negative assets when property prices fell by 50 percent or more. Sustaining Hong Kong's economic recovery and growth has been high on the list of Hong Kong government priorities ever since.

RECENT CHALLENGES

Although no longer in crisis mode, Hong Kong over the last two and a half years has faced a broad and complex set of challenges. These include recovery from the Asian financial crisis and the formulation of plans and programs for sustaining Hong Kong's economic growth in a rapidly changing regional and global environment, especially in preparation for China's entry into the World Trade Organization (WTO). The SAR is also introducing civil service and other reforms intended to improve the government; addressing quality-of-life issues such as affordable housing, environmental protection, and maintenance and expansion of a clean and efficient public transportation system; and safe-guarding Hong Kong's autonomy within the "one country, two systems" framework.

> *Much of the public's discontent in recent years has focused on the government and in particular on Chief Executive Tung Chee-hwa.*

Trade and commerce have been Hong Kong's lifeblood since the beginning of its modern history in 1841 as a British colony and, although drastically transformed over the past 160 years, business is as crucial to Hong Kong today as it was in the past. In the aftermath of the Asian financial crisis, the government introduced reforms in financial and monetary policy, including the promotion of continuous banking sector reform; merging the securities and futures markets and their clearinghouses in March 2000; and implementing new legislation, anticipated by the end of 2001, to improve the regulatory environment. Buoyed in part by improving Asian markets, strong United States and Chinese economies, and rapidly growing United States—China trade, many sectors of Hong Kong's economy recovered relatively quickly from the worst effects of the Asian financial crisis and recession. By late 2000 the stock market had regained its precrisis levels, the unemployment rate had dropped from over 6 percent to 4.8 percent, and Hong Kong's GDP had recorded increases in both 1999 and 2000 after suffering a 5.1 percent decline in 1998. Although a slumping United States economy sharply lowered growth expectations for 2001,

Hong Kong's overall economic outlook in the near term remains positive.

Nevertheless, recognizing that Hong Kong's future economic success will depend on maintaining its competitive edge in an increasingly competitive regional and global environment, the Hong Kong government has undertaken long-term strategic planning that embraces the goal of transforming Hong Kong into a high-tech hub in East and Southeast Asia. The city is committed to establishing a science park and "Cyberport" anchored by several information technology (IT) multinational companies. Hong Kong has also begun to reposition itself by encouraging development of a deregulated telecommunications market, high-speed e-banking and other e-business networks, and state-of-the-art financial services. To cultivate essential human resources, the government is introducing extensive education reforms proposed in May 2000 by a special education commission—reforms seeking to equip Hong Kong citizens with language competency, general education, and work skills that will contribute directly to the city's future advancement. In addition, in March 2001 the Hong Kong government announced it would ease work-permit restrictions to allow mainland Chinese professionals with IT and financial services skills to work in Hong Kong. The government also intends to permit mainland Chinese students who have completed degrees in Hong Kong to work in the SAR.

China's expected entry into the WTO in late 2001 is of special concern. Hong Kong expects to lose its traditional "gateway to China" middleman role with China's entry into the WTO. But Hong Kong's business community hopes it can continue to play a substantial role in China trade and economic growth by capitalizing on Hong Kong's strengths as an international business center and on its extensive and longstanding ties to the mainland China economy. Hong Kong's 2001—2002 budget plan, for example, calls for the expansion of the joint regional infrastructure linking Hong Kong and Guangdong in south China through a sophisticated transport network that will speed the flow of people, goods, and traffic across the border. Accommodating China's WTO membership, however, will also require Hong Kong to make timely and appropriate adjustments to China's rapidly changing economic, social, and political realities—including growing competition from Shanghai, a city determined to reclaim its historic role as China's premier economic powerhouse (a position that it gradually relinquished to Hong Kong in the four decades after 1949).

In addition to Hong Kong's future competitiveness, quality-of-life issues have gained a higher profile in recent years as the demand for adequate and affordable housing has risen, as convenient public transportation has become increasingly important, and as Hong Kong residents have grown more sensitive to the need for improved air and water quality, green areas, and the protection of endangered species. In addressing housing concerns, the Hong Kong government prepared a long-term housing strategy and promised to closely monitor fluctuations in housing supply and demand. But the government's recent performance in the housing sector has been spotty. In the face of a sluggish property market still reeling from the Asian financial crisis, the government equivocated—signaling its intention in 1998 to build 85,000 new residential units annually but subsequently backing away from this commitment. A construction scandal leading to the resignation of Chairman of the Housing Authority Rosanna Wong and an unprecedented no-confidence vote in Legco in June 2000 further married the government's credibility and contributed to public dissatisfaction.

The Hong Kong government has been somewhat more successful in dealing with transportation and environmental concerns. Anticipating continued population growth and rising demand for public transit, the government has developed plans for an extensive railway construction program throughout Hong Kong. In addition, the government endorsed a farreaching strategy to tackle environmental problems. To combat air pollution that has reached record levels in recent years, it is shifting taxi and minibus fuel to cleaner liquid-petroleum gas and establishing strict vehicle-emission controls and fuel-quality standards.

Hong Kong self-rule is likely to experience more continuity than change in the near term.

Despite its achievements, the Hong Kong SAR government remains generally unpopular. The Asian financial crisis and the recession associated with it hit many Hong Kong residents hard in the pocketbook; property that lost half its value in 1997 and 1998 has appreciated slightly since but is unlikely to reach pre-1997 levels anytime soon. Moreover, Hong Kong's postcrisis economic recovery has not benefited everyone, and the income gap is widening. The government's commitment to making Hong Kong a high-tech hub and its embracement of the "new economy" do not reassure those in the workforce who lack the skills or opportunity to make this transition. Civil servants are unhappy with government reforms that promise to trim 10,000 positions from the approximately 185,000-member civil service. Disgruntled teachers of English object to new requirements for mandatory testing of their language skills while many parents and educators accuse the government of emphasizing Chinese-language instruction at the expense of English; doctors and patients criticize the government for doing too little to improve Hong Kong hospitals and the quality of health care; and many employers and employees initially complained about a compulsory deduction of 5 percent from wages, to be matched by a 5 percent employer contribution, for the social security program implemented in December 2000. Whether warranted or not, much of the public's discontent in recent years has focused on the government and in particular on Chief Executive Tung Chee-hwa.

IS "ONE COUNTRY, TWO SYSTEMS" WORKING?

After four years of Hong Kong self-rule, the overriding question remains whether the "one country, two systems" experiment is working. So far the answer is yes—with some qualification. Beijing has honored its pledge to allow Hong Kong to retain a high degree of autonomy and has generally refrained from meddling in Hong Kong SAR internal affairs. The elite People's Liberation Army (PLA) detachment of approximately 8,000 troops stationed in Hong Kong has kept a low profile. In regard to public protests aimed at China, on June 4 every year for the past 12 years tens of thousands of Hong Kong citizens have demonstrated in support of the prodemocracy advocates who died in the 1989 Tiananmen crackdown. While the Chinese government strictly prohibits such demonstrations in mainland China, Beijing has maintained a discreet silence about these protest activities in Hong Kong.

Nevertheless, the complexity of the "one country, two systems" formula in practice has generated some tricky problems. The right-of-abode controversy provides a good illustration. According to the Basic Law, Hong Kong residents of Chinese descent qualify for right of abode (permanent residency) in Hong Kong if at the time of birth at least one parent was a Chinese citizen holding Hong Kong right-of-abode status. New ordinances implemented in July 1997 that made the procedures for proving right of abode more restrictive precipitated a series of court challenges. After the Hong Kong SAR Court of Final Appeal upheld legal challenges to the ordinances in January 1999, the Hong Kong government took its case to the Standing Committee of the National People's Congress in Beijing, which holds the right to interpret the Basic Law. Claiming that the Court of Final Appeal's ruling would extend permanent resident eligibility to 1.6 million potential new immigrants in addition to Hong Kong's current population of 7 million and thereby place "unbearable social and economic burdens on the [Hong Kong] community," the Hong Kong government sought the Standing Committee's interpretation. In June 1999 the committee upheld the position of Hong Kong's government. However, the Standing Committee's action, although legal and binding, led to a massive challenge in Hong Kong courts on behalf of more than 5,000 right-of-abode seekers who claim that the decision denied them the benefit of the January 1999 Court of Final Appeal ruling. The case now awaits a court decision that will presumably clarify who is (or can be) affected when the Hong Kong SAR government seeks Beijing's reinterpretation of the Basic Law. The decision may have important implications for Hong Kong's judicial autonomy under "one country, two systems."

Freedom of expression in Hong Kong remains another crucial issue. While freedom of expression is upheld by law, custom, and an aggressive media reflecting a broad spectrum of opinion, recent statements by Beijing representatives in Hong Kong and by Chinese central government officials, including President Jiang Zemin, have raised questions about Beijing's understanding of and commitment to freedom of the press in Hong Kong. In April 2000, for example, one month after proindependence Democratic Progressive Party (DPP) candidate Chen Shuibian was elected president of Taiwan, Wang Fengchao, deputy director of the Beijing liaison office in Hong Kong, warned Hong Kong news media to refrain from reporting on Taiwan independence activities as "normal" news. Following Wang's comments, Hong Kong Chief Secretary Anson Chan, several Legco members, and the Hong Kong Journalists Association issued public statements in support of press freedom and its legal foundation in Hong Kong. In late October 2000, during a photo session in Beijing concluding a visit by Chief Executive Tung Chee-hwa, President Jiang Zemin lost his temper and scolded Hong Kong journalists for their "too simple, sometimes naive" questions when they pressed him about his personal support for Tung's reelection in 2002. And during a visit to the Macau SAR on December 20, 2000 to commemorate the first anniversary of Macau's return to Chinese sovereignty, Jiang exhorted the local media "not only to value the freedom of the press but also to pay attention to their social responsibilities." Despite statements by Jiang Zemin and other Chinese central government representatives, the Hong Kong media remains generally unrestricted, vocal, and willing to seek news and espouse viewpoints critical of Beijing or the Hong Kong government.

The most delicate current problem related to "one country, two systems" is the Hong Kong government's handling of Hong Kong's Falun Gong, the spiritual organization founded in China in 1992 by the controversial leader Li Hongzhi that combines traditional *qigong* exercises with elements of Buddhism and Li Hongzhi's own beliefs in the pursuit of physical and moral well-being. Falun Gong has been subjected to an intensive crackdown in mainland China since the organization was outlawed in July 1999. In Hong Kong the organization is legally registered under the Societies Ordinance, and Falun Gong practitioners are free to pursue their beliefs. In recent months Hong Kong Falun Gong adherents have organized meetings at city hall, held demonstrations at the Beijing liaison office, and attempted to deliver petitions protesting Falun Gong's suppression in mainland China to high-ranking Chinese central government officials visiting Hong Kong. In turn, Beijing authorities have emphasized that Hong Kong must not become a base of anti-China activities for Falun Gong or any other group.

Internal and external pressures on the Hong Kong government to take action against Falun Gong mounted during the first six months of 2001. In January Hong Kong Secretary for Security Regina lp vowed to keep a "close eye" on Falun Gong activities, claiming that the group was targeting the Chinese central government. Chief Executive Tung stated in February that Falun Gong "more or less bears some characteristics of an evil cult," echoing the "evil cult" terminology Beijing uses to brand Falun Gong. In June Tung termed Falun Gong "without doubt an evil cult" in a question-and-answer session with members of Legco.

Some pro-Beijing legislators suggested that Falun Gong is actually a political, not a spiritual, organization and should be deregistered under the Societies Ordinance. Executive Council member Nellie Fong and others urged the government to enact an antisedition law, as called for under Article 23 of the Basic Law, to control Falun Gong. A Hong Kong government spokesman admitted this May that the government was studying the laws of other countries, including "anti-cult" legislation recently promulgated by the French government, to prepare itself to deal with radical religious groups.

Despite heated rhetoric, conflicting opinions, and pressures from within and outside Hong Kong, cooler heads have continued to prevail, and the Hong Kong SAR government continues to adhere to current laws allowing Falun Gong to conduct its activities in Hong Kong as long as the organization and its members do not violate existing ordinances. Upholding the government's policy in an address to Hong Kong and foreign correspondents in late June, the chief secretary for administration stated emphatically that Hong Kong would enact no anti-Falun Gong legislation and that religious freedom in Hong Kong is non-negotiable.

THE FUTURE OF HONG KONG SELF-RULE

If the present and the recent past provide useful guideposts, Hong Kong self-rule is likely to experience more continuity than change in the near term. The second election for chief executive will take place on March 24, 2002, and, barring unforeseen development, Tung Chee-hwa is almost certain to win reelection to the post. Beijing has made its preference for Tung abundantly clear; an 800-member selection committee that will elect the next chief executive was chosen and empowered in July 2000 in accordance with the Basic Law, and no challengers to Tung have stepped forward. Despite Tung's low public-approval ratings, recently hovering in the 31 to 37 percent range—ratings that reflect Hong Kong's fluctuating economic fortunes as well as the gaffes and blunders that come with his on-the-job training as businessman-turned-politician—Tung Chee-hwa will probably thus continue for a second term as Hong Kong's chief executive.

Similarly, the civil service bureaucracy should undergo few major changes in the near future. The shock waves from Chief Secretary for Administration Anson Chan's unexpected resignation announcement in January 2001, over a year before the expiration of her contract, have subsided. Former Financial Secretary Donald Tsang assumed Chan's position effective May 1, 2001, and Antony Leung gave up chairmanship of the Education Commission to take charge of the financial secretary portfolio. Often described as the "conscience of Hong Kong," Anson Chan was widely viewed as a courageous, outspoken, and talented administrator and will be missed by many, especially those who wish to see the timetable for democratic reforms in Hong Kong accelerated. Others welcome her departure. Tung Chee-hwa now has his first opportunity to work with his own team.

The second post-handover election for the Legislative Council occurred on September 10, 2000. As in 1998, 60 legislators were elected, but in accordance with gradual adjustments stipulated by the Basic Law, 24 (rather than 20) were directly elected by geographic constituencies, 6 were elected by the same 800-person selection committee empowered to elect the next chief executive, and 30 (as in the past) were elected by functional constituencies. While voter participation decreased from 53.3 percent in 1998 to 43.6 percent in 2000, the election results—in terms of the balance between liberal prodemocracy legislators and conservative pro-Beijing legislators—remained about the same as in 1998, with pro-Beijing conservatives holding a slight edge. The Democratic Party headed by Martin Lee—which enjoyed substantial popular support in Legco elections in 1991 and 1995 before the takeover and in 1998—lost ground in the 2000 election; Lee and other Democratic Party leaders cite growing factionalism and a failure to cultivate grassroots support through close attention to constituents and their principal concerns as the reasons for the party's poor performance. Although Legco faces many obstacles—including a powerful executive branch that sets the legislative agenda, a heavy work schedule, few standing committees, and modest public support—the men and women who serve in it play a vital role in Hong Kong's young democracy by drawing public attention to critical issues and demanding accountability from the Hong Kong government. In response to Legco's increasingly vocal role in public debate, the executive branch has had to devise new strategies for gaining public support for its legislative initiatives.

While the next Legco election in 2004 will mark yet another incremental adjustment (when 30 legislators rather than 24 are directly elected by geographic constituencies), it appears unlikely that Legco's composition or relative political strength will change significantly before 2008, when the Basic Law allows Legco to alter the methods of its formation, if it chooses, by a two-thirds vote and the approval of the chief executive. Nevertheless, despite its relative weakness within the current Hong Kong government framework, the Legislative Council may be able to contribute in crucial ways to the evolution of democracy in Hong Kong if it can attract high-caliber legislators who are capable of guiding Legco's role in the process, rising above overly parochial interests and avoiding entrenched ideological positions blindly opposed to "Beijing" or "democracy," and thereby gaining greater public confidence and support.

As for Hong Kong's judiciary and the rule of law, widely perceived as the bedrock of Hong Kong's freedom and economic vitality, the future is likely to bring both continuity and change. Hong Kong's judicial system was carefully transplanted from the British colonial administration into the new Hong Kong SAR government, albeit with some requisite adjustments, such as the creation of a Court of Final Appeal to replace the Judicial Committee of the Privy Council under the British system, and provisions for the coexistence of 11 national (mainland) laws and local Hong Kong laws (or ordinances). The Hong Kong SAR legal structure is now fully in place, and there are no serious indications of a slackening in Hong Kong's longstanding commitment to the rule of law. Change will inevitably occur, however, through the litigation and appeals process as Hong Kong's existing and future laws are applied and tested and as the legal relationship between the Hong Kong SAR and mainland China under the "one country, two systems" model continues to evolve.

NOTES

1. The Basic Law divides the electorate into 5 geographical constituencies and 28 functional (occupational) constituencies representing important economic, social, or professional groups such as labor, business, the legal profession, medical and health services, and education.

CRAIG N. CANNING *is an associate professor of history at the College of William and Mary and the 2000–2001 director of research and development at the Hong Kong–America Center in Hong Kong.*

Keeping Costs Under Control

Indonesia's comparative advantage of being a low-cost, labour-intensive manufacturing base is being eroded. It must change gear to stay competitive

By John McBeth/JAKARTA

DUSK IS DARKENING the windows of his downtown office and the senior executive of an Indonesian textile company is explaining how Chinese manufacturers will eventually lose their competitive edge and be forced to adopt realistic pricing structures. That's the good news. The bad news matches the gloom outside. In the two years before that it is likely to happen, he predicts, a lot more small Indonesian operations will go to the wall.

Slammed at home and abroad by intensified competition, sweeping trade liberalization, low productivity and soaring business costs, two of Indonesia's main export earners—textiles and footwear—are going through their biggest shakeout since labour-intensive manufacturing weighed in behind oil and gas as Indonesia's new engine of growth in the mid-1980s.

Government figures show the number of bankruptcies among large and medium-sized manufacturing companies jumped by 22% in 2002, to 835 from 650 the previous year. Another 767 companies downsized their operations—a 41% increase from the 447 reported in 2001. Worryingly, it's a trend that appears to be nosing back towards 1997-98 levels, at the height of the Asian Crisis, when nearly 3,000 factories were closed.

"I don't think we are competitive any more for labour-intensive manufacturing," says Sofyan Wanandi, the chairman of the private policy group, National Economic Recovery Committee. "That's why we will probably have to go to resource-based industries. That will be our future." But as one Western trade adviser notes: "The shift we are seeing is a function of the demise of one sector rather than the rise of another."

Industry and Trade Minister Rini Soewandi did not respond to requests for an interview, but aides say Indonesia is going to have to look more to downstream processing, particularly of commodities such as palm oil and cocoa and industries capable of providing significant employment. "Agro-industry is where the next growth will come," says Sudarmasto, head of the ministry's research and development agency. "We have to make the value-added process longer."

Higher commodity prices headed off what may otherwise have been another major slump in nonoil exports last year. But economist Hadi Sasaestro, at the Centre for Strategic and International Studies in Jakarta, worries about the next few years and the affects of a dramatic drop-off in foreign direct investment. "The macroeconomic stability we have achieved isn't going to lead to sustained growth," he says. "Why? Because at the micro level, there are so many problems and no one is solving them."

Rising labour and energy costs, higher taxes and other levies are all contributing to the fall-off in the manufacturing sector, boosting the ranks of the unemployed to more than 40 million. Businessman Wanandi now estimates that corruption alone adds as much as 10% to overall business costs. "You can see [corruption] now in every sector of the economy," he says. "During [the] Suharto [years] you could have something done. Now, you're not sure if anything can be done."

What has hit the textile sector in particular has been political decentralization—a success story in some parts of the country, a mess in others. "Labour-intensive industries are a major target for local governments," says the Western trade adviser. "They're using every possible means of extracting revenue." One additional generic problem is the value-added tax and a lengthy restitution system which is causing a blowout in the working capital of many businesses. Another is Jakarta's port-handling charges that far exceed those in Singapore and other Southeast Asian ports.

Nonoil exports earned Indonesia $44.9 billion last year, up $1.3 billion over 2001, but still well below the $47.7 billion recorded in 2000. Last year's electronics exports—mostly audio gear, printers and hand-phone parts—were slightly up on the $6.4 billion recorded in 2001, but still well down on the $7.8 billion they garnered in 2000. Although last year's closure of Sony's factory in West Java has been offset by new investments from Samsung and LG Electronics, that may be more due to the recent removal of luxury taxes aimed at ending rampant smuggling on the domestic market.

Footwear receipts have continued their decline as well—from $2.2 billion in the mid-1990s to $1.1 billion last year. "If we don't improve the market climate, tax and labour policies and strengthen supporting industry, day by day we'll see our market share eroding," says Anton Supit, chairman of the Indonesian Footwear Association.

THE NUMBER OF BANKRUPTCIES AMONG LARGE AND MEDIUM-SIZED MANUFACTURING COMPANIES JUMPED BY 22% IN 2002.

But it is textiles and garments that cause the greatest concern. Although volumes may be up, foreign-exchange earnings for both have been sliding—from $8.1 billion in 2000 to $7.6 billion in 2001 and $6.6 billion last year. With little new investment and machinery that is as much as 20 years old, huge challenges lie ahead, not only in reconciling falling unit values against rising costs, but also in the looming reconfiguration of trade that will usher in a new quota-free era of price and quality-based competition.

Indonesia's relatively low productivity, at least 20% below that of China, is seen by some in the industry as a function of bad management. But the Indonesian textile executive insists that Indonesia's production costs are similar to those in China, pointing to how the Chinese government has spent the past five years offering cheap credit to labour-intensive industry to absorb workers from failing state enterprises. In the next two years, he be-

lieves, China will have to go back to "more realistic pricing" when the cost to the economy from such a policy becomes too great to bear.

Still, China's entry into the World Trade Organization, the emergence of low-cost producers like Bangladesh, Pakistan and Cambodia—as well as Indonesia's continuing exclusion from a recent proliferation of preferential trade agreements—are major problems for a sector whose ability to respond will be constrained by an increasingly highcost economy.

"Indonesia can't compete if it can't sort out its domestic costs," says the trade adviser. "At best it is keeping market share, at worse it is losing market share." Critics blame the government for not being nearly pro-active enough in either going about trade negotiations or in using the political leverage it has to improve market access.

As a result, its textile and apparel exports are vulnerable to trade diversion to higher-cost producers that receive substantial margins of preference. For example, the preferences extended to Pakistan in the European Union and United States textile and apparel markets because of its cooperation in the fight against terrorism could eat into Indonesia's market share as quotas come off altogether in 2005.

Trade experts say that if Indonesia is to deal with the new global order, it will need to rein in costs at both national and local-government levels, develop a strategy to address market-access problems and improve its ability to respond to new forms of protection. Not everyone thinks all is lost. "Logically, I don't believe you'll see textiles and footwear disappear," says a World Bank official. "You'll see a shakeout, but it won't be earthshaking because Indonesia still has a comparative advantage." The next two years will show whether he's right.

THE SACRED WORLD

The Dreamtime is the Aboriginal understanding of the world, of its creation, and its great stories. The Dreamtime is the beginning of knowledge, from which came the laws of existence. For survival these laws must be observed.

The Dreaming world was the old time of the Ancestor Beings. They emerged from the earth at the time of the creation. Time began in the world the moment these supernatural beings were "born out of their own Eternity."

The Earth was a flat surface, in darkness. A dead, silent world. Unknown forms of life were asleep, below the surface of the land. Then the supernatural Ancestor Beings broke through the crust of the earth form below, with tumultuous force.

The sun rose out of the ground. The land received light for the first time.

The supernatural Beings, or Totemic Ancestors, resembled creatures or plants, and were half human. They moved across the barren surface of the world. They travelled, hunted and fought, and changed the form of the land. In their journeys, they created the landscape, the mountains, the rivers, the trees, waterholes, plains and sandhills. They made the people themselves, who are descendants of the Dreamtime ancestors. They made the Ant, Grasshopper, Emu, Eagle, Crow, Parrot, Wallaby, Kangaroo, Lizard, Snake, and all food plants. They made the natural elements: Water, Air, Fire. They made all the celestial bodies: the Sun, the Moon and the Stars. Then, wearied from all their activity, the mythical creatures sank back into the earth and returned to their state of sleep.

Sometimes their spirits turned into rocks or trees or a part of the landscape. These became sacred places, to be seen only by initiated men.

These sites had special qualities.

PHYSICAL WORLD

"OUR LAND OUR LIFE"

'We don't own the land, the land owns us'

'The Land is my mother, my mother is the land'

'Land is the starting point to where it all began. It is like picking up a piece of dirt and saying this is where I started and this is where I will go'

'The land is our food, our culture, our spirit and identity'

'We don't have boundaries like fences, as farmers do. We have spiritual connections'

Land means many things to many people. To a farmer, land is a means of production and the source of a way of life. It is economic sustainability. To a property developer, it is a bargaining chip and the means of financial progress and success. To many Australians, land is something they can own if they work hard enough and save enough money to buy it. To Indigenous people land is not just something that they can own or trade. Land has a spiritual value.

THE HUMAN WORLD

We are the Indigenous people of Australia. Aboriginal people are those traditional cultures and lands lie on the mainland and most of the islands, including Tasmania, Fraser Island, Palm Island, Mornington Island, Groote Eylandt, Bathrust and Melville Islands. The term "Aboriginal" has become one of the most disputed in the Australian language.

The Commonwealth definition is social more than racial, in keeping with the change in Australian attitudes away from racialistic thinking about other people. An Aboriginal person is defined as a person who is a descendant of an Indigenous inhabitant of Australia, identifies as an Aboriginal, and is recognised as Aboriginal by members of the community in which she or he lives.

This definition is preferred by the vast majority of our people over the racial definitions of the assimilation era. Administration of the definition, at least by the Commonwealth for the purposes of providing grants or loans, requires that an applicant present a certificate of Aboriginality issued by an incorporated Aboriginal body under its common seal.

Sometimes non-Aboriginal people get confused by the great range and variety of Aboriginal and Torres Strait people, from the traditional hunter to the Doctor of philosophy; from the dark-skinned to the very fair; from the speaker of traditional languages to the radio announcer who speaks the Queen's English. The lesson to be learned from this is that we should not stereotype people, that people are different, regardless of race.

Our people, of course, did not use the word "Aborigene" (from the Latin ab, origin meaning "from the beginning") to refer to ourselves before the coming of non-Aborigenes. Everyone was simply a person.

A Doomed Reform

North Korea Flirts with the Free Market

Harpal Sandhu

After the fall of the Soviet Union, world politics shifted away from the conflict between communism and capitalism that had characterized much of the 20th century. But in northeast Asia lies one of the last relics of the Cold War and the last Stalinist state, the Democratic People's Republic of Korea, commonly referred to as North Korea. Ruled by a dictatorship that sheltered its citizens from the putative poison of decadent Western culture and influence, North Korea existed in almost complete isolation from the West for decades. However, recent economic reforms and attempts at fostering political discourse with its neighbors mark a conspicuous departure from previous foreign and domestic policy for this international enigma. Once a staunch proponent of communism and national self-reliance, or *juche*, North Korea has devalued its currency 70-fold, allowed prices and wages to be determined by markets, partially eliminated rationing, and announced the creation of a Chinese-style special economic zone (SEZ) open to foreign investment. This arrangement is radically different from traditional North Korean socialism.

On the political front, North Korea has re-opened dialogue with Japan, admitted abducting 13 Japanese citizens in the 1970s and 1980s, begun limited de-mining of the Demilitarized Zone (DMZ), and agreed to the creation of the first railroad link between the two Koreas since World War II. As promising as these developments appear, however, North Korea is unlikely to begin an immediate economic revival like the one China experienced two decades ago. North Korea's unusual

new economic reforms and diplomatic initiatives are misguided attempts to reinvigorate a decaying economy and curry international favor and concessions. In the long run, many of its economic changes and diplomatic maneuvers may prove self-defeating.

Only a serious crisis within the economy could persuade North Korea's government to change its socialist ways. The dramatic decline of the North Korean economy began with the loss of Soviet subsidies and trade following the collapse of the Soviet Union in 1991. North Korea's exports plummeted while its leader, President Kim Il-Sung, refused to reduce imports. Furthermore, military expenditures continued at a level that enabled this country of slightly more than 22 million people to maintain the third largest army in the world. The results were disastrous—gross domestic product (GDP) fell by 50 percent between 1992 and 1997. Worst of all, a devastating famine engulfed North Korea, forcing the reclusive government to appeal to the World Food Programme for aid. While help did come, the humanitarian group Doctors Without Borders estimates that at least 10 percent of the population perished. With a decimated population, a decade of economic recession, and a dearth of communist allies, North Korea has begun to infuse capitalist elements into its supposed workers' paradise.

Getting Ahead

Logic would dictate that North Korea, which is devoid of any substantial capitalist experience in the last half-century,

should begin reform with baby steps. Instead, the government has embarked on several dramatic reforms. Beginning in summer 2002, the North Korean government lifted most price controls and allowed market forces to set prices by supply and demand. Although initial results do not necessarily predict the future, the new policies created huge discrepancies and asymmetries in the economy. While average wages increased 200 times, the price of rice rose 400 times, diesel fuel 40 times, and electricity 60 times. Furthermore, the North Korean government has made no move to integrate private enterprise into the economy, an omission that could cause catastrophic problems in the future. Allowing demand to run rampant while limiting supply by suppressing private enterprise only exacerbates inflation. On the other hand, in a move that foreshadowed the new SEZ and desire for foreign investment, the government slashed the official value of the North Korean currency from the rate of 2.15 won to the US dollar to 150 won to the US dollar. The devaluation of the won to better reflect its market value has been hailed as a wise decision that should help increase trade.

North Korea's efforts at reform is reminiscent of *perestroika*, the late 1980s restructuring that Premier Mikhail Gorbachev implemented in the Soviet Union shortly before its collapse in 1991. With decentralization occurring in a large number of industries, managers were suddenly presented with decisions that they lacked the experience to make. Similarly, the rapid pace of change in North Korea has forced the general public to confront many unfamiliar decisions on a daily basis. Workers without any experience in budgeting will no longer be provided with all their food and shelter by a state rationing system. As demonstrated by the fate of the Soviet Union, these transition problems can prove stubbornly persistent.

North Korean reform has also taken on a decidedly Chinese twist. In September 2002, North Korea announced plans to develop a version of China's SEZs in order to attract foreign investment. This new zone is to be located in Sinuiju, which lies along the Yalu River bordering China on the west coast of the Korean peninsula. The plan calls for 500,000 current Sinuiju residents to be replaced by 200,000 technically skilled new residents from North Korea and China. Estimates place the shipping capacity of the 50-square-mile region at 12 million tons per year, and the nearby Supung hydroelectric power plant will accommodate all of the region's energy needs. The North Korean government has declared that Sinuiju would be a semi-autonomous region with its own legislature and the right to private property. Although the government eventually retracted a promise to allow foreigners to enter the region without visas, there will be other incentives to lure foreign investment. The US dollar will be an official currency and the income tax will be capped at 14 percent. This economic gambit is frequently compared to China's first special economic zone in Shenzhen in 1979. A sleepy fishing village prior to the reforms, Shenzhen ballooned to a population of seven million. Its success convinced China to launch five more zones of the same type. Just over a decade into this capitalist experiment, the five zones in existence in 1991 were responsible for 15 percent of China's US$57 billion in trade that year.

Of course, North Korea seeks to duplicate this kind of success. But there are important distinctions to be made between the situations of Shenzhen and Sinuiju. The first distinction is geographical. Shenzhen, as well the four other zones, is located in the south of China in close proximity to Hong Kong. Already a bustling and successful metropolis, Hong Kong became Shenzhen's primary trading partner. Sinuiju, on the other hand, has no immediate neighbors, aside from South Korea, with the wealth to invest heavily in the region. However, South Koreans are skeptical of investment because of horror stories about bureau-

cratic inertia in the North as well as weariness of risky investments since the 1997 Asian financial crisis.

The potential benefits to international investors of expanding to the general North Korean public do not outweigh the risks of investing in an unpredictable international pariah.

Underlying most Western investment in Shenzhen was the hope of using the SEZ as a gateway to the gigantic consumer market in China. No such motive exists in the case of Sinuiju. Instead of one billion potential CocaCola drinkers in North Korea, there are closer to 22 million. This number is probably not large enough to persuade Western businesses to invest in Sinuiju, especially considering the country's atrocious record of entrepreneurial ventures.

Former Failures

The 1990s featured two spectacular failures of private enterprise and entrepreneurship in North Korea. In fact, the Sinuiju project would not be the first time that the North Korean government has established a special economic zone. In 1996, it created the Rajin-Sonbong Economic and Trade Zone (RSETZ) on the northeastern tip of the peninsula. Surrounded by barbed wire in order to shut out North Korean citizens, the RSETZ occupied a seemingly promising location. The disconnected Sino-Mongolian railroad, if completed, would have made the RSETZ the terminus of a trans-Eurasian railroad. Moreover, the cost of labor in the RSETZ was appreciably lower than in the Chinese coastal cities. Also, it was perennially free of ice and very close to Japan. All that was lacking was additional infrastructure, a problem that was overcome at Shenzhen. To this day, however, Rajin-Sonbong remains a desolate place. The fact that nothing became of the RSETZ suggests that there are inherent problems with North Korea's ability to attract investment. The potential benefits to interna-

tional investors of expanding to the general North Korean public do not outweigh the risks of investing in an unpredictable international pariah.

While the RSETZ was a solely North Korean venture, there also have been joint attempts to open North Korea to the world. In 1998, the Hyundai Corporation of South Korea cooperated with Pyongyang on an ambitious plan to make North Korea's scenic Mount Kumgang a tourist attraction. Located on the eastern Korean coast just north of the DMZ, Mount Kumgang was originally viewed as a beacon of hope in the otherwise dark prospects of reunification.

Unfortunately, the project has been a commercial calamity. The six missile launchers, searchlights, and armed guards littering the mountain hardly make the resort an ideal place to relax. Hyundai reduced the number of ferries to the resort from four per week to one per week, and even that single boat is not filled to capacity. The resort is in such dire straits that Hyundai's hopes of profit have vanished, and the South Korean government must subsidize it in order to keep alive this meager step toward reunification.

Strained Relations

Pyongyang has also emerged from its typical isolationism to engage diplomatically with its neighbors. This newfound openness in diplomacy comes, perhaps not coincidentally, at a time of crippling economic hardship and consequent reform. Indeed, the North Korean government seems to believe that admitting past transgressions gives them some leverage to force concessions to aid its ailing economy. The 1994 nuclear crisis reinforced this pattern of behavior. Pyongyang openly declared that it was in violation of the 1968 Treaty on the Non-Proliferation of Nuclear Weapons and agreed to weapons inspections that have never taken place. In return, they were promised shipments of fuel oil and US assistance in the construction of two light-water nuclear reactors. Thus, when North Korean officials admitted in fall 2002 to a revived nuclear program, they may have been simply repeating previous tactics. However, the adverse Western reaction to that revelation indicates the strong possibility of a diplomatic miscalculation on the part of President Kim Jong-Il.

A similar line of thought can be applied to Pyongyang's surprising approach to Japan. Meeting with Japanese Prime Minister Junichiro Koizumi in September 2002, Kim conceded that a government agency had abducted 13 Japanese citizens during the 1970s and 1980s and subsequently refused to return the children of some of the Japanese nationals. Naturally, the Japanese government was outraged by the news and has ruled out any economic aid so long as the children remain in North Korea.

One can easily understand the deterioration of relations between North Korea and its rivals. However, relations with its ally China have also been tried recently. First and foremost, Pyongyang never consulted China about the special economic zone in Sinuiju, which lies on the Sino-Korean border. Apparently, the Chinese government is concerned that residents of Dandong, China, which is connected by bridge to Sinuiju, will flock to the new capitalist enclave in order to gamble. Furthermore, a statement by China's Foreign Ministry was coldly diplomatic in commenting on the Sinuiju experiment. "We have followed a path of socialism with Chinese characteristics," the Ministry reported. "Such a road is not necessarily suited to other countries' conditions."

But the most cogent evidence that China disapproves of Sinuiju lies in its arrest of Yang Bin, the man hired to govern Sinuiju, just one week after the government announced his appointment. Yang, whose net worth was estimated at US$900 million by *Forbes* magazine, made his fortune in horticulture and took advantage of relaxed immigration laws in the wake of the 1989 student demonstrations of Tiananmen Square in order to gain Dutch citizenship. Although he has confessed to owing US$1.2 million in back taxes to a Chinese provincial government, the timing of his arrest warrants interest. Beijing could have brought the evidence of Yang's criminal activities to Pyongyang's attention before taking action. Instead, his arrest took place just as North Korea was basking in the limelight of attention from the international media.

Clearly, there is friction between North Korea and one of its few remaining allies.

The divergence in results between Soviet *perestroika* and the Chinese reforms is staggering. One state no longer exists, while the other continues to flourish. North Korea's reforms more closely resemble Gorbachev's rash and rapid *perestroika* than the incremental and reserved approach of Chinese leader Deng Xiaoping. Though such a comparison is not encouraging for the future of North Korea's current government, at least Pyongyang has taken the chance of loosening its grip on power in order to try to rejuvenate its crumbling economy. However, North Korea runs the risk of overturning whatever gains it may make from these economic changes. If its odd and mysterious diplomatic initiatives backfire, the state could be deprived of the international backing it so desperately seeks.

Harpal Sandhu, Associate Editor, *Harvard International Review*

Close to Home

*Shunned by Western tourists, the Philippines
rebrands itself as a premier beach resort for Asians*

By James Hookway/BORACAY, THE PHILIPPINES

SHORTLY AFTER the Abu Sayyaf terrorist group started kidnapping Western tourists, among others, two years ago, something unexpected happened on the pristine beaches of Boracay island, a popular Philippine holiday resort. With French and German tourists staying away, pale, northern European faces have been gradually replaced by pale, northern Asian faces—South Korean in particular. Korean-owned restaurants, hotels and scuba-diving centres have multiplied along Boracay's five-kilometre stretch of gleaming white sand to cater to the tourists jetting in from Seoul.

What's more, these new arrivals from Asia have been enough to keep the Philippines' battered tourism industry ticking over. In 2002, South Korean arrivals increased 39% from the previous year, after rising 20% in 2001, according to Department of Tourism figures. Arrivals from China, meanwhile, jumped 47% in 2002. Indeed, nearly half of the 2 million tourists who came to the Philippines last year were from Northeast Asia. They pushed total tourism arrivals 8% higher last year in an industry that employs around 10% of the Philippines' workforce.

Vigorous promotion in Seoul and Beijing has helped raise interest in the Philippines, particularly as a honeymoon and scuba-diving spot. Tourism Secretary Richard Gordon visited South Korea four times last year to assure Korean visitors that they would be taken care of every step of the way in a country that doesn't exactly have the reputation of being the safest destination in Asia. Ads on CNN and BBC World showing lush vistas of sun-kissed beaches and swaying palms also helped put the message across in Asia and beyond.

The success of the campaign has surpassed expectations, demonstrating how the tourism business is changing. Faced with global uncertainty—whether it be war, terrorism or recession—holidaymakers are sticking closer to home. While American travellers are looking again at old standbys such as Mexico, the new rich of China are also checking out their backyard. Thailand has seen an influx of visitors from China, but the Philippines has been a surprise winner too. And its experience could show the way forward for other terror-damaged holiday destinations in Southeast Asia.

Wang Ying stretches back and looks up at the stars while a Filipino artist draws a tattoo around her ankle. The night before, she and her friends had visited the Moon-Dog beachbar and downed 15 shooters—enough to get her name etched on the bar's wall of fame along with thousands of other tourists from across the world. Wang is here to capture the laid-back vibes of Boracay for China's CCTV-2 television channel, where she is a travelogue director. "This place is special," she says.

Further down the beach, a group of South Korean honeymooners dine on grilled seafood, the warm waves lapping gently at their feet as they prise open tiger prawns and crab. A waiter says *kimchi*, the spicy fermented cabbage loved by Koreans, is available on request.

MORE LIKE LATIN AMERICA

Part of the Philippines' attraction to Asians is its cultural uniqueness in Asia. Colonized by Spain and then the United States, the predominantly Catholic islands appear to even casual observers to have more in common with Latin America than its neighbours in Southeast Asia. Korean and Chinese tourists say they find the Philippines' easygoing fiesta atmosphere an intriguing experience compared with the work ethic that dominates in their home countries.

"Many of them have already been to places like Thailand, where once you strip away the food and the climate, the mindset is often the same as they find back home," says Wilson Lee Flores, a Chinese-Filipino journalist. "But when they come to the Philippines they find fiestas, tequila and San Miguel beer. It's like going to Mexico, and for Koreans or Chinese, this is an exotic idea."

Not everybody is pleased with the sudden influx, however. Some Filipinos are still learning to adapt to their new visitors. Religious leaders on Boracay, for example, are campaigning to prevent the opening of a casino there to lure Asian highrollers. "Casinos promote enslavement to the passion of gambling," Bishop Gabriel Reyes told a Manila newspaper. Some dive shops, meanwhile, worry that taking large groups of tourists out to see coral reefs could damage the marine environment around the island.

But many tourists from Northeast Asia so far appear to have little interest in gambling. "I'm on my honeymoon," says 27year-old Park Jung-tae as he looks for sunscreen lotion under his beach towel. "This isn't a time for visiting casinos."

Japan's Homeless Find Their Place in Public Parks

Long Economic Slump, Tolerant Attitudes Let Shantytowns Take Root

Raising Chickens, Vegetables

PHRED DVORAK

OSAKA, Japan—For four months, Osamu Hachiya, an official with the city parks department here, struggled to move 11 homeless people a few hundred feet—from the southeast corner of Nishinari Park to a more crowded spot near the middle.

First, Mr. Hachiya had to inform the squatters that the city planned to renovate the southeast corner of the public park. Then, he went through five rounds of relocation talks with the homeless in the park's junk-filled main square. The final deal: new space for the 11 near the baseball field, three months for them to move and a pledge to "respect the will of the tent-dwellers." At least, Mr. Hachiya says, he rejected demands for free vinyl sheets and stakes for the new tents. "They're there illegally, after all," says the 54-year-old.

Onto the Streets

Once famous for its equitable society, Japan is now suffering from a homeless problem so bad that shantytowns are filling nearly all the big parks in cities such as Tokyo and Osaka. Part of the problem is the economy: 13 years of slump have pushed up unemployment and bankruptcy rates, driving the poorest onto the streets. The number of homeless people has ballooned to about 25,000, according to a recent government estimate, up from almost none in the late 1980s. Activists say the actual figure is probably several times higher.

That's still tiny compared with the U.S., which has an estimated 600,000 homeless. But while the U.S. federal government spent about $2.2 billion last year on homeless-assistance programs, Japan, until recently, hasn't done much. Three years ago, the nation had no national budget for homeless welfare at all. Even now, the country has just four emergency shelters.

Confronted by mounting numbers of squatters and little guidance on how to handle them, local officials such as Mr. Hachiya are taking matters into their own hands. The result is a mish-mash of Band-Aid measures, rule-bending and tolerance that has helped maintain—even foster—orderly shantytowns in public spaces across the nation. Wardens in Tokyo's Shinjuku Chuo Park ask the estimated 130 tent-dwellers to move to another part of the grounds while they trim trees—then direct them back to their spots when they're done. Osaka Castle Park has a shantytown population of 400, with squatters running their own night-time safety patrols.

Nishinari Park, which Mr. Hachiya oversees, has one of the oldest, most-established shantytowns. It even has its own "chairman": Hisakatsu Fujii, who has lived there for a dozen years. City officials let squatters keep chickens, grow vegetables and light cooking fires at dinnertime. Park caretakers shoo away newcomers, hoping the settlement won't expand, but they don't bother the 90 to 100 squatters who have been there for years. Residents of some shantytowns are forming their own support groups or joining with activists to protect the "homeless lifestyle."

"We are recognized," says Shinjuku Chuo Park resident Hideko Asada, pointing to a homemade mailbox near her cardboard hut where the local mailmen deliver letters. "The country has been turning a blind eye to the problem," says Minoru Yamada, a long-time Osaka labor activist who has set up a 300-strong tent city next to the city hall to protest the government's inaction. "Now, they're paying the price with a vengeance."

Japan's welfare system isn't set up to deal with the growing number of homeless people. Handouts tend to be limited to the sick and elderly, on the theory that able men of working age should hold jobs. (Women and children find it easier to get government aid.) For most of Japan's post-World War II history, that worked: Corporations kept their staff employed for life, and a booming economy provided plenty of work for smaller businesses and day laborers.

Now, that system is crumbling, as once-mighty corporate giants are undergoing massive restructuring, moving factories overseas and shedding thousands of workers. Unemployment reached a record 5.5% recently. Retraining programs are scarce or offer few good job prospects. One homeless shelter, for instance, runs a class in fixing bicycles. As a result, an increasing number of people are falling through the welfare cracks. Three-quarters of the homeless in Osaka are men age 50 to 60—too young to collect a pension, but too old to compete for a dwindling pool of jobs in construction and factories. Activists say what's needed is a concerted national overhaul of welfare, labor and housing policy.

Japan is beginning to address the problem. This year, the national budget for homeless-related assistance is $23 million, up from almost nothing in 2000. There are also some local programs, and the national government is coming up with a policy stance, expected within the next few months.

But Japan still has few food-aid programs and there is little coordination among bureaucrats. In Osaka, for example, the parks department is in charge of homeless people in parks, while the construction department oversees those by roads and rivers. But neither can come up with aid programs because that's the responsibility of the welfare department.

The most elaborate shantytowns are in Osaka, the industrial heart of western Japan, where recession and a shrinking market for unskilled labor have put an estimated 10,000 people on the streets. Nishinari Park, tucked in a gritty corner of the city next to a sewage-processing plant, hosts about 100 huts patched together with plywood and blue vinyl sheets. Along the park's main path, a vest-clad resident cuts wood for his stove with a circular saw powered by a portable generator. A few steps away stands a makeshift "barber shop"—a three-sided vinyl-clad box with a chair inside. The park has a new playground, but for now, no one can use it. Park officials have fenced off the play area with barbed wire, in an attempt to keep yet more homeless from streaming in.

Mr. Fujii, the "chairman" of Nishinari Park's shantytown, says there were few homeless people in Japan when he first set up camp more than a decade ago. Now he has a large, vinyl-draped compound in the center of the park, filled with everything from egg cartons to old handbags to a black-and-white television set run off a car battery.

Mr. Fujii, a 79-year-old former day laborer, gathers newspapers and aluminum cans for recycling, bringing in about $250 in yen a month. That's enough to pay for one solid meal a day and a trip to the public bath every three days. At first, he says, park caretakers tried to push him away, making him move his tent once a month for "cleaning." Eight years ago, they forced him and 70 others out. But with nowhere else to stay, he moved back a month later, and there have been no recent attempts to oust him, he says.

"We don't want to take over the park," he says, with a grimace that exposes his one remaining tooth. "But how else are we to live?"

Osaka officials concede they are uneasy about the growing permanence of shantytowns in public parks. But they say they don't have the budget or national backing for a comprehensive relief program. In 2000, the city came up with a partial solution for its three most-squatter-infested parks: Build sparkling-clean temporary shelters inside the park grounds, and move the homeless indoors—and out of sight. According to the plan, the homeless can look for jobs, become "self-supporting" and leave, hopefully within three years. The city has spent about $20 million on the program so far.

About a year ago, Osaka put up temporary shelters in Nishinari Park that can sleep as many as 220 people, just a few hundred feet away from Mr. Fujii and his homeless neighbors. It's the job of Mr. Hachiya, the Osaka official, to direct them to the shelter. His mission is pressing, since the city wants to restart a park-renovation project—stalled for four years because of squatters living on top of the site.

Yet Mr. Hachiya just can't get many people to move in. The shelter remains about two-thirds empty. The city currently frowns upon being too pushy, after activists criticized officials in another park for driving away the homeless. Forcibly removing a tent is a bureaucratic process requiring four different notifications to the owner. So Mr. Hachiya and his team try to persuade park squatters to voluntarily move into the shelters. They're meeting stiff resistance.

In order to enter the shelter, squatters must sign papers promising that even if things don't work out for them, they won't go back to camping in Nishinari Park. But with jobs so few and unstable, residents question whether they can really become self-supporting in three years, as the rules require. At another shelter in Nagai Park, which shut down in March, fewer than 10% of its 200-plus dwellers found jobs and a permanent place to live. Many ended up back on the streets.

"If we leave here there's no guarantee we'll have work," says Akira Kubo, a skinny 55-year-old former construction worker. In his neat shack, he has set up a workshop where he cleans old electrical appliances for resale. He also displays a collection of stone seals that he has carved, traditional Japanese watercolor paintings and lots of books.

Mr. Hachiya concedes the shelter plan is flawed. "There's no exit strategy," he says. "If our positions were switched and I knew I was going to be kicked out in three years, I'd rather stay in the park too."

Though there have been some complaints and even attacks on the homeless, city-dwellers largely tolerate the shantytowns. Japan's homeless settlements tend to be orderly and quiet. Drug problems and violence are rare. Japanese people who wind up homeless "are just like ordinary citizens," says Yohji Morita, a professor who conducted a 1998 survey of homelessness in Osaka.

At the Nishinari shelter, Director Isoji Tanaka bends a few rules. The easy-going former welfare official occasionally doles out medicine when the park squatters get sick, even though the supplies are supposed to be only for people in the shelter. Last November, he put two squatters up for the night when their tents burned down. The two returned to the park the next day. Mr. Tanaka says he is still haunted by the death last year of another one of the park's tent-dwellers. "We said 'hi' every now and then," recalls the 54-year-old Mr. Tanaka, quietly. "It was very sudden."

In the park on a recent day, the square-jawed Mr. Hachiya and a handful of other park officials stood stiffly in blue jackets, preparing for a scheduled, two-hour meeting to discuss moving tent-dwellers out of a proposed renovation area. Mr. Fujii and his neighbors, however, seized the opportunity to lobby for their right to stay put. Among the 50 squatters was a man with a pet rooster tucked into his plaid shirt and an unshaven man in a hooded blue sweatshirt who lives in a cardboard box. They were joined by activists from a group called the Homeless Peoples' Network, which started in 1995 after a homeless man was pushed by youths into a city canal and drowned.

"Do you know why nobody goes to a place like that shelter?" yelled out the man in the blue sweatshirt. "Why don't you tell us?"

"The main reason," Mr. Hachiya ventured gamely, "is that Osaka city has not provided sufficient programs to guarantee that you can become self-supporting after you go in."

There are new programs afoot, he explained, but his voice was drowned out by loud cries of "We've heard all this before." The squatters' demand: A promise to leave their tents alone until the government comes up with a satisfactory welfare program.

Back in his office in a waterfront skyscraper across town, Mr. Hachiya says he doesn't expect to be in this job for too long. Until last year, he handled cultural events for Osaka city, such as puppet shows. In another year, he figures he'll be rotated out into another post. Still, he says he feels the pain of the homeless. "If you were unhappy, you'd take the opportunity to let it out on the officials," he says. "Anyone would."

Wall Street Journal

SOUTH KOREA

Open Education

Seoul prepares to open further the education sector to foreigners, but not too far

By Kim Jung Min/SEOUL

DRASTIC PROBLEMS call for drastic measures, and South Korea is turning to foreign institutions to help revive the country's ailing education system. The country has long been reluctant to open up domestic markets to foreign competition, but in May, as part of its World Trade Organization commitments, it will start negotiations with other WTO member nations on providing greater, if still limited, access to its education sector, officials say. There is significant opposition to the plan among nationalists and teachers, while Education Minister Yoon Deck Hong is against the blanket access that foreign players are seeking. "Opening our education market should be limited to the current level of adult education and other tertiary training," he told a parliamentary meeting on March 18.

So what's wrong with a sector that produces a stream of prize-winning maths and science wunderkinds every year? One of the main criticisms that is levelled against the education system is that it stresses conformity and conservatism over individualism and originality. This in turn means that classes are large, curriculums are rigid and, in the quest for the precious exam passes crucial for higher education, teenagers are put through a gruelling workload and long hours. Stress levels are high—suicide is the third-biggest cause of death among teens, and half kill themselves because of school-related stress, according to the National Statistical Office.

But there's been a backlash over the past decade, and growing numbers of parents have been sending their children to universities, high schools and middle schools abroad as the government has progressively relaxed rules on studying overseas. Favourite destinations are English-speaking countries and, increasingly, China. Parents believe the pressures are lighter, teachers are of a higher quality and the courses are less rigid overseas. In 2001, almost 150,000 Koreans were studying in overseas universities, against around 106,000 in 1995 and some 85,000 in 1993, according to government figures. But it's ex-

pensive studying abroad and many would like to see foreign educational establishments open branches in South Korea.

The government is taking note—up to a point. "The short cut is to promote competition by letting respectable foreign universities set up their operations here," says Lee Man Hee, a policy expert at the state-funded Korean Educational Development Institute. He and other supporters of the move hope that competition will help improve standards and conditions in the system, where about 70% of universities and almost half the number of high schools are private.

"Singapore is a benchmark case for us," adds Lee. The island republic started wooing top overseas universities in 1997, since when elite overseas colleges have tied up with local universities or branched out to offer postgraduate degrees. Eight top-notch U.S. and European universities now have a presence in Singapore, including the Massachusetts Institute of Technology and Johns Hopkins University.

In Korea, however, not everyone was happy when the government announced on March 27 that it planned to discuss further opening of the education sector with WTO members: Almost daily protest marches, mainly in Seoul, have attracted thousands. The country's 390,000 teachers are a powerful lobby group and many oppose a plan that they fear will stress subjects such as English, maths and science over Korean language and history and may put them out of a job. They also argue that small local universities, which offer cheaper tuition fees, could be forced out of business. "Why should foreign investors undertake the role of teaching our children? Their only concern is to earn money," asserts Cho Hee Ju, vice-president of the Teachers and Educational Workers Union, which represents 94,000 teachers.

The government, however, does not seem to be ready to allow wide access just yet. It first opened the higher and tertiary education sector to foreigners in 1995 in a bid to woo investment, but regulatory roadblocks meant that hardly anyone has set up operations in South Korea. That is expected to change, but how far the government is ready to open the door to foreign institutions will be decided in May. At the moment the proposal would allow foreign universities to establish colleges, language institutes and technical colleges in South Korea from 2005. The

negotiations could be tough as some of the countries interested in the Korean education market, including Australia, Britain and the United States, also want access to elementary, middle and high schools, according to a recent report published by Seoul-based Korea Institute for International Economic Policy.

Far Eastern Economic Review

INDONESIA
Going It Alone

Syaiful Bahri Tandjumbulu is a man with a mission: He wants to create a new district that he says will bring wealth and jobs to his people. He's not alone. Each year, Indonesia is being sliced into ever-smaller pieces—a process that's radically reshaping the country

Margot Cohen/AMPANA, CENTRAL SULAWESI

BLUE BLOOD and boundless energy run through the veins of Syaiful Bahri Tandjumbulu. His grandfather was the last raja of the former small kingdom of Tojo; his father was a newspaper publisher and poet. Syaiful's own career has included stints as a student activist, hotel manager, herring-factory worker, pearl-diving supervisor, cacao farmer and convicted ebony smuggler.

Syaiful sleeps about three hours a night. Lately, his waking thoughts have whirred around just one thing: carving out a new 5,721-square-kilometre district, called Tojo Una-Una, from Poso district in central Sulawesi province. The goal, he says, is to give local citizens more political and economic clout.

Syaiful is leading the campaign from Ampana, a coastal town where horse-drawn buggies rattle alongside neat picket fences. White banners for Tojo Una-Una flutter above Syaiful's gate. "Lots of people are smarter or richer. My main asset is charisma. That's what we need in this struggle," says the 43-year-old.

Across Indonesia, citizens like Syaiful are rewriting the map. Since 1999, parliament has approved the creation of four provinces and 106 districts and cities. Many other applications are still being vetted. While the military has pushed for some of these new provinces and districts—hoping to defuse separatist movements in Aceh and Papua—most of the requests have poured in from local lobbyists. In effect, Indonesia is rapidly breaking down into smaller and smaller pieces.

It's a controversial trend. Many analysts, particularly those based in Jakarta, view it all as simply a lucrative game for local and national elites in the run-up to next year's elections. Each district is entitled to one representative in the national parliament, so the major political parties have an interest in cultivating local voters by approving such requests. Significantly, only five applications have been rejected so far.

Indeed, 25 new districts and cities are slated to be announced over the next several months. Tojo Una-Una will be one of them, says Manasse Malo, chairman of the parliament's subcommittee on regional autonomy.

So what's in it for these new entities? Each district gets a locally elected parliament and a district chief, who jointly make crucial decisions on local budget allocations, land-use licences and concessions for some mining and logging operations. Just as important, new autonomy laws allow districts much more leeway in squeezing revenues from natural resources.

Advocates of the process say it will enable local pride to be harnessed to promote democratization. Like others, Syaiful also argues that local communities have a real stake in encouraging sustainable development. He points to his own record of helping to eject two alleged illegal loggers from the area and his commitment to preserving coral reefs in the Togian Islands.

Critics, though, are not convinced. They point to fresh cases of exploitation in some new districts and also worry that new local leaders will spend too much on government

Breaking Up Is Easy To Do

- Since 1999, Indonesia has added four new provinces and 106 districts to its map
- Advocates of the process say it's bringing democracy closer to the people
- Critics say some new areas are economically unsustainable

buildings and civil-service salaries while short-changing health and education for the poor. Even some leading policymakers doubt whether it makes economic sense to create a raft of new districts. "Quite a lot of them won't be able to sustain themselves. Some will have to recombine," says Dorodjatun Kuntjoro-Jakti, coordinating minister for economic affairs.

It's still early days to calculate the burdens or benefits of the country's evolving new map. But Syaiful and his friends are clearly part of an experiment with long-term implications. "It's going to transform this place socially, politically and economically," predicts Sidney Jones, head of the Indonesia office of the International Crisis Group, a European-based advocacy group devoted to conflict resolution. "To ignore it is to turn a blind eye to the biggest change in this country since independence."

THE VIEW IN AMPANA hasn't changed much over the last century. Coconut trees line the coast and crowd the interior. Some of them, at least, are the legacy of Syaiful's grandfather. Aiming to reduce hunger, the raja ordered his subjects to plant the trees and guard them from marauding animals. According to local lore, if the raja discovered any saplings devoured by pigs, he would beat the neglectful farmers with a rattan switch. Syaiful sees himself as a more benign populist. He owns a modest homestead at his 100-hectare cacao plantation, but prefers hanging out in his Ampana home with former student activists, teachers, environmentalists and a cleric or two.

For these people, the idea of going it alone had its immediate roots in the economic and political crisis that hit Indonesia in the late 1990s. Seven months after then-President Suharto fell from power in May 1998, the town of Poso—the district's hub—witnessed violent clashes between Muslims and Christians. Years of riots would eventually claim 542 lives and displace tens of thousands of people. Although the riots never reached Ampana (almost 160 kilometres east of Poso), they strangled commerce throughout the district.

Worried by these developments, Ampana's residents looked to the new regional autonomy laws passed in 1999 and 2000 and decided to revive a request for something first mooted in the 1960s: A district of their own. No one cared that the idea of uniting the two former small kingdoms of Tojo and Una-Una was originally a feat of Dutch colonial mapping. The same was true of many other proposed districts.

Moreover, this new Tojo Una-Una would serve a very modern purpose—locals could declare themselves independent of Poso and the terror that name implied. They would also no longer have to risk the four-hour bus trip to Poso to obtain ordinary documents like identity cards and drivers' licences.

Then, gradually, a bigger idea took hold: A new district with guaranteed security and improved infrastructure would lure both domestic and foreign investors. Expectations soared—not all of them realistic. Fishermen counted on higher prices for their catch. Farmers anticipated higher prices for crops like cacao and copra, or dried coconut. Unemployed students expected new civil service jobs. Business leaders wanted a container port and an airport. "Let's push the car that has stalled," exhorts Syaiful, pointing to a local economy that contracted by 3.8% in 2000. By May 2001, the movement had picked up steam, and Syaiful was selected to head the crusade.

But some members of the Muslim community in the rest of Poso didn't like the idea. If Muslim-dominated Tojo Una-Una broke away, that would leave them living in a district with a 73% Christian majority. Nonetheless, Poso's parliament decided not to veto the creation of the new district (which would slice off almost 40% of Poso's area), partly to stave off a clash with Syaiful's supporters. "We were afraid of an anarchic demonstration," says Poso legislator Firdaus Tato.

Even some of those who would live in the proposed new district had their doubts. Arifin Tutuna, a village headman, wondered how the district would pay for all the benefits its people were expecting, and fretted over a lack of skilled local technocrats. "Natural resources depend on human resources," he says. Indeed, such concerns aren't restricted to Tojo Una-Una: "In Indonesia, we need more people who know how to develop their localities with a sense of entrepreneurship and competition," says P. Agung Pambudhi, executive director of Regional Autonomy Watch, a Jakarta-based business group that's monitoring regional development.

Back in Ampana, Syaiful was hard at work taking care of all the practical details. Under the law, for instance, Tojo Una-Una needed multiple recommendation letters from political leaders and technocrats in Poso, Jakarta, and Palu, the provincial capital. The biggest challenge was raising funds to lobby Jakarta. The town's mayor agreed to redirect funds previously earmarked for new villages. Civil servants' salaries were cut for two months. Forty local businesspeople chipped in. Even Syaiful did his bit by selling 40 cows and a few goats.

Then he hit on another scheme. He asked villagers to comb the forest for ebony shards left over from logging. Half the proceeds would go into the campaign kitty, with the villagers keeping the rest. The scheme backfired. Only licensed operators may remove ebony from the forests, so Syaiful was charged with smuggling and ordered to remain in Poso while awaiting trial. He flouted that order, was jailed for 25 days and was eventually fined 2 million rupiah

($237). Convinced that the charges were trumped up by opponents to the proposed district, Syaiful ignored the fine. "I have the utmost respect for the law," he shrugs, "but I just don't think I did anything wrong."

Syaiful calculates that the Tojo Una-Una campaign required roughly 1.5 billion rupiah. Recommendation letters didn't come cheap. Apart from the steep costs of shuttling back and forth to lobby Jakarta, and bringing technical teams to Ampana, Syaiful says he needed cash to bribe the right people. "This is the situation in Indonesia. We have to use money," he says. Indeed, the Indonesian weekly magazine Tempo has reported that many other lobbyists for new districts are also greasing palms. When Jakarta sent a 17-member inspection team to Ampana in July 2002, Syaiful says his committee spent about 250 million rupiah on air tickets, food and bribes, including 10 million rupiah in "pocket money" for each team member.

In Jakarta, national legislators explain that they ask localities to foot the bill as evidence of grass-roots support. "This is a test case for them. Are they serious, or not?" says Malo, the autonomy subcommittee chairman in Jakarta, who says he sees nothing wrong with "pocket money." (Malo didn't go to Ampana.)

Some critics argue that the bribery and the politicking have obscured the vital question of whether places like Tojo Una-Una are economically viable. Malo rejects those concerns, too. "Budgetary reasons should not be used as a tool to repress people's desire to govern themselves," he argues. As a native of eastern Sumba island, the legislator says he can relate to the yearning of people in remote areas to manage their natural resources for their own benefit.

He compares the situation to a young man asking for his beloved's hand in marriage: "Don't say, 'Oh, you can't get married because you don't have a house, or a car.' Give him a chance." What would Syaiful and his supporters do with that chance? For a start, double the current number of civil servants and security officers. Build new government buildings at a cost of 50 billion rupiah. Develop 200,000 hectares for plantations. Help traditional fisherman upgrade their fleets and quintuple their annual catch. And maybe cut down a few of the old raja's coconut trees to make room for seaside resorts.

But Syaiful is not alone in eyeing the area's bountiful marine resources. Right next door is Fadel Muhammad, governor of the newly formed province of Gorontalo. Fadel wants the fish surrounding the Togian Islands to be exported from his own province. As Indonesia redraws its map, the boundary between neighbour and competitor could become awfully thin.

Macau Sets Its Sights on the High Table

The end of a decades-old monopoly over gambling in Macau has drawn some of the biggest names in the gaming business to the former Portuguese colony. The coming shake-out will likely rewrite the rules of the game but should place Macau in the international big league

By Tim Healy/MACAU

IT'S A TYPICAL weekend night in one of Macau's 65 or 70 VIP gaming rooms where, away from prying eyes and the "grind" players of the main casino tables, the high rollers play baccarat. According to the men who run these VIP rooms, they will face some frightening possibilities on any given night: First, a high roller, playing by himself or with a handful of others, could wager the maximum bet in Macau, HK$4 million ($516,000) on a single hand of cards, and win. And second, the same high roller—a valuable commodity as long as he loses more than he wins—could be enticed by a rival VIP room. Competition among the operators sometimes boils over into violence.

Macau is on the knife-edge of a remarkable transformation. The over-65-year-old monopoly on casino operations—the last 40 under overlord Stanley Ho—has ended. Already a volatile industry, gambling is in a further state of flux. Two years ago Macau awarded gambling concessions to three groups, which in turn are subdividing into an expanding number of casino developers. What has long been a quaint city in the shadow of much-larger Hong Kong is about to be flooded with money and foreign—mostly American—influence. The coming change threatens the existence of a cosy system of oral contracts and extra-legal practices that has evolved over decades. The ensuing overhaul of Macau's

gambling industry will introduce international best practices and cement its position as Asia's gaming capital.

Stanley Ho would not speak to the RE-VIEW for this story. Other key sources were reluctant as well, and many of those who agreed to be interviewed did so only on condition of anonymity. Nevertheless, there was wide agreement among the more than 30 people who the REVIEW spoke to that, while there was likely to be some resistance to such change, the huge influx of money would profoundly alter both Macau's gaming industry and the city itself.

Based on the local government's records, Macau is the world's third-largest gaming market, behind America's Atlantic City and the No. 1, Las Vegas. Ho's casino company, Sociedade de Jogos de Macau, successor to Sociedade de Turismo e Diversoes de Macau as casino operator, recorded $3.5 billion in total gambling revenue in 2003. That was about 27% higher than in 2002, a growth rate that, if it continues, will have Macau nipping at the heels of Atlantic City as the world's No. 2 gaming market by the end of this year. More important, Macau sits on the edge of the world's greatest potential gambling market: China. Estimates of how big the market in China could become range up to the hundreds of billions of dollars, dwarfing even the United States, which recorded just under $70 billion in casino revenue in 2002.

The changes have begun. Macau is beginning to see the first few hundred million dollars in private investment of what could easily be billions by the end of the decade. For a city of only 440,000 people that, according to a government survey at the end of 2002, had accumulated a total of just under $3.5 billion in foreign direct investment in its history, the amount is mind-boggling.

The first wave of investment is coming to a part of Macau close to the ferry terminal that connects the city with what has traditionally been its biggest source of both gamblers and tourists, Hong Kong. Most prominent of the new properties is the Las Vegas Sands, a $240 million casino being developed by one of the new

American competitors, Sheldon Adelson's Venetian group.

It is going up in between a smallish, $60 million hotel casino being developed by Galaxy Resort & Casino, a company started by Hong Kong property tycoon Lui Che-woo, and a $122 million retail-and-entertainment complex being built along the waterfront by Stanley Ho and partner David Chow. Both the Las Vegas Sands and the Galaxy Casino are set to open in the second quarter of 2004. The waterfront project, called Fisherman's Wharf, is supposed to open by the end of the year. Additionally, Galaxy says it will break ground this month for a 600-room hotel casino in the same area.

Just a few blocks away, at the southern tip of the Macau peninsula beyond Ho's iconic flagship hotel casino, the Lisboa, a $1.5 billion land reclamation and property development called Nam Van Lakes is in its final stages. Still to come is a $500 million casino resort from another American, Steve Wynn. Wynn told the REVIEW recently that ground would be broken for the project in June. Now Ho plans to build a 30-floor hotel casino next the Lisboa, to be called Lisboa II, for $187.5 million.

Less concrete but more far-reaching are plans for a strip of reclaimed land connecting Macau's two main outer islands, Taipa and Coloane. Adelson says his company will build a 1,500-room mega-resort and casino on this piece of land and is inviting others to join what he envisions will be a re-creation of the Las Vegas strip with 20 casinos and 60,000 hotel rooms. On a separate parcel of land, the Galaxy group has said it has plans to build a resort casino with 3,000 rooms in 2006.

One early sign of the magnitude of the coming change came with the announcement that the Macau economy grew at a rate of 25.3% in the third quarter of 2003 from the same period a year earlier. For all of 2003, it is expected that Macau will shatter the 9% growth recorded in 2002, which followed six straight years of flat or declining growth.

Tourism growth from the mainland into Macau was already strong in 2003, up almost 23% through the first seven months of the year over the same period in 2002. But after immigration rules changed in late July to allow mainland visitors from selected cities in southern China to enter Macau and Hong Kong without a passport or visa, the number of China arrivals soared. In the last five months of the year, arrivals from mainland China increased 49.3% over 2002.

Macau has a history of gambling that stretches back hundreds of years. What Stanley Ho did was bring professional management and accounting systems to what had been informal, family-run operations. Also, with the opening of the Lisboa in 1970, Ho introduced the most opulent casino gaming hall in the city. In 1987, three years after Britain and China agreed that the transfer of sovereignty over Hong Kong would take place in 1997, Portugal and China came to a similar deal on the return of Macau in 1999. In the run-up to the Macau handover, the city was rocked by murders and kidnappings. The violence was widely attributed to a turf war among triad-connected VIP-room operators battling to see who would control the rooms after Portuguese sovereignty ended and the Chinese arrived.

But after the handover, the violence has subsided. Macau recorded 37 homicides in 1999, three in 2002 and nine in 2003. And Beijing's return as sovereign did much more than just end the violence. Edmond Ho, China's hand-picked chief executive of Macau, accepted 21 applications from companies in Asia and around the world in late 2001 seeking one of three gambling concessions. On January 22, 2002, Macau announced the winners.

As expected, one of the three went to the incumbent monopolist, Stanley Ho. More surprising were the other two winners: One was a joint bid by Galaxy and Adelson, the Las Vegas gaming-and-exhibition impresario, and the second was from a group headed by Wynn, the American credited with helping to transform Las Vegas from a city focused almost entirely on gambling into a broader, family-oriented playground through a series of new casinos in the 1980s and 1990s. The two entrepreneurial swashbucklers had beaten some of the largest and most powerful gambling companies in the U.S. and Asia, including MGM-Mirage, the world's second-largest gaming company, which has been talking to Stanley Ho about opening a casino under his licence for over a year.

In a sense, the January 22 announcement was the city's true transforming event, an economic handover far more consequential than the political handover two years earlier. It has always been assumed that the end of Macau's gaming monopoly would have the greatest negative impact on the incumbent. But anyone who thought that the end of his monopoly would undermine Ho's pre-eminent position doesn't understand the man or appreciate his success in constructing a

company both ubiquitous in, and symbiotic with, the Macau economy.

Not only does Stanley Ho own all or part of 12 fully fledged casinos and control up to 70 VIP gambling rooms, he has majority stakes in the city's dog-racing and horse-racing tracks. Through his own Macau-based companies and Hong Kong-listed Shun Tak Holdings, which he controls, he has key stakes in five-star Macau hotels including the Mandarin Oriental, the Westin Resort and, of course, the Lisboa. He also owns 100% of Macau's second-largest locally based bank, Seng Heng Bank, the city's largest department store, New Yaohan, and numerous restaurants, saunas and property agencies. When the project began in the early 1990s, Ho also owned 25% of the Nam Van Lakes project, which reclaimed an area equivalent to about 20% of mainland Macau's landmass from the South China Sea.

Stanley Ho also dominates Macau's transport industry. Shun Tak owns Far East Hydrofoil, operating between Hong Kong and Macau. And Ho owns one-third of the Macau International Airport and 14% of Air Macau, the main airline serving the territory. In addition, he is said to have deals with all of the major bus companies that on average bring 20,000 mainland Chinese tourists into Macau each day. Now 82, Ho was estimated in 2002 to have a net worth of more than $2 billion by *Forbes* magazine, making him one of the 50 wealthiest people in Asia.

In the 1980s, Ho introduced a system for segregating high rollers in VIP rooms that would prove to be a masterstroke. The high rollers who play in these rooms are the lifeblood of gambling in Macau, generating three-quarters of the city's gaming revenue—which is far more in relative terms than in Las Vegas or Atlantic City, or the biggest casinos in Monte Carlo or Melbourne, Australia.

A common measure of business performance in casinos is win per table per day, which records how much a casino takes in at each table of common casino games like baccarat, blackjack or roulette. In Las Vegas, the largest casinos averaged a win per table of over $2,000 in 2002, according to industry sources. In Macau, the daily average is more than $21,000.

Ho's real innovation was to initiate partnerships for virtually all the VIP rooms. He required that the partner guarantee a monthly income, from which taxes would be paid and then the leftover would be split, usually 70% to Ho and 30% to the partner. The system incorporated the use of junket operators to bring in high rollers, who mostly gambled on credit with the use of "dead chips," or gaming chips purchased by a player that could be gambled or redeemed only in private transactions with another player—often at a discount to the chips' face value.

The beauty of this system, combining junket operators and VIP-room operators, is that it transferred some big risks from ho to subcontractors. Fluctuations in the number of players who come each month or house losses to a player on a hot streak had to be borne by the partners. Also, the responsibility for granting credit to VIP players typically fell to a subcontractor, usually the junket operator who brought them. A high roller from China, for example, might buy a package from a junket operator who acted partly as a travel agent, partly as a gambling facilitator and partly as a financier.

Probably most important was that Ho got out of the business of collecting bad debts, which can be an unsavoury task in China where there is no legal recourse for gambling-debt collection. In fact, the junket system proved to be perfectly suited to China for many reasons. Not only is gambling illegal in the country, even advertising gambling is strictly forbidden. The difficulty, then, is identifying and enticing players to a casino.

Fortunately for Stanley Ho, the junket system evolved into an effective marketing network to identify and connect with potential players. Junket operators also help high rollers circumvent a Beijing prohibition against taking more than 100,000 renminbi ($12,000) out of the country. Often, a player from mainland China will gamble against a credit line set up by the junket operator in Hong Kong and repay with funds back in China.

But much of the junket system built by Ho, so suitable to Macau's unique conditions, may be off-limits for the Americans. The crux of the problem for Wynn, Adelson and any other American casino operator who comes to Macau is that their gaming licences in the state of Nevada, where Las Vegas is located, regulate their activities and associations everywhere in the world. Crucially, the Nevada regulations tell the U.S. licensees that they must associate only with "suitable" partners when they do business overseas—suitability being determined by the Nevada Gaming Commission after an investigation.

How many of these critical junket operators in Macau would be deemed suitable by Nevada? Estimates vary wildly, and it is difficult to find one observer who is both knowledgeable and impartial. One Macau junket operator who spoke to the REVIEW on condition of anonymity says his group is clean but estimates that 80%-90% of Macau's junket operators are not.

Jorge Oliveira, on the other hand, who leads the Macau government's efforts to formulate and rewrite regulations that will control what has traditionally been an unregulated industry, says he doesn't see how the proportion could be so high. "All the biggest junket agents in Macau are already licensed to operate in Las Vegas," he says. "You probably haven't heard this because most of these guys don't necessarily want Stanley Ho to know that they were taking the same players to Las Vegas that they brought to him." Oliveira says that Macau will soon begin scrutinizing and licensing junket agents in much the same way that Las Vegas does.

Wynn, for one, plans to use junket operators, but he recognizes he may have to get new ones. "We use [junket operators] in America, and I intend to use them here. I don't know the people in Guangzhou who want to come to Macau. Or in Dongguan. Or in Shanghai. I need to create an organization of people and remunerate them appropriately."

But Adelson, the first American operator actually confronted with the question of whether to use existing junket operators, seems to have made the decision to mostly steer clear. Neil Ducray, the Hong Kong managing director of global advertising firm TBWA, calls Adelsoris casino plan a *Field of Dreams* strategy—"build it and they will come." TBWA has consulted a number of tourism companies targeting Hong Kong and China, including several involved in Macau. Ducray has been studying the situation. "I think at first you'll see Venetian do very little marketing of any kind. They are going to open their doors and see who comes in."

One gaming-industry analyst who asked not to be identified says that the Venetian told him last year that it expects to bring in $1 billion a year in gambling revenue. If this happens, that would make Macau's Las Vegas Sands one of the highest-grossing casinos in the world in terms of gambling revenue alone. No Las Vegas casino, where revenues tend to be about equally split between gaming and non-gaming sources, reports that much revenue from gambling alone.

Another source familiar with the company's revenue projections says that the low-end goal of the Las Vegas Sands is to bring in an average of 2,000 players a day and see them lose $1,000 each at the tables, which would translate into annual revenue of $730 million, a performance likely to generate enough cash flow to satisfy the money men.

Until now, Macau law prohibited anyone other than banks and credit-card companies from issuing credit to gamblers. In reality, Wynn says, credit had been granted historically by a variety of people—"all of it technically illegal." Wynn was afraid of running afoul of Nevada regulators who demanded that its companies follow the letter of the law where they operate, not the common practice.

EXPANDING CREDIT

Late last month, the Macau government, after consulting closely with representatives of all the casinos, introduced a law that allows casinos and junket operators to grant credit for gambling and spells out rules and conditions. Oliveira says that he expects the law to be passed without any significant changes, though it may be the middle of the year before it is finally approved. Wynn told the REVIEW recently that the new legislation gives him the confidence he needs to start building his casino by the middle of the year.

David Chow, who is developing Fishermaris Wharf, is a Macau legislator and its highest-profile junket operator. He says that the new legislation is a masterpiece of legal draftsmanship in that it satisfies both Wynn's desire to have the matter of issuing credit formalized but still affords flexibility in terms of allowing junket-and-VIP-room operators to grant credit.

Despite the enormous potential for casino operators, the risks are high. If three-quarters of Macau's market, currently attributed to high rollers, might be difficult for the Americans to crack because of its domination by Ho and the network of junket and VIP-room operators, that still leaves $875 million worth of annual business to fight over. Capturing a third of that would give a casino operator almost $300 million in revenue, which, though substantial in Las Vegas, might not be enough to generate the kind of cash flow that banks and investors want to see in Macau.

There is also substantial and growing competitive risk. At one time, it might have been reasonable for a newcomer to assume he could capture a third of the lower-end market without a major marketing push. But if Wynn builds as promised, and MGM cements a long-rumoured deal with Stanley Ho to open a new casino, that will leave three substantial American competitors and Hong Kong's Galaxy in the race to grab Ho's existing business and attract new players within a very few years.

How will the new competitors succeed? They all have the same basic options: They can increase their share of the lower-end business, increase the overall volume of that business, or begin to capture a significant amount of the high-roller turnover. The first option is possible, but it obviously becomes more difficult as competition grows. The second option is probably the most preferable and probably attainable, but the number and characteristics of new potential gamblers heading to Macau, mostly from the mainland, remain a big unknown. And the third option will be very difficult if, as many people believe, Ho and his network of junket operators have a stranglehold on the high rollers.

Oliveira has no doubt that the new competitors know exactly what they're doing. "At first I thought it was strange that these guys, coming into Macau between the 23rd and 24th hour before we granted licences, were so anxious to invest $1 billion. How could they know if they would be successful?" he says. "But let me tell you, they have numbers the Macau government doesn't have. They're not crazy. I think these guys know what they're doing."

Says Robert Broadfoot, managing director of the Political and Economic Risk Consultancy in Hong Kong, "You can debate all you want about whether the change is good for Macau, whether it will be resisted or embraced. The fact is, change is coming. There's too much money at stake for Macau not to change."

The Decline of Japan's Farmers

Farmers have long formed one of Japan's most politically powerful groups. Now they are ageing, losing their clout and facing competition from imports as politicians look for a new support base

Martin Fackler/KUSHIRA

AS TOKYO SEEKS free-trade agreements with several Asian nations and Mexico, Japanese farmers like Koji Suemitsu are getting an uncomfortable lesson in just how much their political clout has diminished.

Last year, when the pig farmer in the rural town of Kushira, population 13,000, in southern Kagoshima prefecture tried to organize a rally at a government office to oppose cheap pork imports, just 12 people showed up. That's compared with almost 100 who routinely came to protests in the 1980s. After the rally, he helped send off a group of local farmers to Tokyo, 950 kilometres away, to lobby lawmakers, only for them to get the cold shoulder from key Diet members who once courted the farm vote.

Meanwhile, Suemitsu has watched helplessly as Japanese negotiators have proposed doubling imports of pork from Mexico costing just ¥250 (about $2.50) per kilogram, ¥100 less than he can produce it for.

"Tokyo isn't listening any more," says Suemitsu, 50, who's already seen pork prices drop in half since he started farming 23 years ago. "Lawmakers don't see any merit any more in protecting agriculture."

In a shift that is radically altering the landscape of Japanese politics, the country's once-powerful farmers are finding themselves in the unaccustomed position of struggling to get Tokyo's ear.

Japanese policymaking has long been dominated by special interests, from small-shop owners to surgeons. Among them, farmers have been one of the mightiest, and most coddled, groups.

Farming has long held a special place in the national psyche. In a nation dominated by factories and urban sprawl, the countryside is seen as a repository of pristine national values, like harmony and community spirit. These emotional associations have helped make agriculture sacred in Japanese politics.

So has the reliability of rural votes, which have helped guarantee the conservative Liberal Democratic Party (LDP) a half-century of rule. In exchange, the party showered the countryside with subsidies, generous public-works projects—and trade barriers that kept out cheaper imports.

This has created the unreal situation where Japan's 3.9 million farmers have been able to sustain middle-class annual incomes of ¥7.6 million from plots that average a mere 1.6 hectares. That compares with average annual incomes of about $60,000 from farms averaging about 200 hectares in the United States.

But in recent years, the farm lobby has begun to lose its grip on Tokyo as demographic change shrinks the base of rural voters.

Japan's declining population is a national problem, but the countryside has been hardest hit. Rural areas are literally draining as young people leave for the bright lights of cities like Tokyo or Osaka. Over the last decade, the number of Japanese living on farms has dropped by a quarter to 13.5 million. Forecasters predict it will drop by half again by 2025. Those left behind are also getting older. At the time of the 2000 national census, more than half of all full-time farmers—53%—were aged 65 or older.

In the long run, farmers and officials say the decline will likely alter the fabric of rural society. As family farms dwindle, larger, more cost-efficient corporate-run operations could rise to take their place.

A DYING BREED

But the more immediate effect has been political. The falling number of rural voters is forcing the LDP to change tactics and make more efforts to court city dwellers. Until the 1990s, most of the party's back-room dealers and kingmakers were from farming districts and got their power by ensuring that

FARMERS LOSE THEIR CLOUT

Japan's political base is moving to the city as:

- **Farmers are ageing**
- **Free-trade agreements bring competition**
- **New regulations support corporate farming over small-scale farms**
- **A new generation of leaders is taking over**

rural interests were protected. Now, a new generation of more photogenic leaders like Prime Minister Junichiro Koizumi and LDP Secretary-General Shinzo Abe—at 49 the youngest person ever to hold that post—has risen by appealing to urban voters.

This also means less influence within the party for pro-farm lawmakers—known as the "agricultural tribe." These politicians fiercely fought efforts to open agricultural markets, shielding farmers behind outright import bans and tariffs as high as 1,000%. But over the past decade, Japan has been steadily lowering trade hurdles and allowing in a growing number of farm imports—a move welcomed by long-suffering consumers, who can still pay as much as $100 for a melon.

"The agricultural tribes are a dying breed," says Ichizo Ohara, who served as agriculture minister in the mid-1990s.

At no time has the waning political influence of farmers been as apparent as during the current round of negotiations for free-trade agreements. With an eye to wooing urban voters, Koizumi has vowed to "make whatever concessions are necessary" to make negotiations succeed, according to news reports.

Such agreements have broad appeal—outside rural areas—as sources of export jobs and lower consumer prices. Despite their larger numbers, urban and suburban residents have been too poorly organized and passive to be a political force. But now, as rural populations shrink and electoral reforms reduce inequalities between rural and urban votes, city dwellers have become a crucial constituency for the LDP.

"The [agreements] are an interesting test of whether the LDP values consumers above farmers," says Richard Jerram, chief economist at ING securities in Tokyo. "Five years ago, the answer would have been no. This will show how priorities have shifted."

Japan is a relative latecomer to the free-trade game, signing its first agreement just last year, with Singapore. But fears of being upstaged by Beijing have prompted Tokyo to make a sudden push across Asia. In the last two years, Japan has made a broad effort to sign separate agreements with Mexico and a halfdozen Asian nations. It is in formal talks with Mexico and South Korea, has begun negotiations with Thailand, and has completed a round of talks with the Association of Southeast Asian Nations.

Talks with Mexico stalled on the issues of pork and orange juice in October, though Mexican President Vicente Fox was already in Tokyo to preside over the signing of a deal. Negotiations with Thailand, a major producer of agricultural products ranging from rice and sugar to chicken, may hit similar snags, say Japanese officials.

But political momentum is behind the agreements and passage is only a matter of time, say farmers and Agriculture Ministry bureaucrats. "The [agreements] are inevitable," says Tatsuya Kajishima, the ministry's director of trade policy. Desperate to limit damage from the deals, Japan's farm lobby has chosen to make what might be its last stand.

In Kagoshima, an area of rolling hills and active volcanoes on Japan's southernmost main island of Kyushu, the trade pacts are feared as a threat to one of the pillars of the local economy: pork production. The prefectoral assembly has twice passed formal resolutions asking Tokyo not to sacrifice farm interests in negotiations.

As the young depart for the cities, the number of Japanese living on farms has fallen; more than half of Japan's farmers are 65 or older.

The Kagoshima agricultural cooperative, to which most the prefecture's farmers belong, has spent ¥3 million to air an hour-long TV documentary showing bucolic images of earnest farmers planting rice and other traditional crops. It's part of a campaign to convince city dwellers that Japan needs a healthy agricultural economy.

"We're drawing the line here, at the agreements," says co-op president Kouichi Kawaida. "If we open to more free trade, it will wipe out agriculture in Japan."

Indeed, more open trade is itself one of the reasons why so many people are leaving rural areas, speeding their political decline. When Torao Kitamura started raising pigs 35 years ago in Nokata, a village in Kagoshima, it was a ticket to a solidly middle-class standard of living. But over the past decade, he says, cheaper imports have crunched prices. Last year, when the price of pork hit ¥370 per kilogram, the lowest he'd ever seen, his income fell to ¥6 million, down from ¥10 million in the early 1990 s." Last year was hell," he says, his usually smiling face turning grave for a moment.

Kitamura says dozens of nearby farmers have given up. When he started farming there were more than 300 pig farmers in Nokata and neighbouring villages, he says. Today, just 70 are left. Now 65 years old, Kitamura says he may retire soon. Despite his family having farmed in Nokata for five generations, he says he's glad his four children have no interest in succeeding him. He hopes his nine grandchildren won't follow in his footsteps, either.

"I worry about being the last farming generation, but I don't want them to succeed me," Kitamura says. "Agriculture is just too risky now. I can't recommend it to young people."

Signs of rural decline are evident across Nokata. Empty homes and weedfilled former rice paddies are now a common sight in the village of 2,500 people, where fields dotted with ancient burial mounds attest to the fact this region has been farmed continuously for more than two millennia. The local elementary school now has just 30 children in its six grades, and may have to be closed soon, say town officials.

Nationwide, agricultural prices have slid 16.4% over the past 10 years, according to Agriculture Ministry figures. Over the same period, farm incomes have fallen 29%. Behind the declines has been Japan's slow but steady opening of its agricultural markets.

Long a major importer of soy beans and cereals like corn, Japan remained self-sufficient in most fruits, vegetables and meats until the 1980s. Pressure from trade partners and global treaties pushed Tokyo to open up. Japan now produces just 40% of the food it consumes, the lowest level in the developed world.

Japanese farmers say they've responded by cutting costs and boosting productivity. Suemitsu says he's tried to increase economies of scale by cramming as many pigs as will fit onto his tiny, 1.5-hectare farm; last year he sold 2,160, making him one of the largest pig farmers in Kushira. He's also reduced disease by feeding them anti-bacterial drugs.

Still, he says, the costs of labour and environmental regulations like building tanks to dispose of pig faeces and urine make raising pigs in Japan far more expensive than in Mexico and Thailand. "We can't compete on an equal basis," says Suemitsu, a strong, compact man who is shyly friendly.

As more farmers drop out, and others retire, organizing political activities has become increasingly difficult, Suemitsu says. Membership in his local pigfarmers' association has dropped to 10 from 70 when he started farming in 1981. "We don't have much influence any more," he says.

As family farms close, a new presence may be starting to take their place in the countryside: Companies are buying empty plots to build larger-scale farms.

A few kilometres away from Suemitsu's farm is what may be the shape of things to come. A subsidiary of Daiei Inc., one of Japan's largest supermarkets, has opened a factory-like pig farm that now produces about 20,000 pigs for slaughter each year, all to be sold at Daiei stores, says a company spokeswoman.

The government is starting to encourage farms like these to make Japan more competitive in an increasingly globalized market place. In 2005, it plans to cut subsidies to smaller farms to spur mergers. Regulations are also being rewritten to facilitate creating corporate farms.

Still, only 1,271 corporate farms existed in Japan in 2000, 0.34% of all the nation's farms. "We need to push forward the commercialization of Japanese agriculture," says Ohara, the former agriculture minister. "The development of internationally competitive farms in Japan has been retarded too long by special-interest groups."

Japan's English Lessons

North Korea's nuclear threat is forcing Japan to choose between two strategic options: draw closer to the United States, or develop a more autonomous and assertive foreign policy. How to balance these competing visions? Look to another island nation, Britain, which has spent the last 60 years strongly allied to the United States yet maintaining its ability to act independently on the global stage.

By Bill Emmott

As political figures, Tony Blair and Junichiro Koizumi have much in common, even though Great Britain's prime minister leads a nominally left-wing party and Japan's leads a conservative one. Both men have risen to be prime ministers on the back of personal popularity; both are good on television and skilled at retail politics; and both have put much of their foreign-policy store in a close relationship with the United States, leading both to support U.S. President George W. Bush during the 2003 Iraq war. Yet this last shared approach is now what divides the two leaders. Where Blair's stance on Iraq is hurting him at home and abroad (except in Washington), Koizumi has, if anything, been helped by his support for the war—even at home, in famously pacifist Japan.

This difference raises an intriguing possibility. The controversy surrounding Britain's involvement in the Iraq campaign, coupled with the war's messy aftermath, could prove a turning point in British foreign policy, discouraging future prime ministers (even Blair himself) from again backing a U.S. military venture so ardently. But this period could also mark the beginning of a new, more active phase in Japanese foreign policy, in which Japan, shaking off some of the legacy of the Second World War, strengthens its own military and diplomatic stance, but does so in a way that aligns it even more closely to the United States.

Admittedly, Iraq is a frail reed on which to base such a proposition, and it is unlikely that Britain and Japan are genuinely going to swap foreign policies. Their starting points are too far apart for that: While Britain, a nuclear power and permanent member of the United Nations Security Council, contributed the largest number of troops to the Iraq war of any U.S. ally, Japan has a six-decade allergy against aggressive action overseas. Initially, the nation merely provided moral support; more recently, it pledged $5 billion for Iraqi reconstruction and offered up to 1,000 peacekeepers once the situation becomes more stable.

But the British and Japanese positions could begin to converge, and as they do, Japan, both in its dealings with the United States and its broader foreign policy, may have much to learn from the British experience since 1945. Britain offers lessons likely to come in handy as Japan grapples with the growing danger posed by North Korea.

DEFEAT AND DEPENDENCE

Visitors to Japan can sometimes be forgiven for thinking it is less the land of the rising sun than the land of the tedious cliché. How often has one heard, in polite but uncomfortable small talk, that Japan is unique because of its four seasons, that its people have a deep love of nature, and—if you are British—that ours are two island nations united by a common love of fish and tea?

> **In a deep sense, Britain and Japan's special relationships with the United States represent a defeat for old aspirations and self-images.**

Superficially, there are indeed points of commonality between Britain and Japan, beyond a liking for tannin-flavored hot water. Each is part of a nearby continent, but not really of it. For hundreds of years, a central aim of British foreign policy has been to keep France and Germany from dominating Europe and threatening Britain's interests, while Britain has sought to pursue those interests around the world as an independent entity. At least since its opening to the outside world in 1853, Japan has followed a similar path, keen to prevent China, Russia, or the European imperial powers from dominating Asia while mostly wishing to act alone or be left alone.

Since 1945, each has also forged what the British call a "special relationship" with the United States. Yasuhiro Nakasone, a Japanese prime minister in the 1980s, said that Japan, with its huge U.S. bases at Okinawa and on the principal island of Honshu, should be "an unsinkable aircraft carrier" for U.S. forces off the Asian mainland. Britain, home to scores of U.S. bombers and fighters, serves a similar function off the European mainland.

In a deep sense, these special relationships represent a defeat for old aspirations and self-images. In Japan's case, a real defeat; in Britain's, a moral and political defeat, for World War II brought an end to the country's vast global empire, to its global network of military bases, to its capacity for independent action against other major powers, and to the belief that Britain was a shaper of world events.

But while postwar Britain and Japan both relied on the United States in many ways, they differed in their degrees of dependence. Britain never seriously contemplated an independent policy toward the Soviet Union, for example, but has shown a willingness to go it alone on regional or minor matters. It has sought to integrate its economy and even political identity more closely with continental Europe, while also pursuing its own military and diplomatic activities in Africa, Latin America, and parts of Asia.

Japan, by contrast, has been much more servile in its relationship with the giant across the ocean, reflecting an almost total lack of an independent foreign policy, even for regional issues. Commentators differ over whether this subservience reflects a deliberate effort to take a free ride on U.S. defense spending and diplomacy, or whether it has simply occurred by default. The choice of explanation doesn't matter much; the answer may well be a bit of both.

Post-1945 Japan, forbidden by its U.S.-imposed constitution from developing military capability beyond basic self-defense needs, cuddled up to the United States for security during the Cold War. The Cold War threats came from the Soviet Union and perhaps China, and in both cases, the United States could be depended upon to provide the protection Japan needed. In turn, successive Japanese governments permitted U.S. forces to do more or less what they liked on their Japanese bases and took a blind-eye approach to the use of Japanese ports by nuclear-armed U.S. naval ships, an obvious sore point with the Japanese public.

To be sure, almost from the day the ink dried on Japan's pacifist constitution, governments sought to bend its rules in order to build up some military strength, and with annual defense spending of about 1 percent of Japan's gross domestic product, that strength became considerable. But it still would have been inadequate to thwart a Soviet invasion during the Cold War. Even to-

day, Japan's armed forces suffer from a chronic lack of operational experience.

Likewise, in recent decades, Japanese defense and industrial bureaucrats pursued a technological insurance policy of sorts—making sure, for instance, that Japanese firms had the ability to build modern fighter planes or even construct nuclear weapons. However, these efforts have been made without any sense that such armaments might actually be needed someday.

HIROSHI GET YOUR GUN

But that attitude is changing, a transition prompted by the growing threat North Korea poses to Japan. While North Korea is pushing the Japanese even closer to the United States, it is also pushing them to develop independent military and diplomatic capabilities. Increasingly, right-wing Japanese figures have become willing to declare openly that if North Korea should become a nuclear state, Japan should do so as well. Such views are still on the margins, but hawkish talk is approaching the mainstream.

Early in 2003, Japan's defense minister suggested that Japan should be willing to contemplate a preemptive strike on North Korea if it saw evidence that a devastating attack against Japan was being prepared. The notion of preemption, made fashionable, of course, by the Bush administration, would have been unmentionable in the past but is now widely discussed in Tokyo.

The change in attitude has been fueled by three specific events: the 1994 crisis, when evidence first emerged of a North Korean nuclear program and war on the Korean peninsula looked like a real possibility; North Korea's testing in 1998 of a medium-range ballistic missile, which flew over Japan en route to a landing in the Pacific Ocean; and North Korea's admission in 2002 that it had a covert nuclear weapons program.

These developments, coupled with confirmation in 2002 that North Korea had, as suspected, kidnapped a number of Japanese citizens over the years, have hardened Japanese public opinion. There is now a desire to see the government take a tougher line with North Korea, and this has made it possible for politicians to openly contemplate preemptive strikes and nuclear deterrence.

From the Japanese perspective, North Korea is no longer merely a nasty irritant but a potentially lethal threat. This realization has not yet produced a clear strategy. It may never, given that Japanese politics tends to mitigate against the production of clear strategies. But it has shifted the balance of Japanese politics away from pacifism, enabling a slow but steady dismantling of some of the legal obstacles preventing the country from having a normal defense stance.

Those obstacles are of a tedious, technical nature and obstruct Japanese participation in U.N. peacekeeping activities and U.S.-led military operations. In 2001, for example, Japan's parliament passed a bill that enabled a supply ship and escorts to go to the Indian Ocean to provide indirect support to the U.S. campaign in Afghanistan. As such laws are passed, Japanese defense forces are given greater freedom of action. Arguably more important in the immediate term, however, is new Japanese investment, totaling more than $2 billion, in a surveillance program that included the launching of two spy satellites in March 2003, with two more scheduled for orbit before March 2004.

Such surveillance could be done on Japan's behalf by the United States, and of course it is. So why launch the satellites? One reason is that Japan simply wants to pull more of its own weight in intelligence-gathering. But this program also indicates that Japan is no longer sure close ties with the United States are enough and is beginning to cover itself for the possibility that Washington might be too preoccupied elsewhere in the world to pay sufficient attention to Japan and the Koreas.

Koizumi's firm support for the war in Iraq, and his offer of peacekeeping troops and financial support for reconstruction efforts, are intended to help ensure that the United States does keep a clear eye on East Asia, even as it is engaged elsewhere. In its effort to make progress with North Korea through multilateral talks involving China, Russia, South Korea, and Japan, the Bush administration is now standing shoulder to shoulder with Koizumi's government on this issue in a way that former U.S. president Bill Clinton never did with the Japanese governments of the 1990s. Koizumi hopes to keep it this way.

One reason for this strategy is that Japan still relies on the United States for many defense needs. For instance, those in Japan who are keen to put preemption on the table readily acknowledge that the country is not yet capable of carrying out a successful preemptive strike. The United States could do it, of course. Indeed, it is now well known that in 1994, the Clinton administration drew up plans for just such a strike on North Korea's nuclear installations, even involving some Japanese security and intelligence officials in the discussions, but ultimately rejected the plan as too dangerous.

In the long term, Japan's best hope of protection against missile attack from the Korean peninsula will be participation in the theater and national missile defense systems currently being developed in the United States. For Japan, a close relationship with Washington is thus made more important than ever by Kim Jong Il's nuclear saber-rattling.

The question, though, is whether close U.S. ties remain enough. Can Japan entirely depend on the United States? Might there be circumstances in the future in which the United States might not intervene to prevent a Korean attack on Japan? Or in which the United States might provoke one? Or where the United States no longer has such an extensive deployment of forces in East Asia?

Such questions have always existed in theory, but they were of little concern during the Cold War. It is hard to envision any of these scenarios playing out now, but they

have become less implausible. And the North Korean threat has become distinctly un-theoretical.

Japanese political and bureaucratic leaders are realizing that preparations ought to be made for other, perhaps less favorable outcomes, both long and short term. What effect might unification of North and South Korea have on U.S. policy? What deal might Washington make to keep China on its side against Kim Jong Il, or in the aftermath of Korean unification? Might the United States pull back from East Asia following a debacle in Iraq? There are imponderables galore, and no responsible Japanese government should ignore them.

THINK BRITISH, ACT BRITISH

Japan needs to develop a capacity for independent knowledge, thought, and action that does not cut across or undermine its close alliance with the United States. The challenge facing Japan is a seemingly contradictory one: Be dependent but also more independent. Yet, during most of the last 60 years, that has been precisely Britain's position—or, if you like, predicament.

The title of a thoughtful new book by Peter Riddell, political commentator of the *Times* of London, sums it up well: *Hug Them Close: Blair, Clinton, Bush, and the "Special Relationship"* (London: Politico's, 2003). Riddell sets out to explore how a Labour Party prime minister, Blair, sought to be a close ally to both Clinton and Bush and sent British armed forces into battle alongside U.S. troops under these two very different U.S. presidents. In the more leftwing parts of the British media, Blair is described as an American "poodle," and there has been much hand-wringing over his failure to develop an independent foreign policy.

Japan needs to develop a capacity for independent knowledge, thought, and action that does not cut across or undermine its alliance with the United States.

Yet, as Riddell points out, this embrace is far from new. Every British prime minister since Churchill, with the exception of Sir Edward Heath in the early 1970s, has made a close relationship with the United States the centerpiece of his foreign policy. That has not always meant participation in U.S. wars; during the 1960s, another Labour prime minister, Harold Wilson, refused American requests to send British forces to Vietnam. But even then, Britain offered moral support.

However, while maintaining the strong bond with the United States, successive prime ministers have also sought to uphold and enhance Britain's capacity for independent military action and to bolster the country's dip-

Want to Know More?

Ian Buruma's *Inventing Japan, 1853-1964* (New York: The Modern Library, 2003) traces modern Japan's journey from preindustrial isolation to imperialism to pacified, prosperous democracy. *Japan: A Reinterpretation* (New York: Pantheon, 1997) by Patrick Smith, challenges the U.S. view of Japan as a liberal democratic ally. Bill Emmott's *20/21 Vision: Twentieth-Century Lessons for the Twenty-First Century* (New York: Farrar, Straus and Giroux, 2003) analyses the challenges Japan will face in the coming century. Frank B. Gibney predicts where Japan will be in 2050 in **"Reinventing Japan… Again"** (FOREIGN POLICY, Summer 2000).

The U.S. ambassador to Japan, Howard H. Baker Jr., offers his vision of what Japan's role in the world should be in the October 7, 2003 **"Doshisa-Yomiuri Ambassadorial Lecture,"** available on the Web site of the U.S. embassy in Tokyo. Charles Krauthammer suggests the United States should make Japan a nuclear power in **"The Japan Card"** (*Washington Post*, January 3, 2003). *Reinventing the Alliance: U.S.-Japan Security Partnership in an Era of Change* (New York: Palgrave Macmillan, 2003), edited by G. John Ikenberry and Takashi Inoguchi, argues that the U.S.-German alliance is a better model for the United States and Japan than the U.S.-Britain-style relationship envisaged by the so-called Armitage-Nye report, **"The United States and Japan: Advancing Toward a Mature Partnership"** (Washington: Institute for National Strategic Studies, 2000).

Roy Jenkins' *Churchill: A Biography* (New York: Farrar, Straus and Giroux, 2001) chronicles the life of the greatest exponent of the "special relationship" between Britain and the United States. In **"Imbalance of Power"** (FOREIGN POLICY, March/April 2002), Robert Skidelsky details the limits of this relationship. A trans-Atlantic debate over the role of multilateral institutions, **"When Worlds Collide"** (FOREIGN POLICY, March/April 2001), reveals contrasting U.S. and British views on global governance. Peter Riddell's *Hug Them Close: Blair, Clinton, Bush, and the "Special Relationship"* (London: Politico's, 2003) explains current British Prime Minister Tony Blair's approach to his country's relationship with the United States. Blair laid out his vision for the special relationship in the 21st century in his speech to a joint session of the U.S. Congress in July 2003, available on the prime minister's Web site.

≫ For links to relevant Web sites, access to the *FP* Archive, and a comprehensive index of related *Foreign Policy* articles, go to www.foreignpolicy.com.

---

lomatic and economic ties within its region, chiefly though membership in the European Union (EU).

This independence has been of practical importance at times. Prime Minister Margaret Thatcher could not have recaptured the Falkland Islands from Argentina in 1982 without it, for her close ally, U.S. President Ronald Reagan, limited his support to offering intelligence cooperation and expediting the delivery of sidewinder missiles. Likewise, Britain's role in peacekeeping and peacemaking operations in the Balkans and in Africa has probably conferred diplomatic benefits. But perhaps the biggest benefit is the symbolic one—the notion that Britain has some choice in whether it supports or depends upon the United States, and that it can, if pressed, deliver its own punches.

Japan, with no permanent seat on the Security Council, no membership in a robust regional economic or diplomatic entity comparable to the EU, no nuclear deterrent, and no reputation for independent action, is in a very different position. If it is to respond to the threat posed by North Korea and to possible changes in East Asia and U.S. policy in the region, it would do well to move closer to Britain's position.

Japan's Ground Self-Defense Force is almost one and a half times the size of the British army. More hardware and software investment, as well as a declared intention to establish a nuclear deterrent, if necessary, would bring Japan's military posture more in line with Britain's. But Japanese troops are completely untested in warfare, and the country lacks the ability to project military power far from its shores. Tokyo needs to demonstrate it is willing to deploy forces in the cause of local or regional security and that those forces can be effective.

To do so, Japan will need to overcome historical enmities in the region and pacifist opposition at home. Offering forces to help sort out international crises such as the one in East Timor several years ago will seem like a huge step. In fact, though, it will simply be an indication that Japan has truly cast aside its imperialist past, is now willing to take responsibility for regional problems, and is dedicated to peace, democracy, and human rights.

Will Japan be willing to act this way? Probably the more appropriate question: Will it have to act this way? The answer lies among the many mysteries of Pyongyang.

Bill Emmott is editor of *The Economist*. He is the author of three books on Japan and of *20/21 Vision: Twentieth-Century Lessons for the Twenty-First Century* (New York: Farrar, Straus and Giroux, 2003).

Women's Universities Struggle in Japan

Shrinking enrollments bring closures, mergers, and reforms

By Alan Brender

TOKYO

Julie Tanaka has spent 17 of her 22 years at Japan Women's University, in Tokyo. She entered the university's kindergarten program at the age of 5, progressed through its elementary school, middle school, and high school, and is now a senior at the university. Next year she will leave the sheltered all-female environment for the first time, to attend graduate school at coeducational Keio University.

"To have guys in the same class is going to be a weird experience," she says. "I guess I'm a little bit worried and excited at the same time. I'm going to miss this school."

Students like Ms. Tanaka are the lifeblood of women's universities in Japan, which have historically recruited many of their students from this university-school "tunnel" system, as it is known. But now that system is ailing. Fewer families are enrolling their daughters in elementary and secondary schools affiliated with women's universities. And fewer teenagers are interested in spending four years at an institution for women.

Adding to the strain, the country's 91 women's universities, along with the rest of Japanese higher education, are struggling to cope with a decade-long recession and a steadily shrinking pool of college-age students. The combined pressures of tight budgets and fierce competition for undergraduates have forced a number of women's universities to convert to coed institutions or close. Others are trying to find ways to become more relevant to the needs of modern women in Japan.

"As competition for students and consolidation takes place, women's universities will have to fight hard to maintain their positions," says a spokesman for Mukogawa

Women's University, in Nishinomiya, a suburb of Osaka.

CODDLING STUDENTS?

Women's universities, locally called *joshidai*, form a broad spectrum: from chic, prestigious institutions that rival Smith and Wellesley to financially strapped, academically mediocre colleges struggling to survive.

Those less-prestigious institutions had, in the past, operated primarily as finishing schools, offering majors in such subjects as home economics. Many are overhauling their academic programs, and some have started admitting men. But even that is no guarantee of success.

Risshikan University, in Hiroshima, had the dubious distinction earlier this year of being the first university in postwar Japan to shut down without graduating a single student. For many years the institution operated as a two-year women's college under the name Hiroshima Women's Commercial Junior College. Two years ago administrators changed its status to a four-year women's university. Last year it became coed. This year, with just 176 students, the university closed, and most of the remaining students transferred to nearby Kure University.

Some female undergraduates say they avoided single-sex colleges because the institutions have a reputation for treating their students as if they were in high school. One university, for example, has a homeroom system and offers its students sleepovers at other campuses.

In an informal survey of 20 female students at Kyorin University, in Hachioji, near Tokyo, only one indicated even a remote interest in attending a women's uni-

versity, and even she was not enthusiastic. "I'd like to go to a women's university, but only for a day because I don't want to meet with only women on a regular basis," says Akiko Ishii, a senior.

Students at most women's universities in Japan actually have a fair amount of exposure to men through social events and cross-registration agreements with coeducational institutions. Male students converge on the campuses of all-female universities to recruit students to their clubs. "Japan Women's University is filled with boys during orientation week," says Masako Matsui, head of student affairs.

Showa Women's University, however, hires private plainclothes officers to deter male students not only from its Tokyo campus but also from the main road leading to the campus from the nearest subway stop.

Anri Uchida, a junior at Waseda University, went to girls' schools from the age of 5 to 18 but chose not to attend a women's college. "I mainly remember the girls didn't want to do anything alone; they always wanted to be in groups," she says of her early school years. She chose Waseda, she says, because she worried that a women's college would perpetuate that group mentality.

> **"As competition for students and consolidation takes place, women's universities will have to fight hard to maintain their positions."**

Ms. Tanaka, the Japan Women's University student, spent a year at Wellesley College and says she was impressed with the individualism of the women there. The biggest difference between Wellesley students and those at Japanese women's universities, she says, is that those in America seemed "proud of their individual successes. Japanese students care too much about how they appear to other people."

Brian J. McVeigh, former head of the women's studies department at Tokyo Jogakkan University, a women's college, and author of the book *Japanese Higher Education as Myth* (M.E. Sharpe, 2002), also sees a great difference between American and Japanese women's universities. "Elite, well-established schools like Smith and Wellesley offer feminist philosophy. Firebrand feminism hasn't penetrated Japanese society," says Mr. McVeigh, who now teaches east-Asian studies at the University of Arizona.

But some students prefer the essentially conservative nature of women's colleges.

"Japanese women's universities may be old-fashioned, but women can learn manners and culture," says Junko Matusawa, a senior at Showa Women's University. "Here in Japan, even young people are still thinking in the traditional Japanese way."

FROM HOME ECONOMICS TO COMPUTERS

For decades, the tunnel system ensured a steady flow of students into women's universities. Many families preferred a guaranteed pass from the affiliated primary school all the way to the university, skipping arduous entrance examinations.

For the elite institutions, the tunnel system still works. At Japan Women's University, 85 percent of the high-school seniors from its affiliated school enter the university. Students at Showa Women's University are enticed to stay in the system by being allowed to start their university careers at the age of 17 after spending just three years at the institution's high school. Kaori Taguchi, now a third-year student at the university, said the tunnel system was her major reason for attending the university.

But for many women's universities, the tunnel is collapsing. Female students at the affiliated high schools no longer assume that the women's university is their only option. And some guidance counselors may even steer students away from attending the university in hopes of getting students into more prestigious colleges.

Since they can no longer rely on the tunnel for students, women's institutions are revamping themselves in an effort to attract undergraduates. In the past, the universities mainly provided training in home economics, foreign languages, and teaching, but now they have expanded their course offerings to include such subjects as computer science, economics, nutrition, and the hard sciences. Many of the institutions also are greatly expanding their graduate programs.

Women's universities are also putting a "feminine" spin on many traditional courses, especially in architecture and economics. For example, architecture courses emphasize interior design over structural design, while economics departments offer courses that focus on gender or female entrepreneurship.

"Since childhood, I wanted to be an architect," said Natsuko Tamai, a student at Showa Women's University. "I chose this school because it teaches architecture more femininely with more emphasis on interior design and color coordination."

Although most women's universities are trying to be progressive, some of their efforts can seem outdated, especially by Western standards. For example, one university proudly notes that its home-economics professors now teach women how to make budgets and plan nutritious meals instead of how to sew.

Some institutions are trying other strategies to attract students. Showa Women's University set up a branch campus in Boston for its Japanese students to study English and to teach Japanese culture and language to local residents. "The students go over in groups, stay in dorms and are quite safe," said Gordon Robson, coordinator of language studies at the Tokyo campus.

HAMPERED BY RULES

Until recently, the ability of women's universities to reform and update their academic programs was severely restricted by the Japanese education ministry. The ministry required all universities to conform to a massive number of rules that controlled everything from the hiring of a new faculty member to the number of students allowed to enroll in specific departments, such as mechanical engineering.

But a vast program to revamp higher education has allowed for more creativity in course design and has released universities from stringent rules that had hampered creative solutions to enrollment problems.

At the same time, the ministry's plan to overhaul the system has also placed new pressure on women's universities to succeed. It calls for the withdrawal of key financial subsidies, for example, from all institutions that do not reach enrollment quotas. Without such subsidies, many universities will be forced to close.

Not all women's universities are suffering. A handful of elite institutions have seen their enrollments steadily increase in recent years. "We have seen a 50-percent growth in applications," says Shoko Goto, president of Japan Women's University, considered the best women's institution in the country. The university is home to 5,900 students and has campuses in Tokyo and in Kanagawa prefecture.

Kiyosi Hirai, president of Showa Women's University, another of the elite institutions, says his institution has also seen the number of applicants rise, especially among women interested in studying nutrition, architecture, and the sciences.

In a society where women are just beginning to be appointed to management positions in companies, students who attend women's universities say that they give them the confidence to compete.

"One of the most important principles of Japan Women's University is 'self-actualization,'" says Ms. Goto, the president. "We teach women to stand alone and to be productive members of society. We provide a place where women are able to establish self-reliance and identity."

In a recent national survey of women attending such universities, most said that they chose to enroll in a women's institution in order to express their abilities and personalities more freely, to take leadership roles they would be reluctant to take in coed institutions, to undertake special majors, and to study without the distraction of men.

Mao Kamakura, a senior at Japan Women's University who plans to pursue a doctorate in physics, says she chose a women's university because she feared being vastly outnumbered by male students at a coeducational college. "The boys tend to lead the research, but here anyone can lead," she says.

Administrators at women's universities also argue that they have a leg up when it comes time for their graduates to look for jobs. Japanese companies view women who attended women's universities as good prospective employees because they see those women as self-confident and poised but not overly aggressive, says Mr. Robson of Showa Women's University.

Showa and Japan Women's University, for example, both boast job-placement rates of more than 95 percent. Ms. Goto says that women at coeducational institutions carefully monitor the companies that recruit at women's universities so that they know which companies to approach for jobs.

Although attitudes toward women in the workplace are rapidly changing in Japan, there is still a strong sense of male and female roles. Women's universities sometimes see that as a benefit.

"Companies usually can't say they want men or women, because it is illegal, so they apply to women's universities," notes Mr. Hirai.

In a survey released recently by the Japanese government, 68 percent of the women questioned believed that men are given preference at work. The same survey also pointed out that only 20 percent of Japanese civil servants are women, compared with 53 percent in the Philippines.

Women are no longer reduced to serving tea to the male employees and company guests, but they face a very low glass ceiling. It is highly unusual to find female executives at major corporations. Some companies still resort to the practice of classifying women as part-time workers, even though they are hired to work 40 or

50 hours per week, in order to avoid giving them the benefits of full-time employees.

Many companies encourage women to quit when they marry or have children. Company personnel departments also encourage employees to marry, going so far as to present pictures of workers of the opposite sex to employees and urging them to date.

This continued sex discrimination is one reason that proponents of women's universities hope the institutions survive and help break down barriers for women. But whether they are agents of change or relics from a past era remains to be seen.

SOUTH KOREA
Take Three

A trio of new films shows how South Korea's young, confident movie directors explore their culture and past

By Kimberly Song/PUSAN

THE PAST few years have seen an extraordinary explosion in South Korean film, finally freeing directors to follow their own personal interests and artistic inclinations and releasing them from the burden of tackling sombre social issues and patriotically correct content.

The result? Film-makers still revisit the country's poignant history, but with more style, flair and grace. Some of this year's most critically acclaimed and commercially successful films aren't modern romantic comedies, but films that treat viewers to a personal and emotive look into the Korean cultural mirror.

Take, for instance, E J-yong. "I've been wanting to do a traditional costume drama for a while," says the 38-year-old director, who took five years to make his unusual, ambitious and highly successful *Untold Scandal*.

"Korean culture is splendid, with its 5,000-year history, but in school we are first taught about Western art," says the soft-spoken E. "I wondered about Korean art, what Korean beauty is."

But in many ways, there's nothing traditional about E's *Untold Scandal*, a visually sumptuous period film set in the conservative, Confucian 18th-century Chosun dynasty.

The movie opened earlier this month at the Pusan International Film Festival and, in a first for costume dramas in South Korea, broke box-office records,

attracting more than 1 million viewers in its first three days on release.

Unlike traditional period dramas, usually about Korean kings and their concubines, *Untold Scandal* takes its cues from the 18th-century French novel *Les Liaisons Dangereuses*. E laboured for a year and a half on the screenplay, carefully tailoring the French script to his Korean setting.

The idea of melding cultures has been with E since he was a student at the Hankuk University of Foreign Studies, where he majored in Turkish language and culture. At 22, his travels through Turkey and across Europe planted the seed of his fascination with other cultures and led to the film's inspiration.

"I wanted to crush two different cultures together and see what happens," says E, who in *Untold Scandals* unflinchingly brings to life intrigue and sexual gamesmanship among Korea's conservative nobility that's more readily associated with 18th century France.

CJ Entertainment, one of South Korea's two major distribution and investment companies, shelled out more than $4.5 million to fund and promote this untried idea.

To ensure the historical authenticity of his movie, E visited museums and consulted historical texts and experts. Then, with production designer Jung Ku Ho, he spent nearly half the film's budget on recreating a splendid, more colourful version of

Chosun Korea. While Chosun nobility wore subdued hues, E's version sees them dressed in fuchsia, azure, deep gold and blood red, while their dramatic hairdos are accented with jade and coral. More than 120 unique, hand-made, hand-embroidered, and hand-dyed *hanboks*—traditional Korean costumes—were especially crafted for the film.

The film opens with a scene of an aristocratic family paying their respects to ancestors in an annual ritual. The camera then switches from this solemn scene to erotic close-ups of a couple making love. Characters use the classical, melodic way of speaking Korean. And the viewer is further lured into the story not by the traditional Korean *kayagum*, a stringed instrument, but by the subtle strums of European baroque music.

Untold Scandal will likely be snatched up by film festivals to delight viewers worldwide. But, says E, "I did not make this films for foreigners, this film is for Koreans."

MURDER REMEMBERED

Director Bong Joon Ho, 34, also looked for a Korean flavour in a genre traditionally dominated by Hollywood. "I wanted to make a crime movie," says the tall, shaggy-haired director. "Not a cold thriller, but something with a Korean feel."

Memories of Murder, his second film, is based on the 1986 serial murders of 10 women in a small country town in Gyeonggi province. Distinctively Korean, it still has all the trappings of typical crime dramas—the good-cop, bad-cop banter and a series of false leads. The movie was a blockbuster at the box office, selling more than 5 million tickets, even if the provincial government of Gyeonggi and families of the victims would sooner forget these murders than relive them on screen.

The movie shows a country shackled by an authoritarian military government and a clash between modernity and rurality, which partly leads to the bungling of murder investigations. "This is a true Korean crime story," says Bong. "The history of Korea in the 1980s, the politics, the economy, runs throughout the story."

Shown at Spain's 51st Donostia-San Sebastian International Film Festival in September, *Memories* bagged the Silver Shell award for best director, the Altadis-New Directors' award and the prestigious Fipresci prize.

Bong grew up reading the works of thriller writers like Agatha Christie, Ed McBain and Raymond Chandler. He took Kim Kwang Rim's 1996 play *Come See Me*, based on the same murder case, and spent a year further researching newspaper clippings, police reports and government documents to write the screenplay.

"While researching the screenplay, I actually thought I might be able to solve this case myself," says Bong. "But the more I got into the story, the more I became confused like the real detectives did." At a special screening for police detectives in Seoul, Bong says a few of them came up to him afterwards and said they would like to see the case reopened in the hope of finally solving it.

FOUR SEASONS

Kim Ki Duk also touches on the theme of murder in his highly original *Spring, Summer, Fall, Winter ... and Spring*, an exploitation of a mystical Korea drawn from the depths of the director's imagination.

In a departure from Kim's previous eight films—mostly gritty, bloody affairs—this lyrical film unfolds quietly at a pristine valley lake reflecting surrounding mountains. "The location of the film, in the mountains at a temple, shows the quiet, cultural, inner side of Korea," Kim says.

With minimal action and dialogue, the film tells the story of a man's journey through the seasons of his life. In it, a young monk falls in love, leaves his monastery and commits a murder. Later, in the autumn, he returns to accept responsibility for his crime, but is first given a task by an older monk to help release his anger. "I wrote the synopsis of the film in about 30 minutes," says the 43-year-old director.

With that, and about $1 million, he made the film, which was later picked as Korea's nomination for the Best Foreign Language Film category at this year's Academy Awards. Sony Pictures Classics has also picked it up for distribution in North America.

Kim used four different actors for each season, taking one of the parts himself. "I played the character in winter because I personally feel like I am in winter," says Kim. "Filming winter was more like a documentary, it was real for me."

Marriage in Japan

Yuka Ocura

"I don't feel any need to get married." "I don't want to loose that easy-going lifestyle." So say an increasing number of single people who are at what has traditionally been seen as marriageable age. Even if they intend to get married at some stage in the future, more than half of unmarried men and women say that they "are happy to stay single until the right person comes along," meaning that the overall outlook towards marriage is less positive than in the past.

If we compare the results of surveys of unmarried males and females from 1950 and 2000, we see that the figure for males has jumped from 1.5% to 12.6% and from 1.4% to 5.8% for females. There is an increasing trend among both men and women to opt not to marry. Similarly, there is a move to marry later: the average age for first marriage has risen for both men and women. The length of time that couples spend together before getting married is increasing, and this one of the reasons for the increase in the average age for first marriage. These are, of course, many other factors also at work here.

At the same time, there has been a remarkable rise in the divorce rate. The divorce rate is not high compared to many Western countries, but the 70,000 cases of divorce in 1960 had doubled to 140,000 by 1980, and by 2001 had reached a record figure of 290,000. Despite the fact that couples are spending longer together before getting married—and, one would assume, getting to know each other better—some 40% of couples getting divorced do so within five years of having got married, and an increase in divorce among older people is also noticeable. Cases of wives losing interest in their husbands and divorcing them as a result are becoming increasingly common.

Surveys suggest that one wife in three is unhappy with her husband, that more than half regret getting married and that as many as 70% have considered divorce. Also, divorce was once seen as socially unacceptable, but attitudes are changing: there has been a 10% increase in the number who think people should get divorced if they feel dissatisfied with their marriage partner. A contributory factor is that women are participating in society more than before, and have greater economic independence.

Be that as it may, while people in Japan appear to be satisfied with less lavish weddings than before, the average cost of the ceremony and the reception is still 2.78 million yen. It may not be the best way to think of it, but that is a lot to pay when the chances of divorce are increasing...

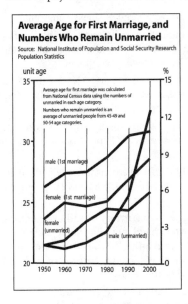

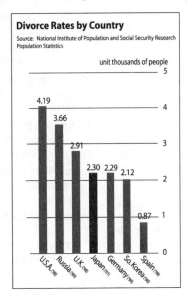

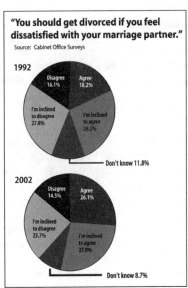

Glossary of Terms and Abbreviations

Animism The belief that all objects, including plants, animals, rocks, and other matter, contain spirits. This belief figures prominently in early Japanese religious thought and in the various indigenous religions of the South Pacific.

Anti-Fascist People's Freedom League (AFPFL) An anti-Japanese resistance movement organized by Burmese students and intellectuals.

ANZUS The name of a joint military-security agreement originally among Australia, New Zealand, and the United States. New Zealand is no longer a member.

Asia Pacific Economic Cooperation Council (APEC) Organized in 1989, this body is becoming increasingly visible as a major forum for plans about regional economic cooperation and growth in the Pacific Rim.

Asian Development Bank (ADB) With contributions from industrialized nations, the ADB provides loans to Pacific Rim countries in order to foster economic development.

Association of Southeast Asian Nations (ASEAN) Established in 1967 to promote economic cooperation among the countries of Indonesia, Malaysia, the Philippines, Singapore, Thailand, and Brunei.

British Commonwealth of Nations A voluntary association of nations formerly included in the British Empire. Officials meet regularly in member countries to discuss issues of common economic, military, and political concern.

Buddhism A religious and ethical philosophy of life that originated in India in the fifth and sixth centuries B.C., partly in reaction to the caste system. Buddhism holds that people's souls are endlessly reborn and that one's standing with each rebirth depends on one's behavior in the previous life.

Capitalism An economic system in which productive property is owned by individuals or corporations, rather than by the government, and the proceeds of which belong to the owner rather than to the workers or the state.

Chaebol A Korean term for a large business conglomerate. Similar to the Japanese *keiretsu.*

Chinese Communist Party (CCP) Founded in 1921 by Mao Zedong and others, the CCP became the ruling party of the People's Republic of China in 1949 upon the defeat of the Nationalist Party and the army of Chiang Kai-shek.

Cold War The intense rivalry, short of direct "hot-war" military conflict, between the Soviet Union and the United States, which began at the end of World War II and continued until approximately 1990.

Communism An economic system in which land and businesses are owned collectively by everyone in the society rather than by individuals. Modern communism is founded on the teachings of the German intellectuals Marx and Engels.

Confucianism A system of ethical guidelines for managing one's personal relationships with others and with the state. Confucianism stresses filial piety and obligation to one's superiors. It is based on the teachings of the Chinese intellectuals Confucius and Mencius.

Cultural Revolution A period between 1966 and 1976 in China when, urged on by Mao, students attempted to revive a revolutionary spirit in China. Intellectuals and even Chinese Communist Party leaders who were not zealously communist were violently attacked or purged from office.

Demilitarized Zone (DMZ) A heavily guarded border zone separating North and South Korea.

European Union (EU) An umbrella organization of numerous Western European nations working toward the establishment of a single economic and political European entity. Formerly known as the European Community (EC) and European Economic Community (EEC).

Extraterritoriality The practice whereby the home country exercises jurisdiction over its diplomats and other citizens living in a foreign country, effectively freeing them from the authority of the host government.

Feudalism A social and economic system of premodern Europe, Japan, China, and other countries, characterized by a strict division of the populace into social classes, an agricultural economy, and governance by lords controlling vast parcels of land and the people thereon.

Greater East Asia Co-Prosperity Sphere The Japanese description of the empire they created in the 1940s by military conquest.

Gross Domestic Product (GDP) A statistic describing the entire output of goods and services produced by a country in a year, less income earned on foreign investments.

Hinduism A 5,000-year-old religion, especially of India, that advocates a social caste system but anticipates the eventual merging of all individuals into one universal world soul.

Indochina The name of the colony in Southeast Asia controlled by France and consisting of the countries of Laos, Cambodia, and Vietnam. The colony ceased to exist after 1954, but the term still is often applied to the region.

International Monetary Fund (IMF) An agency of the United Nations whose goal it is to promote freer world trade by assisting nations in economic development.

Islam The religion founded by Mohammed and codified in the Koran. Believers, called Muslims, submit to Allah (Arabic for God) and venerate his name in daily prayer.

Keiretsu A Japanese word for a large business conglomerate.

Khmer Rouge The communist guerrilla army, led by Pol Pot, that controlled Cambodia in the 1970s and subsequently attempted to overthrow the UN–sanctioned government.

Kuomintang The National People's Party (Nationalists), which, under Chiang Kai-shek, governed China until Mao Zedong's revolution in 1949; it continues to dominate politics in Taiwan.

Laogai A Mandarin Chinese word for a prison or concentration camp where political prisoners are kept. It is similar in concept to the Russian word *gulag.*

Liberal Democratic Party (LDP) The conservative party that ruled Japan almost continuously between 1955 and 1993 and oversaw Japan's rapid economic development.

Martial Law The law applied to a territory by military authorities in a time of emergency when regular civilian authorities are unable to maintain order. Under martial law, residents are usually restricted in their movement and in their exercise of such rights as freedom of speech and of the press.

Meiji Restoration The restoration of the Japanese emperor to his throne in 1868. The period is important as the beginning of the modern era in Japan and the opening of Japan to the West after centuries of isolation.

Monsoons Winds that bring exceptionally heavy rainfall to parts of Southeast Asia and elsewhere. Monsoon rains are essential to the production of rice.

National League for Democracy An opposition party in Myanmar that was elected to head the government in 1990 but that has since been forbidden by the current military leaders to take office.

New Economic Policy (NEP) An economic plan advanced in the 1970s to restructure the Malaysian economy and foster industrialization and ethnic equality.

Newly Industrializing Country (NIC) A designation for those countries of the developing world, particularly Taiwan, South Korea, and other Asian nations, whose economies have undergone rapid growth; sometimes also referred to as newly industrialized countries.

Non-Aligned Movement A loose association of mostly non-Western developing nations, many of which had been colonies of Western powers but during the cold war chose to remain detached from either the U.S. or Soviet bloc. Initially Indonesia and India, among others, were enthusiastic promoters of the movement.

Opium Wars Conflicts between Britain and China in 1839–1842 and 1856–1866 in which England used China's destruction of opium shipments and other issues as a pretext to attack China and force the government to sign trade agreements.

Pacific War The name frequently used by the Japanese to refer to that portion of World War II in which they were involved and which took place in Asia and the Pacific.

SARS (Severe Acute Respiratory Syndrome) A pneumonia-like virus that first appeared in China in 2002 and spread rapidly in 2003 to China, Hong Kong, and other countries, causing hundreds of deaths and adversely affecting the economy of the region.

Shintoism An ancient indigenous religion of Japan that stresses the role of *kami,* or supernatural gods, in the lives of people. For a time during the 1930s, Shinto was the state religion of Japan and the emperor was honored as its high priest.

Smokestack Industries Heavy industries such as steel mills that are basic to an economy but produce objectionable levels of air, water, or land pollution.

Socialism An economic system in which productive property is owned by the government as are the proceeds from the productive labor. Most socialist systems today are actually mixed economies in which individuals as well as the government own property.

South Pacific Forum An organization established by Australia and other South Pacific nations to provide a forum for discussion of common problems and opportunities in the region.

Southeast Asia Treaty Organization (SEATO) This is a collective-defense treaty signed by the United States and several European and Southeast Asian nations. It was dissolved in 1977.

Subsistence Farming Farming that meets the immediate needs of the farming family but that does not yield a surplus sufficient for export.

Taoism An ancient religion of China inspired by Lao-tze that stresses the need for mystical contemplation to free one from the desires and sensations of the materialistic and physical world.

Tiananmen Square Massacre The violent suppression by the Chinese Army of a prodemocracy movement that had been organized in Beijing by thousands of Chinese students in 1989 and that had become an international embarrassment to the Chinese regime.

Tsunami Japanese word composed of two Chinese characters meaning "harbor wave." Tsunamis or tidal waves are created when earthquakes, underwater volcanic activity, or undersea mudslides upseat ocean stability and produce unusually high and fast-traveling ocean waves.

United Nations (UN) An international organization established immediately after World War II to replace the League of Nations. The organization includes most of the countries of the world and works for international understanding and world peace.

World Health Organization (WHO) Established in 1948 as an advisory and technical-assistance organization to improve the health of peoples around the world.

World Trade Organization (WTO) Successor organization to the General Agreement on Trade and Tariffs (GATT) treaties. WTO attempts to standardize the rules of free trade throughout the world.

Bibliography

GENERAL WORKS

Amitav Acharya, *The Quest for Identity: International Relations of Southeast Asia* (New York/Oxford: Oxford University Press, 2000).

Explores the complex relationship between intense feelings of nationalism and the international-order post–Cold War.

Mark Borthwick, *East Asian Civilizations: A Dialogue in Five Stages* (Cambridge, MA: Harvard University Press, 1988).

The development of philosophical and religious thought in China, Korea, Japan, and other regions of East Asia.

Richard Bowring and Peter Kornicki, *Encyclopedia of Japan* (New York: Cambridge University Press, 1993).

Barbara K. Bundy, Stephen D. Burns, and Kimberly V. Weichel, *The Future of the Pacific Rim: Scenarios for Regional Cooperation* (Westport, CT: Praeger, 1994).

Commission on U.S.–Japan Relations for the Twenty-First Century, *Preparing for a Pacific Century: Exploring the Potential for Pacific Basin Cooperation* (Washington, D.C.: November 1991).

Transcription of an international conference on the Pacific with commentary by representatives from the United States, Malaysia, Japan, Thailand, Indonesia, and others.

Susanna Cuyler, *A Companion to Japanese Literature, Culture, and Language* (Highland Park, NJ: B. Rugged, 1992).

William Theodore de Bary, *East Asian Civilizations: A Dialogue in Five Stages* (Cambridge, MA: Harvard University Press, 1988).

An examination of religions and philosophical thought in several regions of East Asia.

Richard E. Feinberg, ed., *APEC as an Institution: Multilateral Governance in the Asia-Pacific* (Singapore: Institute of Southeast Asian Studies, 2003).

Explores the strategy of the Asia Pacific Economic Cooperation Council and analyzes its accomplishments and failures.

Syed N. Hossain, *Japan: Not in the West* (Boston: Vikas II, 1995).

James W. McGuire, ed., *Rethinking Development in East Asia and Latin America* (Los Angeles: Pacific Council on International Policy, 1997).

Charles E. Morrison, ed., *Asia Pacific Security Outlook 1997* (Honolulu: East-West Center, 1997).

Seijiu Naya and Stephen Browne, eds., *Development Challenges in Asia and the Pacific in the 1990s* (Honolulu: East-West Center, 1991).

A collection of speeches made at the 1990 Symposium on Cooperation in Asia and the Pacific. The articles cover development issues in East, Southeast, and South Asia and the Pacific.

Edwin O. Reischauer and Marius B. Jansen, *The Japanese Today: Change and Continuity* (Cambridge: Belknap Press, 1995).

A description of the basic geography and historical background of Japan.

Leo Suryadinata, ed., *Nationalism and Globalization: East and West* (Singapore: Institute of Southeast Asian Studies, 2000).

Case studies of six Asian countries and six European countries, showing how each has handled the pressures of the modern era.

NATIONAL HISTORIES AND ANALYSES

Australia

Boris Frankel, *From the Prophets Deserts Come: The Struggle to Reshape Australian Political Culture* (New York: Deakin University [St. Mut.], 1994).

Australia's government and political aspects are described in this essay.

Herman J. Hiery, *The Neglected War: The German South Pacific and the Influence of WW I* (Honolulu: University of Hawaii Press, 1995).

David Alistair Kemp, *Society and Electoral Behaviors in Australia: A Study of Three Decades* (St. Lucia: University of Queensland Press, 1978).

Elections, political parties, and social problems in Australia since 1945.

David Meredith and Barrie Dyster, *Australia in the International Economy in the Twentieth Century* (New York: Cambridge University Press, 1990).

Examines the international aspects of Australia's economy.

Brunei

Wendy Hutton, *East Malaysia and Brunei* (Berkeley, CA: Periplus, 1993).

Graham Saunders, *A History of Brunei* (New York: Oxford University Press, 1995).

Nicholas Tarling, *Britain, the Brookes, and Brunei* (Kuala Lumpur: Oxford University Press, 1971).

A history of the sultanate of Brunei and its neighbors.

Cambodia

David P. Chandler, *The Tragedy of Cambodian History, War, and Revolution since 1945* (New Haven, CT: Yale University Press, 1993).

A short history of Cambodia.

Michael W. Doyle, *UN Peacekeeping in Cambodia: UNTAC's Civil Mandate* (Boulder, CO: Lynne Rienner, 1995).

A review of the current status of Cambodia's government and political parties.

Craig Etcheson, *The Rise and Demise of Democratic Kampuchea* (Boulder, CO: Westview Press, 1984).

A history of the rise of the Communist government in Cambodia.

William Shawcross, *The Quality of Mercy: Cambodia, Holocaust, and Modern Conscience; with a report from Ethiopia* (New York: Simon & Schuster, 1985).

A report on political atrocities, relief programs, and refugees in Cambodia and Ethiopia.

Usha Welaratna, ed., *Beyond the Killing Fields: Voices of Nine Cambodian Survivors* (Stanford, CA: Stanford University Press, 1993).

A collection of nine narratives by Cambodian refugees in the United States and their adjustments into American society.

China

Julia F. Andrews, *Painters and Politics in the People's Republic of China, 1949–1979* (Berkeley, CA: University of California Press, 1994).
A fascinating presentation of the relationship between politics and art from the beginning of the Communist period until the eve of major liberalization in 1979.

Ma Bo, *Blood Red Sunset* (New York: Viking, 1995).
A compelling autobiographical account by a Red Guard during the Cultural Revolution.

Jung Chang, *Wild Swans: Three Daughters of China* (New York: Simon and Shuster, 1992).
An autobiographical/biographical account that illuminates what China was like for one family for three generations.

Kwang-chih Chang, *The Archaeology of China*, 4th ed. (New Haven, CT: Yale University Press, 1986).

___, *Shang Civilization* (New Haven, CT: Yale University Press, 1980).
Two works by an eminent archaeologist on the origins of Chinese civilization.

Nien Cheng, *Life and Death in Shanghai* (New York: Penguin Books, 1988). A view of the Cultural Revolution by one of its victims.

Qing Dai, *Yangtze! Yangtze!* (Toronto: Probe International, 1994).
Collection of documents concerning the debate over building the Three Gorges Dam on the upper Yangtze River in order to harness energy for China.

John King Fairbank, *China: A New History* (Cambridge, MA: Harvard University Press, 1992).
An examination of the motivating forces in China's history that define it as a coherent culture from its earliest recorded history to 1991.

David S. G. Goodman and Beverly Hooper, eds., *China's Quiet Revolution: New Interactions between State and Society* (New York: St. Martin's Press, 1994).
Articles examine the impact of economic reforms since the early 1980s on the social structure and society generally, with focus on changes in wealth, status, power, and newly emerging social forces.

Richard Madsen, *China and the American Dream: A Moral Inquiry* (Berkeley, CA: University of California Press, 1995).
A history on the emotional and unpredictable relationship the United States has had with China from the nineteenth century to the present.

Jim Mann, *Beijing Jeep: A Case Study of Western Business in China* (Boulder, CO: Westview Press, 1997).
A crisp view of what it takes for a Westerner to do business in China.

Suzanne Ogden, *China's Unresolved Issues: Politics, Development, and Culture* (Englewood Cliffs, NJ: Prentice-Hall, 1992).
A complete review of economic and cultural issues in modern China.

John Wong and Lu Ding, *China's Economy into the New Century* (London: Imperial College Press, 2002).
A policy-oriented and fact-based analysis of the new Chinese economy.

John Wong and Nan Seok Ling, *China's Emerging New Economy* (London: Imperial College Press, 2000).
This book is intended for readers interested in China's internet and e-commerce sectors.

Li Zhisui, *The Private Life of Chairman Mao* (New York: Random House, 1994). Memoirs of Mao's personal physician.

East Timor

Hal Hill and Joao M. Saldanha, eds., *East Timor: Development Challenges for the World's Newest Nation* (ISEAS/Palgrave Publishers, 2001).
A summary of economic information about East Timor and an analysis of the pitfalls ahead as the country attempts to recover from its difficult birth as a new nation.

Hong Kong

"Basic Law of Hong Kong Special Administrative Region of the People's Republic of China," *Beijing Review*, Vol. 33, No. 18 (April 30–May 6, 1990), supplement.

Ming K. Chan and Gerard A. Postiglione, *The Hong Kong Reader: Passage to Chinese Sovereignty* (Armonk, NY: M. E. Sharpe, 1996).
A collection of articles about the issues facing Hong Kong during the transition to Chinese rule after July 1, 1997.

Walter Hatch and Kozo Yamamura, *Asia in Japan's Embrace: Building a Regional Production Alliance* (Cambridge: Cambridge University Press, 1996).
Discusses the future likelihood of Japan building an exclusive trading zone in Asia.

Berry Hsu, ed., *The Common Law in Chinese Context* in the series entitled *Hong Kong Becoming China: The Transition to 1997* (Armonk, NY: M. E. Sharpe, Inc., 1992).
An examination of the common law aspects of the "Basic Law," the mini-constitution that will govern Hong Kong after 1997.

Benjamin K. P. Leung, ed., *Social Issues in Hong Kong* (New York: Oxford University Press, 1990).
A collection of essays on select issues in Hong Kong, such as aging, poverty, women, pornography, and mental illness.

Jan Morris, *Hong Kong: Epilogue to an Empire* (New York: Vintage, 1997).
A detailed portrait of Hong Kong that gives the reader the sense of actually being on the scene in a vibrant Hong Kong.

Mark Roberti, *The Fall of Hong Kong: China's Triumph and Britain's Betrayal* (New York: John Wiley & Sons, Inc., 1994).
An account on the decisions Britain and China made about Hong Kong's fate since the early 1980s.

Frank Welsh, *A Borrowed Place: The History of Hong Kong* (New York: Kodansha International, 1996).
A presentation on Hong Kong's history from the time of the British East India Company in the eighteenth century through the Opium Wars of the nineteenth century to the present.

Indonesia

Amarendra Bhattacharya and Mari Pangestu, *Indonesia: Development, Transformation, and Public Policy* (Washington, D.C.: World Bank, 1993).
An examination of Indonesia's economic policy.

Frederica M. Bunge, *Indonesia: A Country Study* (Washington, D.C.: U.S. Government, 1983).
An excellent review of the outlines of Indonesian history and culture, including politics and national security.

Freek Colombjin and J. Thomas Lindblad, eds., *Roots of Violence in Indonesia* (Singapore: ISEAS/KITLV Press, 2002).
The authors explain the violence in Aceh, East Timore, and other places by exploring the long history of Indonesia.

East Asia Institute, *Indonesia in Transition* (New York: Columbia University, 2000).
An analysis of the revolution in Indonesian politics since the overthrow of Suharto.

Philip J. Eldridge, *Non-government Organizations and Political Participation in Indonesia* (New York: Oxford University Press, 1995).
An examination of Indonesia's nongovernment agencies (NGOs).

Audrey R. Kahin, ed., *Regional Dynamics of the Indonesian Revolution: Unity from Diversity* (Honolulu: University of Hawaii Press, 1985).
A history of Indonesia since the end of World War II, with separate chapters on selected islands.

Hamish McConald, *Suharto's Indonesia* (Australia: The Dominion Press, 1980).

The story of the rise of Suharto and the manner in which he controlled the political and military life of the country, beginning in 1965.

Susan Rodgers, ed., *Telling Lives, Telling Histories: Autobiography and Historical Immigration in Modern Indonesia* (Berkeley, CA: University of California Press, 1995).

Reviews the history of Indonesia's immigration.

David Wigg, *In a Class of Their Own: A Look at the Campaign against Female Illiteracy* (Washington, D.C.: World Bank, 1994).

Looks at the work that is being done by various groups to advance women's literacy in Indonesia.

Japan

David Arase, *Buying Power: The Political Economy of Japan's Foreign Aid* (Boulder CO: Lynne Rienner Publishers, Inc., 1995).

An attempt to explain the complexities of Japan foreign-aid programs.

Michael Barnhart, *Japan and the World since 1868* (New York: Routledge, Chapman, and Hall, 1994).

An essay that addresses commerce in Japan from 1868 to the present.

Marjorie Wall Bingham and Susan Hill Gross, *Women in Japan* (Minnesota: Glenhurst Publications, Inc., 1987).

An historical review of Japanese women's roles in Japan.

Roger W. Bowen, *Japan's Dysfunctional Democracy: The Liberal Democratic Party and Structural Corruption* (New York: M. E. Sharpe, 2003).

Explores the Japanese public's weakness in the face of major government corruption.

John Clammer, *Difference and Modernity: Social Theory and Contemporary Japanese Society* (New York: Routledge, Chapman, and Hall, 1995).

Dean W. Collinwood, "Japan," in Michael Sodaro, ed., *Comparative Politics* (New York: McGraw-Hill, 2000).

An analysis of Japan's government structure and history, electoral process, and some of the issues and pressure points affecting Japanese government.

Dennis J. Encarnation, *Rivals beyond Trade: America Versus Japan in Global Competition* (Ithaca NY: Cornell University Press, 1993).

Explains how the economic rivalry that was once bilateral has turned into an intense global competition.

Mark Gauthier, *Making It in Japan* (Upland, PA: Diane Publishers, 1994).

An examination of how success can be attained in Japan's marketplace.

Walter Hatch and Kozo Yamamura, *Asia in Japan's Embrace: Building a Regional Production Alliance* (Cambridge: Cambridge University Press, 1996).

Discusses the future likelihood of Japan building an exclusive trading zone in Asia.

Paul Herbig, *Innovation Japanese Style: A Cultural and Historical Perspective* (Glenview, IL: Greenwood, 1995).

A review of the implications for international competition.

Ronald J. Hrebenar, *Japan's New Party System* (Boulder, CO: Westview Press, 2000).

An analysis of the political structure in Japan since the end of complete LDP dominance.

Harold R. Kerbo and John McKinstry, *Who Rules Japan? The Inner-Circle of Economic and Political Power* (Glenview, IL: Greenwood, 1995).

The effect of Japan's politics on its economy is evaluated in this essay.

Hiroshi Komai, *Migrant Workers in Japan* (New York: Routledge, Chapman, and Hall, 1994).

An examination of the migrant labor supply in Japan.

Makoto Kumazawa, *Portraits of the Japanese Workplace: Labor Movements, Workers, and Managers* (Boulder, CO: Westview Press, 1996).

Translated into English from Japanese, the book includes reviews of the workplace lifestyle of bankers, women, steel workers, and others.

Solomon B. Levine and Koji Taira, eds., *Japan's External Economic Relations: Japanese Perspectives,* special issue of *The Annals of the American Academy of Political and Social Science* (January 1991).

An excellent overview of the origin and future of Japan's economic relations with the rest of the world, especially Asia.

John Lie, *Multi-Ethnic Japan* (Cambridge, MA: Harvard University Press, 2001).

The existence of Ainu, Koreans, Chinese, Taiwanese, Burakumin, and Okinawans challenges the widely held belief that Japan is a monoethnic society.

E. Wayne Nafziger, *Learning from the Japanese: Japan's Pre-War Development and the Third World* (Armonk, NY: M. E. Sharpe, 1995).

Presents Japan as a model of "guided capitalism," and what it did by way of policies designed to promote and accelerate development.

Nippon Steel Corporation, *Nippon: The Land and Its People* (Japan: Gakuseisha Publishing, 1984).

An overview of modern Japan in both English and Japanese.

Asahi Simbun, *Japan Almanac 1998* (Tokyo: Asahi Shimbun Publishing Company, 1997).

Charts, maps, statistical data about Japan in both English and Japanese.

Patrick Smith, *Japan: A Reinterpretation* (New York: Pantheon Books, 1997).

A discussion of the rapidly changing Japanese national character.

Korea: North and South Korea

Chai-Sik Chung, *A Korean Confucian Encounter with the Modern World* (Berkeley, CA: IEAS, 1995).

Korea's history and the effectiveness of Confucianism are addressed.

Donald Clark et al., *U.S.–Korean Relations* (Farmingdale, NY: Regina Books, 1995).

A review on the history of Korea's relationship with the United States.

James Cotton, *Politics and Policy in the New Korean State: From Rah Tae-Woo to Kim Young-Sam* (New York: St. Martin's, 1995).

The power and influence of politics in Korea are examined.

James Hoare, *North Korea* (New York: Oxford University Press, 1995).

An essay that addresses commerce in Japan between 1868 and the present.

Dae-Jung Kim, *Mass Participatory Economy: Korea's Road to World Economic Power* (Landham, MD: University Press of America, 1995).

Korean Overseas Information Service, *A Handbook of Korea* (Seoul: Seoul International Publishing House, 1987).

A description of modern South Korea, including social welfare, foreign relations, and culture. The early history of the entire Korean Peninsula is also discussed.

___, *Korean Arts and Culture* (Seoul: Seoul International Publishing House, 1986).

A beautifully illustrated introduction to the rich cultural life of modern South Korea.

Callus A. MacDonald, *Korea: The War before Vietnam* (New York: The Free Press, 1986).

A detailed account of the military events in Korea between 1950 and 1953, including a careful analysis of the U.S. decision to send troops to the peninsula.

Christopher J. Sigur, ed., *Continuity and Change in Contemporary Korea* (New York: Carnegie Ethics and International Affairs, 1994).

A review of the numerous stages of change that Korea has experienced.

Joseph A. B. Winder, ed., *Korea's Economy 1999* (Washington, D.C.: Korea Economic Institute, 1999).

A review of the economic impact of the Asian financial crisis on South Korea.

Yonhap News Agency, ed., *North Korea Handbook* (New York: M. E. Sharpe, 2002).

A comprehensive guide to North Korea's military, economy, and culture.

Laos

Sucheng Chan, ed., *Hmong: Means Free Life in Laos and America* (Philadelphia: Temple University Press, 1994).

Arthur J. Dommen, *Laos: Keystone of Indochina* (Boulder, CO: Westview Press, 1985).

A short history and review of current events in Laos.

Grant Evans, ed., *Laos: Culture and Society* (Silkworm Books, 2000).

A comprehensive social and cultural analysis of the nation and people of Laos.

Joel M. Halpern, *The Natural Economy of Laos* (Christiansburg, VA: Dalley Book Service, 1990).

___, *Government, Politics, and South Structures of Laos: Study of Traditions and Innovations* (Christiansburg, VA: Dalley Book Service, 1990).

Macau

Charles Ralph Boxer, *The Portuguese Seaborne Empire, 1415–1825* (New York: A. A. Knopf, 1969). A history of Portugal's colonies, including Macau.

W. G. Clarence-Smith, *The Third Portuguese Empire, 1825–1975* (Manchester: Manchester University Press, 1985). A history of Portugal's colonies, including Macau.

Malaysia

Mohammed Ariff, *The Malaysian Economy: Pacific Connections* (New York: Oxford University Press, 1991).

The report on Malaysia examines Malaysia's development and its vulnerability in world trade.

Richard Clutterbuck, *Conflict and Violence in Singapore and Malaysia, 1945–1983* (Boulder, CO: Westview Press, 1985).

The Communist challenge to the stability of Singapore and Malaysia in the early years of their independence from Great Britain is presented.

Virginia Hooker and Norani Othman, eds., *Malaysia: Islam, Society and Politics* (Singapore: Institute of Southeast Asian Studies, 2003).

Collection of essays written by sociologist Clive Kessler on modern Malaysian religion and society.

K. S. Jomo, ed., *Japan and Malaysian Development: In the Shadow of the Rising Sun* (New York: Routledge, 1995).

A review of the relationship between Japan and Malaysia's economy.

Gordon Means, *Malaysian Politics: The Second Generation* (New York: Oxford University Press, 1991).

R. S. Milne, *Malaysia: Tradition, Modernity, and Islam* (Boulder, CO: Westview Press, 1986).

A general overview of the nature of modern Malaysian society.

A.B. Shamsul, *From British to Bumiputera Rule: Local Politics and Rural Development in Peninsular Malaysia* (Singapore: Institute of Southeast Asian Studies, 2004).

Originally published in 1986, this book nevertheless provides keen insight into the mindset of the Malay people as they deal with domestic and international pressures.

Myanmar (Burma)

Michael Aung-Thwin, *Pagan: The Origins of Modern Burma* (Honolulu: University of Hawaii Press, 1985).

A treatment of the religious and political ideology of the Burmese people and the effect of ideology on the economy and politics of the modern state.

Aye Kyaw, *The Voice of Young Burma* (Ithaca, NY: Cornell SE Asia, 1993).

The political history of Burma is presented in this report.

Chi-Shad Liang, *Burma's Foreign Relations: Neutralism in Theory and Practice* (Glenview, IL: Greenwood, 1990).

Mya Maung, *The Burma Road to Poverty* (Glenview, IL: Greenwood, 1991).

Myat Thein, *Economic Development of Myanmar* (Singapore: Institute of Southeast Asian Studies, 2004).

A compilation of studies on the economy of Myanmar from 1948 to 2000.

New Zealand

Bev James and Kay Saville-Smith, *Gender, Culture, and Power: Challenging New Zealand's Gendered Culture* (New York: Oxford University Press, 1995).

Patrick Massey, *New Zealand: Market Liberalization in a Developed Economy* (New York: St. Martin, 1995).

Analyzes New Zealand's market-oriented reform programs since the Labour government came into power in 1984.

Stephen Rainbow, *Green Politics* (New York: Oxford University Press, 1994).

A review of current New Zealand politics.

Geoffrey W. Rice, *The Oxford History of New Zealand* (New York: Oxford University Press, 1993).

Papua New Guinea

Robert J. Gordon and Mervyn J. Meggitt, *Law and Order in the New Guinea Highlands: Encounters with Enga* (Hanover, NH: University Press of New England, 1985).

Tribal law and warfare in Papua New Guinea.

David Hyndman, *Ancestral Rainforests and the Mountain of Gold: Indigenous Peoples and Mining in New Guinea* (Boulder, CO: Westview Press, 1994).

Bruce W. Knauft, *South Coast New Guinea Cultures: History, Comparison, Dialectic* (New York: Cambridge University Press, 1993).

The Philippines

Frederica M. Bunge, ed., *Philippines: A Country Study* (Washington, D.C.: U.S. Government, 1984).

Description and analysis of the economic, security, political, and social systems of the Philippines, including maps, statistical charts, and reproduction of important documents. An extensive bibliography is included.

Manual B. Dy, *Values in Philippine Culture and Education* (Washington, D.C.: Council for Research in Values and Philosophy, 1994).

James F. Eder and Robert L. Youngblood, eds., *Patterns of Power and Politics in the Philippines: Implications for Development* (Tempe, AZ: ASU Program, SE Asian, 1994).

A review of the impact of politics and its power over development in the Philippines.

Singapore

Lai A. Eng, *Meanings of Multiethnicity: A Case Study of Ethnicity and Ethnic Relations in Singapore* (New York: Oxford University Press, 1995).

Paul Leppert, *Doing Business with Singapore* (Fremont, CA: Jain Publishing, 1995).

Singapore's economic status is examined in this report.

Hafiz Mirza, *Multinationals and the Growth of the Singapore Economy* (New York: St. Martin's Press, 1986).

An essay on foreign companies and their impact on modern Singapore.

Nilavu Mohdx et al., *New Place, Old Ways: Essays on Indian Society and Culture in Modern Singapore* (Columbia, MO: South Asia, 1994).

South Pacific

C. Beeby and N. Fyfe, "The South Pacific Nuclear Free Zone Treaty," Victoria University of Wellington *Law Review,* Vol. 17, No. 1, pp. 33–51 (February 1987).

A good review of nuclear issues in the Pacific.

William S. Livingston and William Roger Louis, eds., *Australia, New Zealand, and the Pacific Islands since the First World War* (Austin, TX: University of Texas Press, 1979).

An assessment of significant historical and political developments in Australia, New Zealand, and the Pacific Islands since 1917.

Leo Suryadinata, *Ethnic Relations and Nation-Building in Southeast: The Case of the Ethnic Chinese* (Singapore: Institute of Southeast Asian Studies, 2004).

A review of the ethnic tensions that have long been a feature of Southeast Asian life.

Lim Chong Yah, *Southeast Asia: The Long Road Ahead* (London: Imperial College Press, 2001).

A cross-country discussion of the problems that will be encountered as Southeast Asia develops.

Taiwan

Joel Aberbach et al., eds., *The Role of the State in Taiwan's Development* (Armonk, NY: M. E. Sharpe, 1994).

Articles address technology, international trade, state policy toward the development of local industries, and the effect of economic development on society, including women and farmers.

Bih-er Chou, Clark Cal, and Janet Clark, *Women in Taiwan Politics: Overcoming Barriers to Women's Participation in a Modernizing Society* (Boulder, CO: Lynne Rienner, 1990).

Examines the political underrepresentation of women in Taiwan and how Chinese culture on the one hand and modernization and development on the other are affecting women's status.

Stevan Harrell and Chun-chieh Huang, eds., *Cultural Change in Postwar Taiwan* (Boulder, CO: Westview Press, 1994).

A collection of essays that analyzes the tensions in Taiwan's society as modernization erodes many of its old values and traditions.

Dennis Hickey, *United States–Taiwan Security Ties: From Cold War to beyond Containment* (Westport, CT: Praeger, 1994).

Examines U.S.–Taiwan security ties from the Cold War to the present and what Taiwan is doing to ensure its own military preparedness.

Chin-chuan Lee, "Sparking a Fire: The Press and the Ferment of Democratic Change in Taiwan," in Chin-chuan Lee, ed., *China's Media, Media China* (Boulder, CO: Westview Press, 1994), pp. 179–193.

Sheng Lijun, *China and Taiwan: Cross-Strait Relations Under Chen Shui-bian* (Singapore: Institute of Southeast Asian Studies, 2002).

With the electoral defeat of the Kuomintang (KMT), Taiwan's relations with China are undergoing important changes.

Robert M. Marsh, *The Great Transformation: Social Change in Taipei, Taiwan, since the 1960s* (Armonk, NY: M. E. Sharpe, 1996).

An investigation of how Taiwan's society has changed since the 1960s when its economic transformation began.

Robert G. Sutter and William R. Johnson, *Taiwan in World Affairs* (Boulder, CO: Westview Press, 1994).

Articles give comprehensive coverage of Taiwan's involvement in foreign affairs.

Wei-Bin Zhang, *Taiwan's Modernization* (London: Imperial College Press, 2003).

The impact of Confucianism on the modernization of Taiwan and other Confucian-based societies in East Asia.

Thailand

Medhi Krongkaew, *Thailand's Industrialization and Its Consequences* (New York: St. Martin, 1995).

A discussion of events surround the development of Thailand since the mid-1980s with a focus on the nature and characteristics of Thai industrialization.

Ross Prizzia, *Thailand in Transition: The Role of Oppositional Forces* (Honolulu: University of Hawaii Press, 1985).

Government management of political opposition in Thailand.

Susan Wells and Steve Van Beek, *A Day in the Life of Thailand* (San Francisco: Collins SF, 1995).

Vietnam

Chris Brazier, *The Price of Peace* (New York: Okfam Pubs. U.K. [St. Mut.], 1992).

Ronald J. Cima, ed., *Vietnam: A Country Study* (Washington, D.C.: U.S. Government, 1989).

An overview of modern Vietnam, with emphasis on the origins, values, and lifestyles of the Vietnamese people.

Chris Ellsbury et al., *Vietnam: Perspectives and Performance* (Cedar Falls, IA: Assn. Text Study, 1994).

A review of Vietnam's history.

Hy Van Luong, *Postwar Vietnam: Dynamics of a Transforming Society* (Lanham, MD: Rowman and Littlefield Publishers, 2003).

The Vietnam War is not the most important thing about Vietnam. This book shows how rich are Vietnam's culture and history.

Hy V Luong, ed., *Postwar Vietnam: Dynamics of a Transforming Society* (ISEAS/Rowman & Littlefield Publishers, 2003).

The dynamics of economic reform, socioeconomic inequality, and other social conditions are covered in this book, which attempts to go beyond the usual focus on the Vietnam War.

D.R. SarDeSai, *Vietnam: The Struggle for National Identity* (Boulder, CO: Westview Press, 1992).

A good treatment of ethnicity in Vietnam and a national history up to the involvement in Cambodia.

PERIODICALS AND CURRENT EVENTS

The Annals of the American Academy of Political and Social Science
  c/o Sage Publications, Inc.
  2455 Teller Rd.
  Newbury Park, CA 91320
  Selected issues focus on the Pacific Rim; there is an extensive book-review section. Special issues are as follows:
  "The Pacific Region: Challenges to Policy and Theory" (September 1989).
  "China's Foreign Relations" (January 1992).
  "Japan's External Economic Relations: Japanese Perspectives" (January 1991).

Asian Affairs: An American Review
  Helen Dwight Reid Educational Foundation
  1319 Eighteenth St., NW
  Washington, D.C. 20036-1802
  Publishes articles on political, economic, and security policy.

The Asian Wall Street Journal,
  Dow Jones & Company, Inc.
  A daily business newspaper focusing on Asian markets.

Asia-Pacific Issues
  East-West Center
  1601 East-West Rd.
  Burns Hall, Rm. 1079
  Honolulu, HI 96848-1601
  Each contains one article on an issue of the day in Asia and the Pacific.

The Asia-Pacific Magazine
  Research School of Pacific and Asian Studies
  The Australian National University
  Canberra ACT 0200, Australia
  General coverage of all of Asia and the Pacific, including book reviews and excellent color photographs.

Asia-Pacific Population Journal
  Economic and Social Commission for Asia and the Pacific
  United Nations Building
  Rajdamnern Nok Ave.
  Bangkok 10200, Thailand
  A quarterly publication of the United Nations.

Australia Report
  1601 Massachusetts Ave., NW
  Washington, D.C. 20036
  A monthly publication of the Embassy of Australia, Public Diplomacy Office, with a focus on U.S. relations.

Canada and Hong Kong Update
  Joint Centre for Asia Pacific Studies
  Suite 270, York Lanes
  York University
  4700 Keele St.
  North York, Ontario M3J 1P3, Canada
  A source of information about Hong Kong emigration.

Courier
  The Stanley Foundation
  209 Iowa Ave.
  Muscatine, IA 52761
  Published three times a year, the *Courier* carries interviews of leaders in Asian and other world conflicts.

Current History: A World Affairs Journal
  Focuses on one country or region in each issue; the emphasis is on international and domestic politics.

The Economist
  25 St. James's St.
  London, England
  A newsmagazine with insightful commentary on international issues affecting the Pacific Rim.

Education About Asia
  1 Lane Hall
  The University of Michigan
  Ann Arbor, MI 48109
  Published 3 times a year, it contains useful tips for teachers of Asian Studies. The Spring 1998 issue (Vol. 3, No. 1) focuses on teaching the geography of Asia.

Indochina Interchange
  Suite 1801
  220 West 42nd St.
  New York, NY 10036
  A publication of the U.S.–Indochina Reconciliation Project. An excellent source of information about assistance programs for Laos, Cambodia, and Vietnam.

Japan Echo
  Maruzen Co., Ltd.
  P.O. Box 5050
  Tokyo 100-3199, Japan
  Bimonthly translation of selected articles from the Japanese press on culture, government, environment, and other topics.

Japan Economic Currents
  Keizai Koho Center
  1900 K Street NW
  Suite 1075
  Washington, D.C. 20006
  A commentary on selected economic and business trends. Available online at www.kkc-usa.org.

The Japan Foundation Newsletter
  The Japan Foundation
  Park Building
  3-6 Kioi-cho
  Chiyoda-ku
  Tokyo 102, Japan
  A quarterly with research reports, book reviews, and announcements of interest to Japan specialists.

Japan Quarterly
  Asahi Shimbun
  5-3-2 Tsukiji
  Chuo-ku
  Tokyo 104, Japan
  A quarterly journal, in English, covering political, cultural, and sociological aspects of modern Japanese life.

The Japan Times
  The Japan Times Ltd.
  C.P.O. Box 144
  Tokyo 100-91, Japan
  Excellent coverage, in English, of news reported in the Japanese press.

The Journal of Asian Studies
  Association for Asian Studies
  1 Lane Hall
  University of Michigan
  Ann Arbor, MI 48109
  Formerly *The Far Eastern Quarterly;* scholarly articles on Asia, South Asia, and Southeast Asia.

Journal of Japanese Trade & Industry
  11th Floor, Fukoku Seimei Bldg., 2-2-2 Uchisaiwai-cho Chiyoda Ku
  Tokyo 100-0011, Japan
  A bimonthly publication of the Japan Economic Foundation, with a focus on trade but including articles on Japanese culture and other topics.

Journal of Southeast Asian Studies
  Singapore University Press
  Singapore Formerly
  The *Journal of Southeast Asian History;* scholarly articles on all aspects of modern Southeast Asia.

Korea Economic Report
Yoido
P.O. Box 963
Seoul 150-609
South Korea
An economic magazine for people doing business in Korea.

The Korea Herald
2-12, 3-ga Hoehyon-dong
Chung-gu
Seoul, South Korea
World news coverage, in English, with focus on events affecting the Korean Peninsula.

The Korea Times
The Korea Times Hankook Ilbo
Seoul, South Korea
Coverage of world news, with emphasis on events affecting Asia and the Korean Peninsula.

Malaysia Industrial Digest
Malaysian Industrial Development Authority (MIDA)
6th Floor
Industrial Promotion Division
Wisma Damansara, Jalan Semantan
50490 Kuala Kumpur, Malaysia
A source of statistics on manufacturing in Malaysia; of interest to those wishing to become more knowledgeable in the business and industry of the Pacific Rim.

The New York Times
229 West 43rd St.
New York, NY 10036
A daily newspaper with excellent coverage of world events.

News From Japan
Embassy of Japan
Japan–U.S. News and Communication
Suite 520
900 17th St., NW
Washington, D.C. 20006
A twice-monthly newsletter with news briefs from the Embassy of Japan on issues affecting Japan–U.S. relations.

Newsweek
444 Madison Ave.
New York, NY 10022
A weekly magazine with news and commentary on national and world events.

The Oriental Economist
380 Lexington Ave.
New York, NY 10168
A monthly review of political and economic news in Japan by Toyo Keizai America Inc.

Pacific Affairs
The University of British Columbia
Vancouver, BC V6T 1W5
Canada
An international journal on Asia and the Pacific, including reviews of recent books about the region.

Pacific Basin Quarterly
c/o Thomas Y. Miracle
1421 Lakeview Dr.
Virginia Beach, VA 23455-4147
Newsletter of the Pacific Basin Center Foundation. Sometimes provides instructor's guides for included articles.

South China Morning Post
Tong Chong Street
Hong Kong
Daily coverage of world news, with emphasis on Hong Kong, China, Taiwan, and other Asian countries.

Time
Time-Life Building
Rockefeller Center
New York, NY 10020
A weekly newsmagazine with news and commentary on national and world events.

U.S. News & World Report
2400 N St., NW
Washington, D.C. 20037
A weekly newsmagazine with news and commentary on national and world events.

The US-Korea Review
950 Third Ave.
New York, NY 10022
Bimonthly magazine reviewing cultural, economic, political, and other activities of The Korea Society.

Vietnam Economic Times
175 Nguyen Thai Hoc
Hanoi, Vietnam
An English-language monthly publication of the Vienam Economic Association, with articles on business and culture.

The World & I: A Chronicle of Our Changing Era
2800 New York Ave., NE
Washington, D.C. 20002
A monthly review of current events plus excellent articles on various regions of the world.

Index

Index

I

identity confusion, in Pacific Rim, 5
India, influence in Cambodia and, 44
Indians, among Pacific Rim populations, 17
Indonesia, 1, 5; Borneo crisis in, 68; East Timor and, 57–58, 67–68; Japanese control of, 65; impact of Asian financial crisis, 66; national debt and, 66; overview, 64–68
interisland migration, in Pacific Islands
International Action Against Genocide (Kuper), 45
International Atomic Energy Agency (IAEA), 94
International Monetary Fund (IMF), 10, 71, 104, 112; Indonesia and, 67; Thailand and, 123
Islam, 54
Islamic militants, in Indonesia, 66–67
Islamism, radical, Philippines and, 102

J

Japan International Development Organization (JIDO), 10
Japan: business in, 27–28; barriers to foreign businesses in, 33; Cambodia and, 44;early Chinese influence in, 20; cultural characteristics of, 25–27, 28; defense policy and, 31; economy, ten commandments of post–war success of, 24–25; empire building and, 22; influence among Pacific Islands, 8–10; "internationalization" and, 23; Pacific Rim investments and, 1, 8; Korean influence on, 110; Laos and, 71; limitations on military in, 23, 30–31; Meiji restoration in, 21–22; opening to the West of, 21; overview, 20–33; rearmament of, 3; social classes in, 21; trade with U.S. and, 22; U.S. foreign aid and, 24; U.S. occupation of, 22; World War II and, 2, 22
Jemaah Islamiyah, 66
Jiang, Zemin, 53

K

Keating, Paul, 37
Khmer Empire, 44
Khmer Rouge, 45
Kim Il Song, 92, 94, 95
Korean War, 92–93; New Zealand and, 88
Kowloon Peninsula, 60
Kuomintang, 50, 115

L

laogai, 53
Laos 1, 2; economy, 70–71; ethnic diversity of, 70; France and, 71; Japan and, 71; opium and, 71; overview of, 69–72; Vietnam War and, 72
Law of the Sea, 16
Lee Kuan Yew, 108
Legalism, in China, 50
less developed countries (LDCs), 6
Liberal Democratic Party, in Japan, 25, 31
Lon Nol, 44

M

Macapagal–Arroyo, Floria, 104
MacArthur, Douglas, 92
Macau: overview, 73–75
Madura, 68
Maisin, 98
Malaysia, 1; economic development of, 78–79; environmental issues in, 79; ethnic diversity of, 77; financial crisis of 1998 and, 78; "Operation Get Out" and, 78; overview, 76–79; politics of, 77–78, 79; social problems in, 79
Malinowski, Bronislaw, 99
Mao Zedong, 49, 50–52, 93, 115
Maori, 86, 89–90
Marcos, Ferdinand, 5, 103
Marquesas Islands, 13
Maylay people, 77, 79
Meiji Restoration, 21–22
Melanesia, characterized, 13
Mendana, Alvaro de, 13
Micronesia, characterized, 13
Muslims: in China, 49; in Indonesia, 65, 66
Myanmar (Burma) 1, 2, 11; Buddhism in, 84; ethnic conflict in, 81–82; Great Britain and, 81; heroin and, 82; overview, 80–84; tightly controlled society of, 81; Western sanctions against, 82–84

N

Nagasaki, 31
natural disasters, Pacific Rim and, 6–7
Nauru, 15; refugees and, 37; Australia and, 38
nerve gas disposal and, 88
New Guinea, 14
New Zealand: overview of, 85–90
newly industrializing countries (NICs), 5
Nihonjin–ron, 26

Norodom Sihamoni, 46
Norodom Sihanouk, 44–45
North Korea, 2, 3, 111; overview of, 91–95; China and, 93; reunification of, 7, 94; United States and, 94
nuclear free zone, 88
nuclear weapons, 15, 88

O

one child policy, in China, 49
"Operation Get Out," in Malaysia, 2, 78
Opium Wars, 60
opium, Hong Kong and, 60
"overseas Chinese," 116
overseas development assistance (ODA), by Japan, 1; by U.S., 1

P

Pacific Islands, three groupings of, 13–14; geographic distinctions of, 14; atomic weapon testing and, 15; toxic waste and, 15; political independence of, 16;
Pacific Rim: armies in, 3; defined, 1; forces at work in, 1; overview of, 1–11; piracy in, 2; population imbalances and, 3–4
Papua New Guinea, overview, 96–99; Australia and, 38
"parasite singles," 4
Pathet Lao, 71, 72
Philippines, 2, 3, 4, 5; Chinese community in, 102; disputes over Sabah and, 78; overview 100–104; social problems, 104; women in, 104
Phumiphon Adunyadet, 122
pirate attacks, in Pacific Rim, 2
Pol Pot, 45
political instability, of Pacific Islands, 4
Polynesia, characterized, 13
population densities in Pacific Rim, 4
population imbalances among Pacific Islands, 3–4
Portugal: East Timor and, 57–58; Macau and, 74

R

Raffles, Stamford, 106
Recruit Scandal, 24
religion, in Japan, 30
religious diversity in Pacific Rim, 5
Russia, 3
Russo–Japanese War, 110–111

S

samurai code, 21

SARS, 6; epidemic, China and, 53

savings rate: in Japan, 25, 26; in West, 27

Sea of Japan, Soviet nuclear waste and, 6

Senkaku Islands, 2, 15

Shanghai, 48, 49

Shinto, 11, 20

shoguns, 20

Singapore, 1, 3; Asian financial crisis and, 107; overview of, 105–108;

Sino–Japanese War, 115

SLORC (State Law and Order Restoration Council), 83–84

Souphanouvong, Prince, 71

Southeast Asian Treaty Organization, 87–88

South Korea, 2, 3, 9; overview of, 109–113; reunification of, 7, 94; social problems of, 112–113

South Pacific Commission, 15

South Pacific Forum, 15

South Pacific Islands Fisheries Development Agency, 15

South Pacific Nuclear Free Zone Treaty, 15–16

South Pacific Regional Trade and Economic Agency, 15

Soviet Union, and Laos, 71–72; North Korea and, 11, 92, 95

Spanish–American War, 102–103

Special Economic Zones, in China, 52

Spratly Islands, disputes over, 2, 42, 104

Suharto, 5, 67

Sukarno, 67

Sun–Yat-sen, 50–51

syngman Rhee, 111

T

Tahiti, 14

Taiwan, 1, 51; economic success of, 9, 116–117; official recognition of, 116; overview of, 115–118; politics in, 117–118; trade via Hong Kong and, 117; United Nations and, 115–116;

Taoism, 54, 61, 77

Thailand, 1, 2, 3, 119–125; demographic changes in, 124: Japan and, 123, 124; military role in, 122; shift to export–economy by, 123–124; social problems in, 124–125

"The Emergency," in Malaysia, 78

Three Gorges Dam, 55

Tiananmen Square massacre, 52–53, 63

Tibet, China and, 49

Tojo, Hideki, 22

Tokugawa Era, 21

toxic waste disposal, 15, 16, 88

Treaty of Nanking (Nanjing), 60

Truman, Harry, 92

U

United Maylay National Organization (UMNO), 77, 78

United Nations: in East Timor, 57–58; peacekeeping in Cambodia, 46; Khmer Rouge and, 46; Taiwan and, 115–116

United States: Korean War and, 92–93; occupation of Japan, 22, 24; Vietnam and, 128–130

V

Vietnam War: Cambodia and, 45; Laos and, 72; New Zealand and, 88; containment and, 93; Thailand and, 121; United States and, 128–129

Vietnam, 1, 2, 3; Chinese in, 127; economy of, 130; isolation of, 46; overview of, 126–130; religions in, 129

W

warfare, among Pacific Islands, 2–3

Western Samoa, 16

"White Australia" policy, 36; One Nation Party and, 37

wildfires of Pacific Rim, 1997 and 1998, 6, 67

women, in Japan, 28, 29–30

World Bank, 67

World Health Organization, 4, 5, 17

World Trade Organization, 5; China and, 52; Taiwan and, 118

World War II, 2, 8–9, 22, 38–39, 41, 87, 88, 103, 123

Y

Yamato clan, 20

"yen bloc," 24, 37

Z

zaibatsu, 24